Y0-BZF-577

DISCARDED

Evolution's Destiny

Co-evolving Chemistry of the Environment and Life

Evolution's Destiny

Co-evolving Chemistry of the Environment and Life

R. J. P. Williams
Inorganic Chemistry Laboratory, University of Oxford, Oxford, OX1 3QR UK
Email: bob.williams@chem.ox.ac.uk

R. E. M. Rickaby
Department of Earth Sciences, University of Oxford, Oxford, OX1 3AN UK
Email: rosr@earth.ox.ac.uk

RSC Publishing

ISBN: 978-1-84973-558-2

A catalogue record for this book is available from the British Library

Published by The Royal Society of Chemistry,
Thomas Graham House, Science Park, Milton Road,
Cambridge CB4 0WF, UK

Registered Charity Number 207890

For further information see our web site at www.rsc.org

Printed and bound in the United States of America

Preface

This book is written as an addition both to Darwin's work and that of molecular biologists on evolution, so as to include views from the point of view of chemistry rather than just from our knowledge of the biology and genes of organisms. By concentrating on a wide range of chemical elements, not just those in traditional organic compounds, we show that there is a close relationship between the geological or environmental chemical changes from the formation of Earth and those of organisms from the time of their origin. These are considerations that Darwin or other scientists could not have explored until very recent times because sufficient analytical data were not available. They lead us to suggest that there is a combined geo- and biochemical evolution, that of an ecosystem, which has had a systematic chemical development. In this development the arrival of new very similar species is shown to be by random Darwinian competitive selection processes such that a huge variety of species coexist with only minor differences in chemistry and advantages, which is in agreement with previous studies. By way of contrast, we observe that on a large scale, groups of such species have special, different energy and chemical features and functions so that in fair part they support one another. It is more difficult to understand how they evolved and therefore we examine their energy and chemical development in detail. Overall we know that there is a cooperative evolution of a chemical system driven by capture of energy, mainly from the Sun, and its degradation, in which the chemistry of both the environment and organisms are facilitating intermediates. We will suggest that the overall drive of the whole joint system is to optimise the rate of this energy degradation. The living part of the system, the organisms, is under the influence of inevitable inorganic environmental change which moves rapidly to equilibrium conditions, though much of it was forced by the different chemicals added to it by organisms at different times. We are also able to explore some ways in which the organic chemicals of

Evolution's Destiny: Co-evolving Chemistry of the Environment and Life
R. J. P. Williams and R. E. M. Rickaby

Published by the Royal Society of Chemistry, www.rsc.org

organisms evolved. Such evolution was dependent on the inevitably changing environment for all its chemicals and therefore much novel organic chemistry followed a determined path. We recognise that as complexity of the chemistry of organisms increased, the organisms had to become part of a cooperative overall activity and could not remain as isolated species. Prokaryotes and bacteria managed by long-distance exchange between different cells; eukaryotes evolved by incorporating some bacteria – the organelles. The eukaryotes also had increasing numbers of other compartments found in both animals and plants. Later division of essential activities was by direct combination of differentiated cells and by further different forms of symbiosis. Only in the last chapter do we attempt to make a connection between the changing chemistry of organisms with the coded molecules, DNA, of each cell which have to exist to explain reproduction.

This book has to encompass the full spectrum of chemistry, from the extreme of Earth Sciences to those of Biological Sciences. No author can claim to cover all these disciplines to an appropriate depth. The authors of this book apologise for misjudgements of any particular topics and trust that readers will inform them of any errors.

Acknowledgements

We are indebted to the University of Oxford and to the Royal Society for support. We are also grateful to Susie Compton for her help in typing the documents and David Sansom for the illustrations. RR acknowledges financial support from the ERC, grant SP2-GA-2008-200915.

Evolution's Destiny: Co-evolving Chemistry of the Environment and Life
R. J. P. Williams and R. E. M. Rickaby

Published by the Royal Society of Chemistry, www.rsc.org

Contents

Evolution's Destiny: Co-evolving Chemistry of the Environment and Life
R. J. P. Williams and R. E. M. Rickaby

Published by the Royal Society of Chemistry, www.rsc.org

Glossary

Biological

Anaerobic The condition of the environment and organisms in the absence of oxygen.

Aerobic The condition of the environment and organisms in the presence of oxygen which they utilise.

Micro-aerobic The condition of the environment and organisms in the presence of extremely low levels of oxygen but sufficient to produce some sulfate or ferric ions which assist in prokaryote metabolism.

Prokaryotes The earliest cellular life including both bacteria and Archaea. Earliest date 3.5 Ga. They have a simple rigid membrane enclosing all chemical activities.

Eukaryotes (a) Single cell. These cells evolved at about 2.5 Ga, from prokaryotes? They are large cells with a flexible outer membrane and contain many vesicles including organelles, mitochondria and chloroplasts. They are strictly aerobic.

(b) Multicellular. These organisms are the immediate precursors of the plant, fungal and animal organisms of today. They developed after cellular clusters with an outer containing membrane or 'skin' with many types of differentiated cells or groups of cells (organs) internally. They have an extracellular, but internal to the organisms, connective structure and fluids.

Cytoplasm The main internal solution of all cells.

Periplasm The outermost region of a prokaryote cell between its inner membrane and its outer wall or membrane.

Nucleus The central unit of DNA. It is relatively simple in prokaryotes but much more complex in eukaryotes in a separate compartment.

Vesicle Any membrane-enclosed space with no nucleus.

Plasmids Small units of DNA separate from the nucleus.

Mitochondria see Organelles

Chloroplasts see Organelles

Organelles Eukaryotes capture bacteria and modify them so that they remain internal and partly functional. This may be to reduce complexity in a single

Evolution's Destiny: Co-evolving Chemistry of the Environment and Life
R. J. P. Williams and R. E. M. Rickaby

Published by the Royal Society of Chemistry, www.rsc.org

enclosed volume. They are of two kinds: mitochondria, which can obtain energy from oxygen and reduced organic matter, and chloroplasts which obtain energy directly from the Sun. The chloroplasts produce reductive metabolism and as a consequence liberate free oxygen.

Endoplasmic reticulum The large internal membrane enclosed structures which give rise to multiple forms of internal vesicles such as lysozomes for lysing proteins and peroxysomes containing oxidising enzymes.

Differentiated cells Cells of different activity derived from an 'initial' single stem cell of multicellular eukaryotes which carry out different functions although they have the same DNA. They are the original source of organs.

Species and genes The separation of organisms (species), each with different DNA (genes) (see text for details).

Geological

Ocean/hydrothermal vents Fissures of the upper crust, often near divergent plate boundaries or mid-ocean ridges which allow high-temperature reducing liquid circulating through the hot new ocean crust to enter the sea.

Black smokers Typically high-temperature hydrothermal vents with waters emanating at 350 °C. The black smoke is largely FeMn oxyhydroxides which precipitate on cooling of the fluid due to admixing with ambient seawater.

Igneous rock Rock formed from melts and distinguished from sedimentary rock formed from aqueous solutions.

Upper and lower mantle The mantle is the compositionally different layer immediately beneath the crust which passes seismic waves at a higher velocity than the crust. It is separated into two layers, upper and lower mantle, by a seismic discontinuity at ~660 km.

Crust (continental and oceanic) The outer layer of the Earth which is high in silica, more granitic, (continental crust, ~30 km thick) or lower in silica, more basaltic (oceanic crust, ~7 km thick).

Pegmatite A very coarsely crystalline intrusive rock of granitic composition.

Banded iron formation (BIF) Distinctive units of sedimentary rock that are almost always of Precambrian age. A typical BIF consists of repeated, thin layers of iron oxides, either magnetite (Fe_3O_4) or hematite (Fe_2O_3), alternating with bands of iron-poor silicates, shale and chert.

Snowball Earth Periods when there is geological evidence to suggest that glacial conditions extended to equatorial latitudes.

Subduction The process by which one tectonic plate moves under another tectonic plate, sinking as the plates converge.

Cambrian Explosion The rapid (quasi-simultaneous) emergence of multicellular and biomineralised life across multiple phyla at 542 Ma seen in the fossil record.

Mass independent isotopic fractionation Any process that acts to separate isotopes, where the amount of separation does not scale in proportion with the difference in the masses of the isotopes.

Aragonite A metastable calcium carbonate mineral, one of the two common, naturally occurring, crystal forms of $CaCO_3$ (the other being the mineral calcite). Aragonite has a low magnesium content.

Abbreviations

Ma	Millions of years ago
Ga	Gigayears ago (billions of years ago)
‰	In percentages of units per mil.
BIFs	Banded Iron Formations
REE	rare earth element

Evolution's Destiny: Co-evolving Chemistry of the Environment and Life
R. J. P. Williams and R. E. M. Rickaby

Published by the Royal Society of Chemistry, www.rsc.org

About the Authors

Professor R. J. P. Williams F.R.S., Emeritus Napier Royal Society Professor at Oxford and Fellow of Wadham College, Oxford. Professor Williams is often called the Grandfather of Biological Inorganic Chemistry, a subject which he started in 1953. He has published over 700 papers on this and related subjects and he is the coauthor of several related books, including The Biological Chemistry of the Elements. He is a medallist of several academic societies and has been awarded Honorary Degrees from Universities in the UK and abroad.

Rosalind E. M. Rickaby is a Professor in Biogeochemistry at the University of Oxford and a Fellow of Wolfson College. Her research explores the interactions of biology and chemistry in the carbon cycle throughout the geological history of the Earth. She has authored over 40 peer-reviewed articles, been awarded Outstanding Young Scientist by the EGU, and the Philip Leverhulme Prize, 2008, The Rosenstiel Medal, 2009 and the James B. Macelwane Award of the AGU, 2010.

Evolution's Destiny: Co-evolving Chemistry of the Environment and Life
R. J. P. Williams and R. E. M. Rickaby

Published by the Royal Society of Chemistry, www.rsc.org

CHAPTER 1

Outline of the Main Chemical Factors in Evolution

1.1 Introduction to the Chemistry of the Ecosystem

This chapter contains a general introduction to the multidisciplinary subject that includes chemistry, geochemistry, biochemistry and biology of the evolution of and on Earth, *i.e.* both the environment and its organisms. The book does depend heavily on chemistry so we give an outline of the principles of chemistry in this chapter for a reader who is not familiar with it as a discipline. Chemists may wish to skip quickly over Sections 1.2 to 1.6. In the minds of most scientists the evolution of organisms is based solely on organic chemicals, which quantitatively form by far the largest part of all living systems. In the book we wish to explore an additional part of this evolution, which in the first instance seems to be of little relevance to that of organisms. We refer to the early presence and the evolution of the inorganic surface of Earth, *i.e.* the atmospheric gases, the minerals and their solutions, mainly in the sea which, together, have formed the later changing environment for life. Here we consider these two parts of evolution, inorganic and organic, to be interacting in a common ecosystem. We will show that a major feature of life and its evolution, in addition to developing organic chemistry, is a changing availability and adopted essential use of selected inorganic chemical elements from this environment in cells. Many of these chemicals were dissolved from their minerals into solution (Table 1.1), increasingly by weathering, and then were taken into the cells of organisms.[1] (A cell can be looked upon as an enclosed volume of space, in part permeable to particular chemicals.) Eventually these chemicals were returned to the environment, frequently in a transformed state. These elements perform one essential role in cellular

Evolution's Destiny: Co-evolving Chemistry of the Environment and Life
R. J. P. Williams and R. E. M. Rickaby

Published by the Royal Society of Chemistry, www.rsc.org

Table 1.1 Some Minerals from Weathering and Indirect Biological Causes

Mineral	*Source*
$CaCO_3(Mg)$	Adsorption of original CO_2 by initial ions from the weathering of silicates
Mg_2SiO_4	Weathering of magnesium oxides
Fe_3O_4, Fe_2O_3	The products of oxidation of Fe^{2+} seen in Banded Iron Formations
$BaSO_4$	Due to the oxidation of sulfide

Note. There are many other minerals formed from these two causes in small quantities

catalysis – they are required to activate the small molecules, such as H_2O, H_2 and O_2, and those in some organic metabolic cellular chemical reactions. The need for them follows from the fact that, although all organic chemicals are thermodynamically unstable relative to stable CO_2, especially in the presence of the small molecules H_2O and O_2, they are generally kinetically quite stable at 20 °C. (Virtually all organic chemicals are kinetically unstable at >150 °C, particularly to hydrolysis and oxidation, implying that life has a restricted temperature range, that of liquid water, say from −10° to 150 °C.) At low temperature, 20 °C, they require energy input and catalysed activation in order to bring about synthesis, as well as catalysts for degradation. Therefore both energy and catalysts were required to activate organic chemicals before there could be any coded cellular chemistry, which we call life. The major catalytic inorganic ions are frequently strongly bound and of moderate or slow exchange rate in molecules. They are absolutely required. The essential role of other inorganic elements, which are poor catalysts, lies in their much weaker binding and fast exchange. These properties and the larger available quantities of these elements in the sea make them irreplaceable both in the management of osmotic and electrical balance of cells and in fast transfer of information, *i.e.* in message transmission necessary for balance between the several restricted paths of organic chemical change in cells. Later their fast transfer from outside to inside cells enabled organisms to respond quickly to rapid changes in their environment. The advantage of the exchange of some trace catalytic elements extended to their use in maintaining metabolic homeostasis inside cells. They also acted as controls of genetic expression in transcription factors.

A special chemical interest will be in the controlled biominerals (Table 1.2), produced by, even in, many organisms and giving rise to fossils,[2–5] as well as those made by their decomposition as deposits on the surface of Earth after death, *e.g.* the White Cliffs of Dover in the south of England and the grains of some deserts, called diatomaceous earth.[5] All these features of fossil and general biochemistry provide firm evidence of the coupled evolution of life with that of the surface of the Earth. We shall be led to propose that as well as the Darwinian random search amongst species of organisms for those of greatest survival value, associated with the small advantages of certain of them under given slowly changing environmental conditions,[†] there was and is a systematic larger-scale evolution dependent upon the opportunities which the

Table 1.2 The Major Biominerals

Mineral	*Variations*
SiO_2	$SiO_n(OH)_{2m-n}$ (n <2, m <2)?
$Ca(Mg)CO_3$	Various forms, many with impurities: calcite, aragonite
$Ca_2(OH)PO_4$	Apatite with impurities
$SrSO_4$	In Acantharia (pure)
$BaSO_4$	In a few plants
CaF_2	In shrimps

Many other biominerals are listed and described in specialist publications[5]

large-scale evolving chemical element environment provided. It is, we believe, this strong and faster environmental development, in a given chemical direction, that guided the way to today's organisms in a systematic, overall much slower, chemical evolution.[6] However, the increasing complexity ruled out the possibility that they could manage it all, especially the novel oxidation chemistry and the original reductive chemistry in one compartment. As stress increased from oxidation it became necessary to produce different types of prokaryotes, bacteria, and in succession multicompartment then also multicellular organisms and mutually dependent organisms (symbiosis). Many of their evolving changes are seen in the inorganic chemical content of different organisms.

A particular problem we wish to tackle then is the changing role of the inorganic elements both in solution and in minerals in the evolution of the ecosystem. We shall observe that it is the waste by-products of the cellular organic chemistry, particularly oxygen, which initiated relatively quickly the major changes in environmental inorganic chemistry. The timing of the changes depended on their redox potential. We shall then show that it is the back-reaction of these changes which in turn affected the evolution of organisms. The two are in an interactive feedback system. In summary we have to examine the evolution of environmental and cellular inorganic with that of cellular organic/inorganic chemistry. In doing so it is extremely helpful to follow initially the geological (inorganic) chemical record of all the minerals, especially that of sediments and their impurities. The minerals include fossils, the most clear-cut evidence of organism evolution available (see Chapter 3).[4] To do so we divide the surface minerals of Earth into four classes.[2] (i) Minerals formed without any intervention of solution or biological activity, for example on the solidification of melts, magma. (ii) Mineral sediments, formed later by weathering of rocks (see Table 1.1). (iii) Minerals which have arisen from chemical transformations

† The phrase often used in this context is 'survival of the fittest', which implies competition. All we can observe is the organisms that survived at a given time and it is difficult to know the meaning of fitness, especially as the environment at a given time is unknown. The problem is illustrated by the history of the dinosaurs. We shall observe later that as organisms evolved they became mutually dependent. This indicates that it is a total system that evolves, including the environment and organisms.

of elements in the sea and where it is release of chemicals from organisms, *e.g.* oxygen, which have caused their transformation such as oxidation of iron, giving Fe_3O_4 and Fe_2O_3 precipitates,[5] and oxidation of sulfide to precipitated sulfur or released soluble sulfate and to release trace elements (see Table 1.1). With those chemicals from weathering, they gave the trace elements typical of the sea at a given time. (iv) Biominerals where the mineral remains attached to the cell surface or which grow internally in the organisms and are easily seen in fossils (see Table 1.2).[3] Confusing the issue somewhat is the production of some of the same minerals by more than one of these routes. The history of all these geological deposits has been dated in geological periods (Table 1.3), *i.e.* when a variety of surface rocks and sediments formed (see Sections 2.5 to 2.11). We shall also use this geological table with reference to the timetable of evolution of organisms and related fossils with associated chemistry in the ecosystem. As we have already noted, making the main physical–chemical connection between these minerals and living organisms is the solubility of ions from them, especially in the sea. The limiting possible changes of the inorganic content of the sea at any time arose directly from hydrothermal interaction with basalt, from weathering, or indirectly from chemical reactions of the minerals with chemicals released by cells, and from the death of organisms. We turn to which elements are of importance in the environment and of great influence upon the nature of life and its evolution.

1.1.1 The Involvement of the Elements in Evolution

Not all the elements of the Periodic Table (Figure 1.1) are involved in evolution to any marked degree, certainly to 1900 AD. In addition to hydrogen, carbon and oxygen we shall be concerned with the major ions of

Table 1.3 Geological Periods

Period	*Date x 10^6 (yrs) ago*	
Archaen Eon	4,500 – 2,500	Earth Forms Prokaryotes
Proterozoic Eon	2,500 – 1,000	First Single-cell Eukaryotes Slow Oxygen Rise
Ediacaran	1,000 – 542	First Multi-cell Eukaryotes Next Oxygen Rise
Cambrian	542 – 488	Biominerals Explosion of Species
Ordovician	488 – 443	Vertebrates First Land Plants
Silurian	443 – 415	
Devonian	415 – 358	
Carboniferous	358 – 300	Coal Formation
Permian	300 – 252	First Extinction
Triassic	252 – 200	
Jurassic	200 – 150	Earliest Birds
Cretaceous	150 – 70	Seeds of Plants
Paleogene	70 – 25	
Neogene	25 →	Homo Sapiens

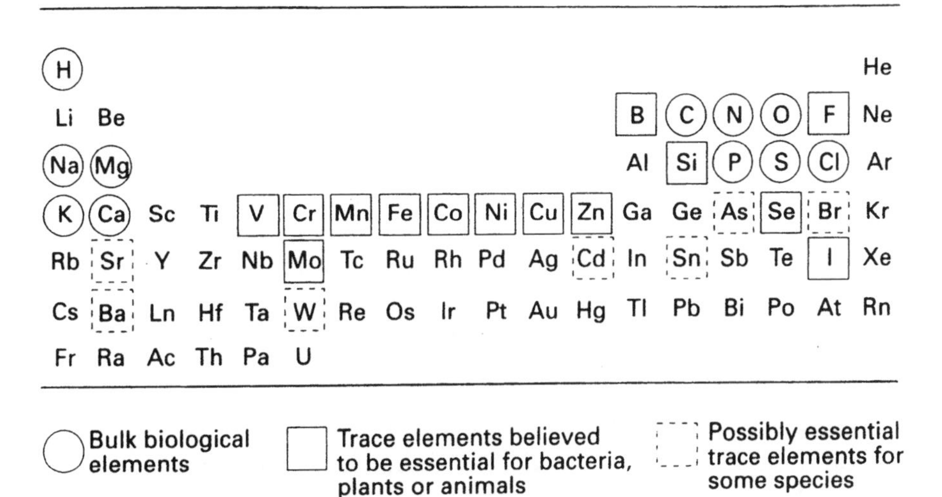

Figure 1.1 The Periodic Table indicating the elements of value in organisms.

sodium, potassium, calcium and magnesium, with the anions carbonate, silicate, sulfide (later sulfate) and phosphate in the sea, all of which are also in organisms in considerable amounts. Of these ions, carbonate and sulfide/sulfate showed the greatest changes in concentration later in time. However biological activity is also generally catalysed and controlled by small amounts of ions of several other elements from the sea such as iron, manganese, cobalt, nickel, copper, zinc, molybdenum and selenium and a few others, in particular organisms, all of which have their geological sources largely in mineral oxides (silicates) and sulfides.[1] The availability of some of these ions, found in many biological catalysts, without which there would be no life, changed with time, as seen in sediments. We know that life today depends on some 20 elements which differ, qualitatively and quantitatively, from those which were required initially, and that they all have aqueous solutions as their biological sources, which are for the most part connected to abiotic minerals. Note that very few other elements, if any, have ever been very available in the sea. But why were so many elements needed both in catalysis and in controls of cellular activity?

As we have already stressed there are two spatial parts of chemical activity of early cells which are of particularly different concern, the zone of the internal metabolism and biopolymers and that of the external surfaces, both of which have to be synthesised with the aid of different metal ions. As we shall show, from the beginning of life the use of internal specific powerful catalysts was required in order to activate in particular oxidation/reduction and hydrolytic reactions of rather inert chemicals, *e.g.* H_2, CH_4 and peptide molecules inside cells, while less powerful catalysts were needed for those reactions which occur relatively easily, *e.g.* hydrolysis of phosphates, inside and outside cells. The different concentrations of ions inside and outside cells then allowed different metal ions, mostly combined with proteins, to both catalyse and control differentially parts of both internal and external metabolism. The requirement for powerful catalysts of both acid/base and redox reactions inside cells is met by the use of some of the above transition metal ions (Fe, Cu, Zn), as they are of high electron affinity and several can change valence state readily (see Figure 1.6). Moreover several can interact with inert small molecules, such as O_2, in a specific, idiosyncratic way, so that we observe specific uses for them. The outside surfaces of cells are of molecules which, later in time, say approaching 0.54 Ga, are often selectively changed differently from those inside, again with the aid of strongly but differently active metal ions. They and/or more weakly active metal ions also stabilised these surface molecules. The weaker catalysis often, of acid/base reactions, was more generally executed by non-transition metal ions of lower electron affinity, for example Mg^{2+} and Ca^{2+}, both inside and outside cells. Lastly, bulk osmotic and electrical balance rested with bare ions of no catalytic activity (Na^+, K^+ and Cl^-), which are in maintained gradients across boundary membranes. To preserve selective action, therefore, cells came to use a considerable variety of metal ions (see Figure 1.1), many of which changed in availability and use with time. Much of this inorganic/organic chemistry is retained in today's cells, but its beginnings are obscure.

Very little if any of the selective catalytic activity of the metal ions was or is due to the bare ions but it arose from active sites, themselves selected, in proteins, enzymes, so that the inorganic chemistry has to be considered with the synthesis of binding proteins as well as with the reactions of organic molecules in cells. One illustrative telling example of the development of external catalysed cell surface reactions, giving rise to biominerals common to later organisms, will be seen to be particularly intriguing during later evolution, because the earliest living cells did not mineralise. The earliest cells left little dependable fossil record, basically only imprints. We shall take it that biomineralisation required particular organic molecules for nucleation, growth and final form.[5] They arose, relatively suddenly, at a particular time of cellular and environmental chemical change. Biomineralisation is then a signature of the evolution not just of organisms but of particular organic chemistry catalysed by special metal ions, with selected binding of other metal ions and of the oxidising strength at particular times. We shall ask what happened to the environment exactly when these special metal ions and biological mineralisation arose.

We shall also need to describe the historical development of message systems used to create and maintain control of organisation in space and in time, because at all stages of evolution both internal and external cellular activities were and are controlled by messengers. Some of these messengers are free inorganic ions (Fe^{2+}, Mg^{2+}, Ca^{2+}, Na^+ and K^+), but many are organic molecules often requiring catalysis for their synthesis. Now synthesis of the mainly metalloenzyme catalysts is under instruction from genes, coded information, controlled by other messengers. We can of course use knowledge of the evolution of genetic molecules (DNA and RNA), and their expression as proteins, to help examine all selective internal changes of organic and inorganic cellular components with time. Some of the controlling free metal ion messengers interact with proteins bound to DNA, so-called transcription factors. However genetic information is poor before 0.54 Ga and it is in the period 3.5 to 0.5 Ga when the knowledge of inorganic element changes both in the environment and in cells is most reliable in providing evolutionary markers.

All cellular activity also depends on energy sources, which undoubtedly changed with time too. Both sources of materials (elements) and energy for very early living systems and their changes are probably directly or indirectly dependent upon the mineral environment and its changes from the earliest times. We shall then describe environmental evolution first from its very inorganic beginnings (Chapter 2). We shall try to keep these observed geochemical changes separate from changes in living systems as far as possible, so as to simplify understanding, but during extensive analysis of each separately we will have to bring them together to examine the whole ecosystem from very early times. To appreciate chemical evolution, therefore, we shall have to follow the analytical, chemical content of the inorganic environment with an examination of the later changing organic chemical content of organisms and its energy capture, including the genome, the proteome, the metabolome, and the metallome.[1,6] Later all energy was from the Sun.

Because our concern is with the environment and organism chemistry and the chemicals which go between them, we need to describe the factors that are important for maintaining the states of both the inorganic and organic chemicals. The constraints on inorganic chemistry are frequently equilibria, thermodynamic relationships which are quantitatively well-defined by constants, solubility products, complex binding constants and redox potentials (see Sections 1.3 to 1.5). The constraints on organic chemistry are quite differently, overwhelmingly, kinetics, rates of reaction, controlled by energy barriers (Section 1.6). Hence many organic chemicals have to be constantly reproduced as they decay. They are energised molecules and react very slowly. They require catalysts and extra energy to change because they are in trapped forms behind energy barriers. We describe next the limiting factors in inorganic chemistry, which give us markers of evolution from geochemistry or studies of the environment and its evolution. These limitations then allow us to make a strong connection to the manner in which organic chemistry and hence organisms could evolve.

Before we begin this analysis of the chemical evolution of an ecosystem we must pay homage to the insight and well-developed theory of life's evolution due to Darwin.[7] Darwin considered that organisms evolved in an ever-branching tree (Figure 1.2a).[8] The modern tree (Figure 1.2b) has become generally accepted as being based on survival of the fittest organisms by chance exploration and exploitation of the environment as it changed.[9] The tree is one of increasing diversity of biological form but must also be in each particular branch, one of increasing chemical complexity. Darwin and more recent biologists describe all this evolution in terms of species, where a species is historically connected by inherited characteristics and today by genes. Their

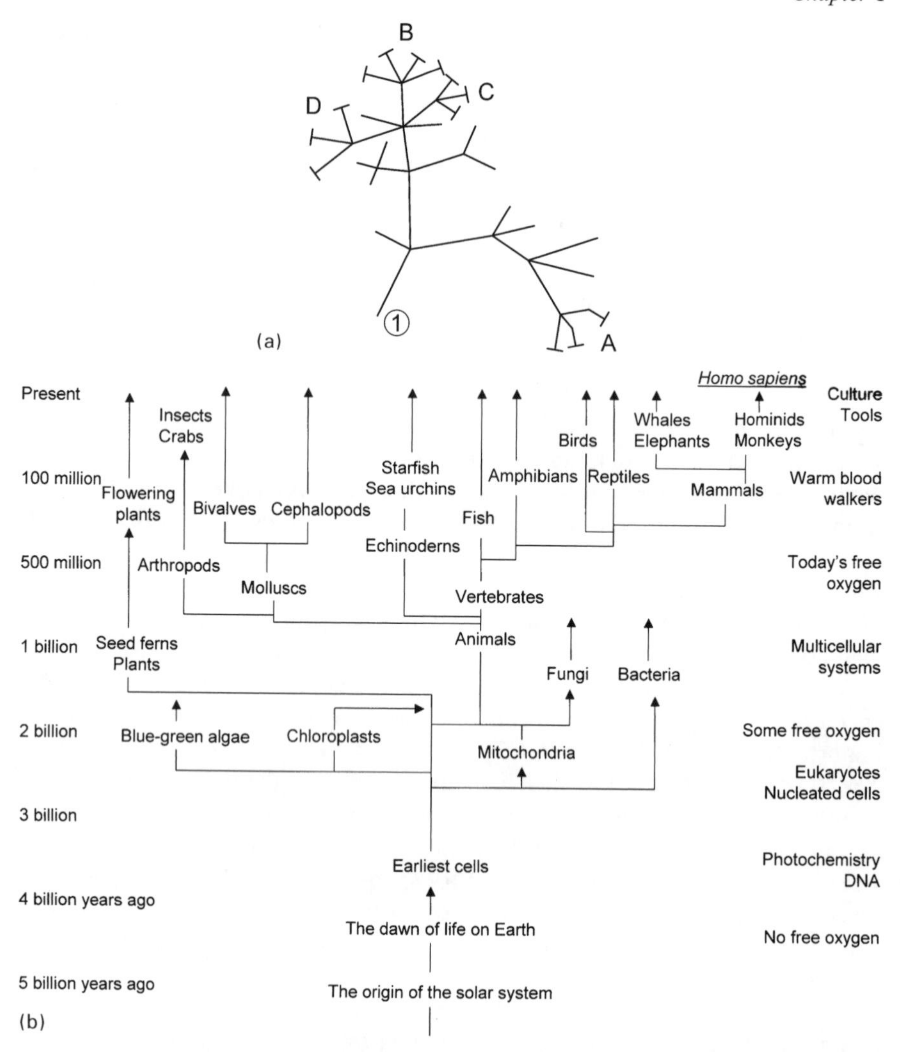

Figure 1.2 (a) Darwin's original musings on a tree of development. (b) A descriptive drawing of the modern "tree" of evolution with an outline of possible dates and of a roughly dated series of events.

discussion has been aided by the observed fossils which could be dated and we will show in Chapter 3 that this study has been greatly extended recently. Now Darwin had virtually no knowledge of the way in which organism or environmental chemistry changed. Indeed in his time there was little knowledge of chemistry. Thus by studying the chemistry of the environment and/or of organisms we can check the idea of random evolution and of an evolutionary tree while examining if it and competitive fitness are correct even in principle.* (The idea of the tree is not an absolute requirement of natural selection.) One very important point Darwin could not have known is that the environment changes which interact with life were systematic, as we will show, and he regarded all evolution to be without system. Survival of the fittest must be defined against the context of the environment, which we agree is changing systematically. When we examine the standing of evolution in the light of our knowledge of chemistry today, we shall class considerable differences between organisms not in terms of species but as one of chemical element differences in large groups of species, chemotypes.[6] Chemotypes, we will say, only arose as a consequence of systematic environmental changes, strongly implying that the dependent chemical evolution of groups of organisms has itself to be systematic, contrary to common belief. A chemotype will include many related species of organisms – genotypes. (Here we must note that a gene is related by molecular biologists to a particular stretch of DNA which is inherited and it is often assumed that therefore a species, called man for example, is completely described in its inherited characteristics by its DNA, for man the human chromosomes. In fact the chromosomes are only viable in any organism, man, with many other inherited chemical factors and even symbiotic organisms subsequent to fertilisation of a species cell. We return to the problem in Chapter 4.)

In passing it is sometimes asked if life could have arisen elsewhere. There are two separate issues. We do not have any clear idea how the great complexity of life on Earth arose. Even the very first forms of life we postulate are very complex, as we have indicated above by reference to general organism chemistry, organic and inorganic. It seems to have arisen once on Earth. Thus we do not know how to estimate the probability of it being found on another planet. Secondly we have no certain knowledge of the environmental requirements for life. What was the environment of Earth 3.5 Ga ago? We do know something of the atmosphere, rocks and sea, including the likely temperature and pressure, but we do not know if they are uniquely suitable for engendering life. Table 1.4 gives the composition of Earth in comparison with that of two other planets. There is no possibility of our kind of life on these planets. Mars is or was a better prospect, but possibly only for very primitive life. The idea that we can detect life resembling life on Earth by analyses of elements in objects, in the residues of meteorites, or on the surfaces of planets may well be misleading when we appreciate the demands of the life which have existed or do now exist on Earth. We will see that life has always required close to 20 elements in selected amounts.

In concluding this introduction we stress that our discussion of evolution is based on systematic chemistry and is quite different from other descriptions. We have indicated that the chemistry can be examined not just descriptively

---

* The phrase often used in this context is 'survival of the fittest', which implies competition. All we can observe is the organisms that survived at a given time and it is difficult to know the meaning of fitness, especially as the environment at a given time is unknown. The problem is illustrated by the history of the dinosaurs. We shall observe later that as organisms evolved they became mutually dependent. This indicates that it is a total system that evolves, including the environment and organisms.

Table 1.4 Composition of Earth, Venus and Mercury

	Earth	*Venus*	*Mercury*
H ppm	33	35	0.4
C ppm	446	468	5.1
N ppm	4.1	4.3	0.05
O %	30.1	30.9	14.4
Mg %	13.9	14.5	6.5
Al %	1.4	1.4	1.1
Si %	15.1	15.4	7.0
S %	2.9	1.6	0.2
Fe %	32.0	31.1	64.4
Ni %	1.8	1.7	3.9
Ca %	1.5	1.6	1.2

Note Mars is similar to Earth but has a smaller atmosphere like Venus. The outer large planets Jupiter, Saturn, Uranus and Neptune are gaseous. From J.W. Morgan and E. Anders Proc. Natl. Acad. Sci. USA, **77**, 1980, 6973–6977.

but within its systematic character by a quantitative approach. By systematic we imply that the chemical changes are in large part predictable, while previous analyses of evolutionary change have been described by the phrase 'random selection'. The distinction comes about through the connection between organism and inorganic chemistry rather than through gene changes. The reactions of inorganic compounds are often fast so that they proceed to the most stable condition, quantitative equilibrium. By contrast organic compounds react very slowly because they are unstable but trapped in long lifetime energised states. Hence we consider first the principles of changes of much of inorganic equilibrium chemistry separately from any approach to organic chemistry. Because the organic chemistry is linked to the inorganic chemistry it follows that any changes in organic chemistry will be led by the fast inorganic changes, especially those producing catalysts of organic reactions.

(Those readers familiar with the general principles of equilibria and kinetics may prefer to go immediately to Section 1.7 or to the last section of this chapter (Section 1.10), which is a summary of the main points of concern in this book.)

1.2 Equilibrium and Steady State Conditions

If the rates of transformation of reactants to products and those of the reverse reaction are fast enough in a given solution then the system does not store energy in C + D and is said to be at equilibrium, which we write:

$$A + B \rightleftarrows C + D$$

The position of balance, equilibrium, is temperature dependent. It cannot be alive or evolve: it is dead. The main equilibria that will concern us are the solubility products of compounds, complex ion stability constants, and the standard oxidation/reduction (redox) potentials of elements and compounds in a solution.[10] All the cases of interest are of inorganic compounds, complexes or ions. All the solution-binding equilibria are set up relatively quickly between inorganic ions and with small organic molecules, but as we have explained, organic molecules are not in equilibria with the most stable state of their simple sources, for example H_2, CO_2, N_2, nor with regard to reaction with H_2O or O_2. Not all the solids of Earth are in equilibrium with their ions either. These sources are only 'stable' in a kinetic description, meaning that they have considerable but limited time of existence. In particular, apart from the large non-equilibrium temperature change from the very centre of Earth to the surface, the rapid original cooling on Earth's formation has left the surface in part in a non-equilibrium energised chemical condition. There are, however, later sediments which we may suppose came to be close to equilibrium with concentrations of components in the sea, which are governed by solubility products, complex ion and redox reaction constants. We shall have to acknowledge that there are exceptions to these generalisations. We outline the nature of the three types of equilibria: solid/solution (solubility products), complex ion formation (stability constants) and reduction/oxidation (standard redox potentials) in the next sections. The concentrations of free individual ions are then mutually dependent on the concentration of partners in these reactions and the redox conditions in the solution.

One general difficulty with both biological and geological systems is that all the material in them is in flow. The flows in biological liquids are not all fast, so that rapid exchanges can reach equilibrium (Table 1.5). Fortunately this is true for many inorganic ionic reactions and reactions of small molecules in solution with one another and with surfaces so that we can apply equilibrium considerations to them, for example incorporation of trace elements in sediments. The flow of other geological systems extends from extremely slow diffusion and movement of such bodies as tectonic plates to the faster motion of materials from volcanic activity and of the mixing of layers of the sea. Again we can select the agents we wish to discuss in these bodies so that we know which have motions fast enough to come to equilibrium locally. Other products are entirely irreversible, *e.g.* initial formation of magma from volcanoes. In many cases in both types of system the flows are strongly, constantly energised, but mixing is fast when the conditions, which are open to analysis though with some difficulty, go towards a steady state, not an equilibrium condition. A biological

Table 1.5 Simplified Classification of Reaction Rates of Bonds

Rate	*Chemical Species*
Very slow	C – H, C – C, C – N, C – O, C – Halide, S = O
Slow	PO_4R^{2-}, – CO_2R, S – R,
Intermediate	Some complex ions e.g. of Cr^{3+} and Co^{3+}, Mn = O, Mo = O
Fast	Mg^{2+}, Ni^{2+}
Very fast	Na^+, K^+, Cl^-, Ca^{2+}, Zn^{2+}, H^+

$$
\begin{array}{l}
R{-}\overset{\overset{\large O}{\|}}{C}{-}O{-}CH_2 \\
R'{-}\underset{\underset{\large O}{\|}}{C}{-}O{-}CH{-}CH_2{-}O \\
\qquad\qquad\qquad\qquad\quad PO_2^- \\
\qquad\qquad\qquad\qquad\quad O{-}CH_2{-}CH_2{-}N^+(CH_3)_3
\end{array}
$$

(a)

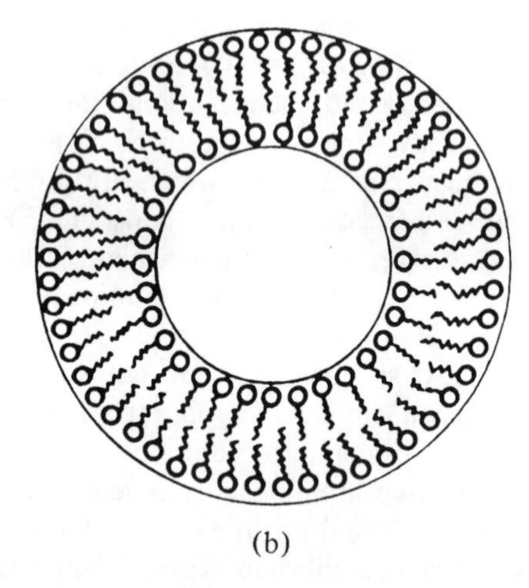

(b)

Figure 1.3 (a) The type of molecule which forms membranes with a polar headgroup and long lipid tails of $(CH_2)_n$, R and R' (b) A diagram of these molecules forming a membrane around a trapped aqueous compartment, a vesicle.

cell is of this kind and can be illustrated by two aqueous phases separated by a membrane (Figure 1.3). When the membrane has pumps for ions or molecules to the inside to which energy is applied and there is an opposed flow outwards through diffusion a steady inside/outside state condition can arise. We consider the general case of a steady state next.

A very different situation from equilibrium arises if the reactants A + B are constantly energised, say by light, and then the excited condition C + D reverts slowly. Here C + D will form disproportionately relative to the equilibrium condition and we can write a final steady condition under fixed radiation by light:

In this case, which is especially relevant to our ecosystem (in particular organisms), we shall therefore need to understand physical and chemical rates

$$\mathrm{A + B} \underset{\text{heat}}{\overset{\text{light}}{\rightleftarrows}} \mathrm{C + D}$$

of change. The system will eventually reach a steady state in which C + D concentration can greatly exceed that at equilibrium. Such a condition stores energy. Any steady state can evolve, say through further reaction between C + D and the environment, and particularly if selective catalysts are added to the system, and which affect rates of A + B and of C + D differently. We will show that the introduction of novel catalysts, which affect rates, to organic chemical reaction systems is in fact a major part of evolution. We also show that this introduction could only occur in a systematic way in evolution. A question which arises is the length of time any such steady state of our environment with life, such as that proposed under the name Gaia,[11] can survive.

There is another way in which a partial steady state can arise, which is in part totally irreversible. Consider a flow in space of A in part to C via B. Provided there is a constant input of A at a given place it will flow steadily to B, which in part gives rise to C, which then may diffuse away and leave some of B, which returns to A, giving a steady concentration of B. B will form in a concentration around the source which does not change with time. C can be looked upon as waste. A cellular system, which is cyclic inside but rejects oxygen to the outside, is of this kind. Flowing chemicals, not said to be living, can also set up such patterns, as we shall discuss. All the flows require the irreversible use of energy and we need to consider disturbances to these flows (Section 1.9).

While organic compounds generally are energised in all their compartments (Chapter 4), inorganic ions normally equilibrate in any compartment but their free ion concentrations have energised flow between compartments (Chapters 5 and 6). It is because of the speed of their reactions in a compartment that they come to quantitative equilibria. This difference makes inorganic ions of particular value in cellular chemistry and in following evolution (see Chapters 5 and 6), because they differ from those of organic compounds in speed of response. We shall also be aware that since the Sun energises the surface of the ocean and life in it there is a continuous gradient of chemicals from the top to the solid surface at the bottom of the ocean. To make discussion simpler we shall often refer to changes of the average property of the whole ocean with time. There are also geological reservoirs of different compounds, which have become frozen or heated in different places, but we shall ignore them very largely. In any energised system of many components several different steady states may be possible but with different life times or survival strengths. Here we include the possibility of self-reproduction or multiplication, which could lead to long-term dominance of particular conditions and steady states.

With this general description of the difference between equilibrium, energised steady states only irreversible in energy, and continuous irreversible flow of energy in some materials, we now expand our descriptions, looking first at the three very important kinds of equilibrium. Those readers who have difficulty with this quantitative approach may wish to go directly to Section 1.10 where the main conclusions are given. Note that a quantitative description is a thermodynamic description and differs from previous quantitative (linear) analysis of previous approaches to evolution.

1.3 Solubility

General restrictions on the availability of elements as free cations, M^+ and M^{2+} in the sea are insolubility and complex ion formation, especially reactions with anions. Insolubility in the very earliest sea may well have been due to silicate as well as sulfide, both of which are variable with temperature, weathering and, in the case of sulfide, oxidising conditions. Carbonates would not have been stable at high temperature of magma formation but only became so after water condensed. The hydroxides (oxides) are open to precipitation too but this is limited mainly to less common states, M^{3+} ion concentrations. The insolubility at equilibrium is described by equilibrium solubility products $K_S = [M^{n+}][A^{n-}]$ where A is an anion. The insolubility products of salts of the abundant metal ions, M^{2+}, is that $Ca^{2+} > Mg^{2+}$ for carbonates and phosphates but $Mg^{2+} > Ca^{2+}$ for silicates while the order is $Ba^{2+} > Sr^{2+} > Ca^{2+} > Mg^{2+}$ for sulfate insolubility. They do not form sulfides. These orders and the abundances of the elements have meant that free Mg^{2+} ions were more concentrated in the sea than Ca^{2+}. Very much later Ca^{2+} formed the major external biominerals. The general order of insolubility of salts in the series of divalent transition metal ions, M^{2+}, is

$$Cu^{2+} > Ni^{2+}, Zn^{2+} > Co^{2+} > Fe^{2+} > Mn^{2+}$$

This order holds for sulfides and oxides (Figure 1.4). A major mineral in the Earth's mantle is olivine, $Fe^{2+}Mg^{2+}SiO_4$, which formed through the abundance of Mg and Fe (see Figure 2.1). It is thought that the early sea had high Mg^{2+} and Fe^{2+} concentrations, as olivine is relatively soluble. The high Fe^{2+} reduced the amount of sulfide somewhat as the iron precipitated it also as pyrite, FeS_2, though it too is not extremely insoluble. Even so the residual sulfide greatly restricted the free ion concentrations of Cu (probably Cu^+ which has a very insoluble sulfide), Zn^{2+} and some heavier metal ions. Whereas the pyrite can enter into weathering reactions, when solubility product considerations are not so useful a guide (see Section 2.4), several of the other cation concentrations in the environment and in the consideration of biomineral formation can be usefully estimated from their solubility products. A particularly interesting feature of Figure 1.4 is the small difference in the nickel sulfide and hydroxide solubility products which is due to so-called ligand field lattice effects (see Section 2.10.2). This makes nickel of particular importance to the early life, especially in Archaea. The trace elements found in sediments are a very useful guide to the composition of seawater at any time (see Section 2.11).

The solubility of organic compounds in water is also of major concern. If we allow that before there was any life saturated C/H compounds could form then these compounds would be insoluble in water as oil, chain hydrocarbons, or gases such as methane. Some of these chain compounds could have polar end groups such as long-chain fatty acids and alcohols. In water they could form bilayers or films, which on agitation could generate bubbles with air inside or

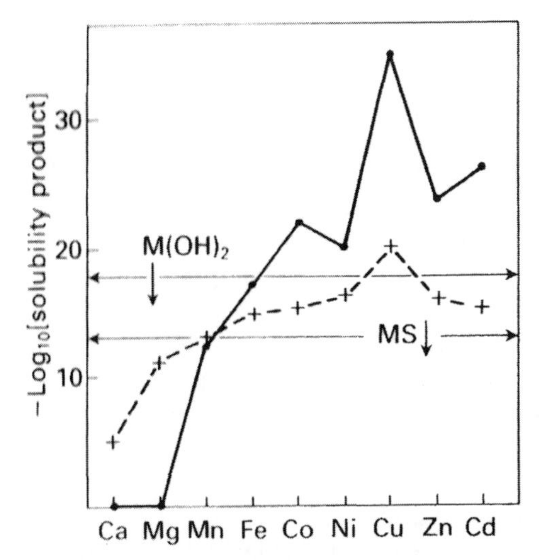

Figure 1.4 The logarithm of the solubility products of hydroxides, broken line, and sulfides, full line, of the divalent ions, M. The horizontal lines give the solubility limits of hydroxides at pH = 7 and for the sulfides at 1.0mM HS^-, for millimolar metal ion concentrations at pH = 7.0 above which the sulfide precipitates. Compare Fig. 1.5 as both reveal a general order of metal ion binding strengths. The low value of nickel sulfide relative to its hydroxide (oxide) probably led to its early relative availability even in sulfidic conditions. From Ref. 1.

vesicles with water inside (see Figure 1.3). It is these vesicles, which we take to be the initial form of the membranes which were the progenitors of cells. The organic molecules form 'liquid' mobile barriers between aqueous phases but unlike inorganic ions they form few solids. In marked contrast the formation of inorganic solids is particularly important in Earth Sciences and also has a considerable influence on biological development of more complex organisms, but somewhat curiously in organisms it is under the influence of organic chemicals (Chapters 4 and 5). There are suggestions that inorganic minerals alone could have formed the earliest membranes, but this is impossible to test.

The insolubility of inorganic ion combinations with organic molecules is also extremely important. Many calcium combinations with particularly organic compounds produce insoluble material. Hence, as we shall see, calcium has to be kept very low in all cells.

1.4 Complex Ion Formation

The hydrated free ion concentrations, availability in solution, and critical for life's evolution, are related to their combinations with ligands in the sea. Here we write the equilibrium $K = [M^{n+}][A^{n-}]/[MA]$ and note that such equilibria will hold in cells as well as in the sea as many ions react rapidly. In cells the anion A is more likely to be an organic molecule while in the sea it is an inorganic anion. Especially in the earliest times the restriction of [M] in the sea would have been especially due to the presence of hydroxide, carbonate, silicate and sulfide but later by oxyanions of stronger acids, for example sulfate. The affinity of complex formation for first four compounds, A, are closely parallel to the above insolubility of their salts. Hydroxide and oxide greatly reduced the free ion concentrations of cations with a charge of more than two. Thus at pH = 7, M^{3+} such as Al^{3+} could be held by OH^- in complexes or precipitates. However in the presence of silica Al^{3+} also forms large soluble aluminosilicates. Acidity, decreasing pH, increases free Al^{3+} concentration for example as in acid rain. The only divalent ion, M^{2+}, which may have been restricted by complex formation with aluminosilicate is nickel (see the solubility of nickel silicates in Section 2.11).[12] The later metal ions of the series Mn^{2+}, Fe^{2+}, Co^{2+}, Ni^{2+}, Cu^{2+}, Zn^{2+} formed sulfide complexes of increasing strength, 1/K, in this order but with $Cu^{2+} > Zn^{2+}$. This Irving–Williams order of binding also holds generally, but not quite universally, with organic ligands in cell compartments (Figure 1.5; compare Figure 1.4).[6] Now some of these elements can exist in more than one ionic cellular compartment. In particular iron is found as Fe^{2+} and Fe^{3+} and copper as Cu^{2+} and Cu^+ in complexes in different compartments of organisms, illustrating how different metal ion redox states in complexes can be present as well as different metal ions can be in separate spaces (in local different equilibrium). The ions Mg^{2+} and Ca^{2+} form few complexes in the sea but they have very selected partners in their complexes in cells while Na^+ and K^+ form hardly any complexes. Anions can also bind to one another but rarely, or to organic surfaces. We shall find stability constants of complexes of metal ions and organic ligands, including

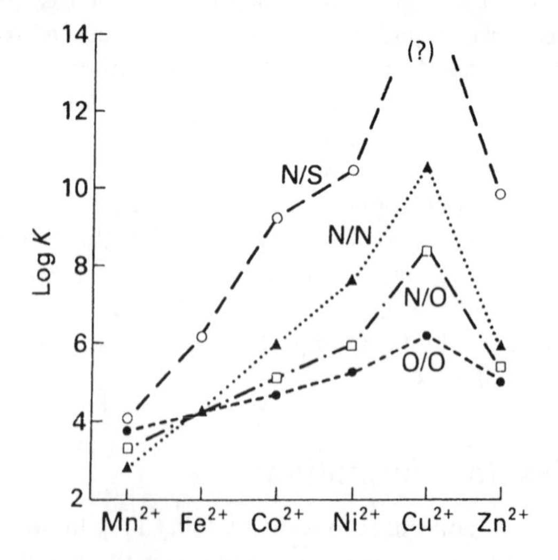

Figure 1.5 The logarithm of the stability constants of the divalent ions in combination with organic molecules which have $-S^-$, $-NH_2$, $-NH_2$ and $-CO^-$–, and $-CO_2^-$ only binding groups. The ordering is very important in general and in cellular chemistry. Compare Fig. 1.4. From Ref. 1.

the major biopolymers, especially proteins, of great value in estimating the free metal ion concentration but only in given compartments of cells. The reasons for the orders of selectivity of M for L in complexes are described fully in reference 1. Limitations due to rates of reaction are described in reference 12. Of particular interest in cell chemistry is that the selectivity of Figure 1.5 holds so that the thermodynamic equilibria of inorganic ions have greatly influenced chemistry, both insolubility and complex ion formation, in environmental and organism systems. As stated above the interaction with organic chemicals has had a great influence on the possibilities of evolution of the environment and life. We must mention too that some metal complex ion formation in cells is irreversible, *e.g.* with porphyrin ligands.

1.5 Standard Oxidation/Reduction Potentials

The major oxidation/reduction chemical changes of the surface of Earth were the uptake of carbon dioxide and nitrogen in combination with hydrogen either as H_2 from hydrogen sulfide or water into organic compounds, that is reduction (in cells), with release of sulfur or oxygen. The second case of O_2 release led to the oxidation of surface minerals and ions and molecules in solution. We shall describe the case of oxidation by 'waste' oxygen reaction, whence much of the environment changed. All this activity was and is driven by light aided by essential catalysts largely based on bound inorganic ions.

$$H_2O + CO_2 \ (+ \text{ elements}) \xrightarrow[\textit{catalysts}]{h\nu} \text{reduced elements in organic compounds} + O_2$$

These changes, small at first, gained systematically as the rate of adsorption of light by chemicals increased. The resultant non-equilibrium systematic changes in organic chemistry will be described in Chapter 4. The changes of oxidation state, of the 'inorganic' elements, at first in the environment and then especially in cell catalysts and message systems, will be described in Chapters 2, 5 and 6. The metal ion oxidation changes are relatively fast and systematic due to the equilibrium thermodynamics of oxidation/reduction even if full equilibrium is not quite achieved. We write the redox potential, E, for the equilibrium between two redox states, say M^+ and M^{2+}

$$E = E^\circ + \frac{RT}{n\Im} \ell n. [M^{2+}]/[M^+]$$

where T is the temperature, $\Im$ is the Faraday and [ ] indicates concentration. E^o, the standard redox potential for $[H^+]/[H_2]$ at pH = 0 is 0.0 V and is −0.42 V at pH = 7.0. The range of possible oxidation/reduction potentials is constrained by the dominant reactions of H_2 at first and O_2 later in water, where the O_2/H_2O standard potential at pH = 7.0 is +0.8 V. We limit ourselves immediately to the elements rather than further consideration of them incorporated into complex chemicals as the first major development from

4.5 Ga to today of the redox states is in the sea throughout the slow switch from H_2 to O_2 limiting conditions and is then readily described (Figure 1.6).[6] This is the inevitable direction of change of element evolution and occurs in a given order in the environment, which is then a control of cell chemistry too in evolution. Some elements were directly affected by change of oxidation conditions of the sea as described while other elements were not affected by the presence of hydrogen or oxygen, for example Na^+, K^-, F^-, Cl^- and HPO_4^{2-}. Further details of redox potentials are to be found in standard inorganic textbooks.[12] Weathering assists oxidation of the environment by exposing reduced compounds. We can observe and date oxidation most easily in the sedimentary precipitates, for example iron oxides and silicate and in sulfate deposits which incorporate radioactive elements or using the isotope fractionations of S (Section 2.7).

Radiation of the surface by the Sun increased the oxidising power of the atmosphere as water disproportionated, H_2 was lost or bound to carbon and O_2 was made available. Its energisation, a major subsequent part of biological chemistry, also slowly changed the oxidation condition of the equilibrated minerals of the surface and the ions in the sea as described. At the same time eruption of reduced material from deep in the Earth did from time to time reverse this process to euxinic conditions but the general oxidation direction has been preserved throughout time. Together they can introduce new elements in solution in the sea. These elements are the major driving force of evolution, they give rise to the essential novel catalysts and then reactions of

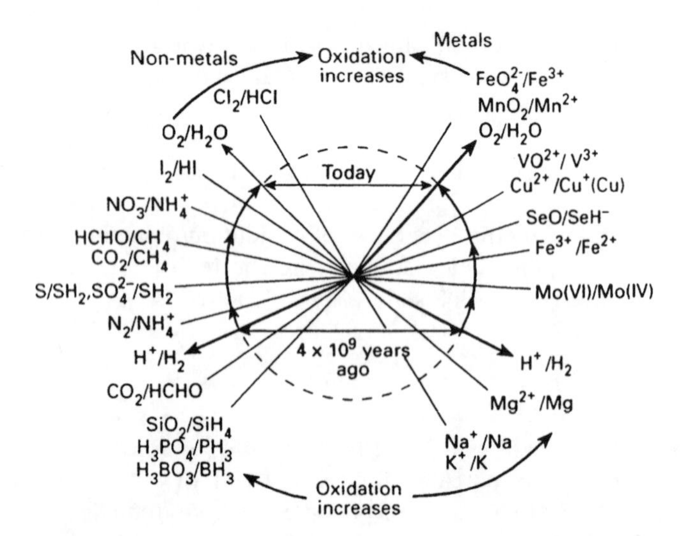

Figure 1.6 Standard redox potentials at pH = 7.0. Also indicated is the changing value of the average redox potential of the sea before life began and today. It is this slow change which drives evolution in an orderly predictable way. From Ref. 1.

cell organic chemistry. The redox behaviour of elements is also affected by insolubility and complex ion formation of different oxidation states. As noted above, perhaps the most important interaction in the environment is that of oxygen with sulfides, giving free metal ions and eventually sulfates. One approach to the control of solubility of the metal ions in the sea with time stressed this changing control exerted by sulfide as it was removed under the influence of oxidation.[1,6] An improved analysis was given by Saito *et al.*,[13] which included sulfide complex formation (see below). The overall picture of concentrations of free ions is not greatly altered but stresses two major oxidation periods (see Chapter 6). We shall have to be careful because, as stated earlier, the sea has never been entirely uniform, that is at equilibrium at all depths with the atmosphere, although it is but one phase and has no physical barriers vertically. (We shall in our account prefer to describe the redox changes as if they were without sharp large steps.) Many cations that do not undergo redox changes themselves are affected by the presence and state of anions of non-metals which are affected by oxidation. In particular we note that amongst the elements most affected little copper can be oxidised, until after iron and sulfide have been transformed, and much zinc will also not be released until these two are removed. One other element important for life that is very sensitive to sulfide oxidation is molybdenum. From the beginning of Earth carbon dioxide was increasingly reduced by release of Ca and Mg by weathering of silicates, giving insoluble $CaMgCO_3$, dolomite. This weathering continued to limit CO_2, even though oxidation of methane to CO_2 would have increased it.

In the following parts of the book we shall find standard redox potentials of considerable value in understanding these changing oxidation conditions of the environment of vesicles and extracellular fluids as opposed to the nearly fixed controlled value of the redox potential in the cytoplasm of cells. Cells have to maintain reduced conditions for synthesis, as we shall see in Chapter 4. The implication is that it is the oxidised states of organic chemicals belonging to cells but outside the cytoplasm where we can expect the greatest evolution. The organic chemicals are never in equilibrated states so that all these three equilibria, solubility, complex formation and redox potentials, make it easy to understand inorganic reactants and products but not organic compounds at given times. Remember that generally, as equilibrium is a state of no change it alone cannot explain evolution or its rate. In the next section we consider, much more qualitatively, the nature of kinetics, that is rate of change, which is critical for evolution. The very big difference is that equilibrium constants such as solubility products, complex ion formation constants and oxidation/reduction potentials are independent of time. They are dependent on temperature but not as strongly as kinetics and we shall rarely have to describe such changes in the Earth's surface. We turn next to a somewhat deeper look at simple steady states, irreversible in energy, and flow systems, irreversible in energy and material. As we explained earlier, this requires a discussion of rates of change.

1.6 Rate Controls and Catalysis

It is necessary for the purposes of this book to make the role of rates of change and catalysis in evolution very clear. It returns us, in part, to the discussion of the controls on reactions of steady states, now with great emphasis on organic chemicals (Section 1.2). We begin with simple principles. Evolution is a time-dependent process which includes all rates of events in the overall physical–chemical system associated with the Sun and the Earth. It is inevitably moving toward some ultimate balanced position, which is extremely difficult to reach, before it dies. The rate of this change is inhibited by energy barriers (Figure 1.7), but is increased by catalysts or higher temperatures. In Figure 1.8, an unstable excited state system goes to the more stable ground state over barriers. Now it can easily happen that progress to the ground state can lead from energisation of an initial condition to one in long-lasting intermediate traps as in Figure 1.8. Here we consider a ground state forced to a higher energy condition by absorption of energy such as light, and on its way back, it is held temporally but often for long periods, in an energised trap. The excited state energy of chlorophyll in Figure 4.3 avoids fluorescence or immediate energy transformation to heat loss by transferring it from an intermediate state to energised chemicals on either side of a membrane. This creates an energy gradient of extreme importance in Chapter 4.An energised trap[3] could also have arisen during rapid cooling of the initial state of very hot matter. For example even some of the distribution of the elements now present on Earth, which were formed at very high temperature in the giant stars, is an unstable state but which can hardly change at all on the very cool Earth (Figure 1.9),[14] as the energy barriers are generally too high. Changes of a few

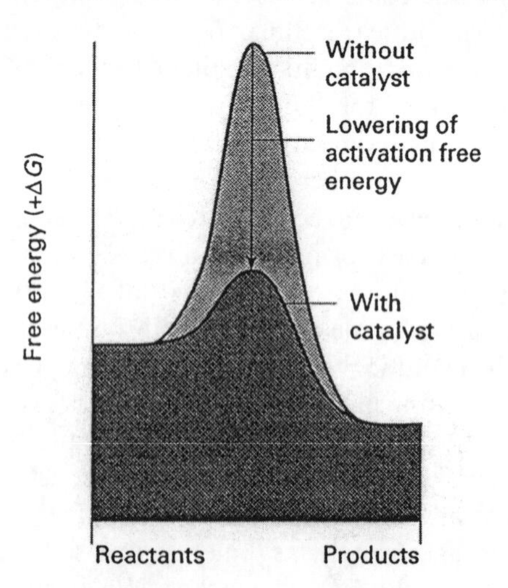

Figure 1.7 An outline of the energy barrier, +ΔG, to a chemical reaction and the reduction of it by a catalyst. Here the initial chemicals are of a higher energy (+ΔG) than the products as when a system moves toward equilibrium. The probability of crossing the barrier is related to temperature, rate α $e^{(-\Delta E/RT)}$.

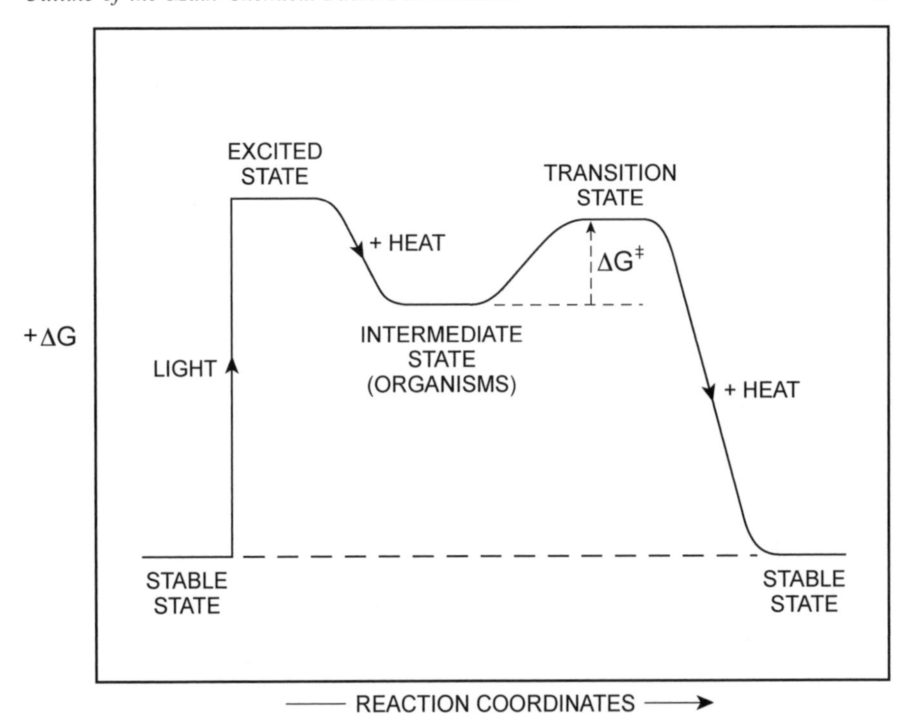

Figure 1.8 The initial (stable) state is energised for example by light and loses some energy as heat before entering an energised trap, labelled intermediate state, which can exist for a long period before crossing an activation energy barrier $\Delta G^{\ddagger}$ and decaying back to the original condition giving out excess energy as heat. The intermediate state may pass through many steps of chemical change and represents an organism.

of the heaviest elements do occur, which we observe as radioactivity. These nuclear radioactive changes are not of concern here except for the fact that the extent of them allows us to date periods of time. We are deeply concerned only with the rate of change of physical and chemical element combinations in the materials around us and in organisms. Many inorganic examples involve cooling of the still hot Earth which led to, and still causes, such events as volcanoes and earthquakes and the slow decomposition of the unstable chemicals or unstable concentrations left when Earth formed. These are processes moving towards equilibrium (Figure 1.7). All such steps require energy to get over barriers and are therefore temperature, or other energisation, dependent. Extreme cold stops organic chemical change and all heat accelerates it. Of great interest in apparent contradiction to this general direction of change toward equilibrium is the production of unstable, especially organic, chemicals so obvious in organisms and the accompanying production of unstable concentrations locally of many elements within their controlled cell volumes (Figure 1.8). This case is exemplified by the intermediate state which we have labelled 'organisms'. It is largely the energy

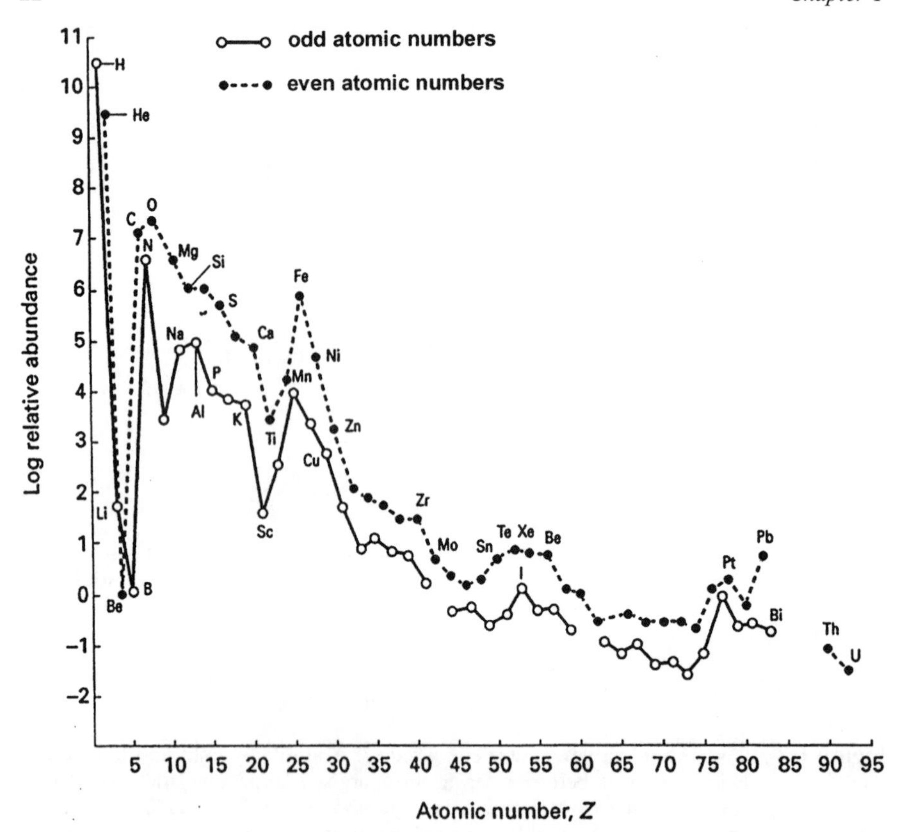

Figure 1.9 The logarithm of the abundance of the elements on Earth. It is element availability, see Chapter 2, which is important for the connection between organisms and the environment.

of the Sun absorbed by Earth's chemicals which brings about these syntheses. Their chemical condition is metastable, inevitably unstable over time. The intermediate state is not isolated and in an environment it can change by moving over barriers before returning to the original condition. This is our basic picture of the chemistry of evolution in which the overall irreversible step is that light finally goes to heat. Weathering is another not so dramatic irreversible process produced by the Sun's energisation of water and giving rise to material in solution and then sediments (Figure 1.9). The state of organic chemicals is understandable when it is remembered that the unstable conditions of chemicals is ultimately short-lived and much is in a cycle of synthesis and degradation of material which increases the overall rate of degradation of the Sun's energy into heat.[§] The energy of erosion leads to

[§] The Sun's radiation is largely of high energy quanta, light. Heat leaving the Earth is in thermal, low energy quanta. The energy of one 'particle' (quantum) of visible light gives very many particles (quanta) of thermal energy at 20°C. The increase in the number of 'particles' gives an entropy gain which is the overall direction toward equilibrium of all spontaneous change.

precipitation and further transformation also yielding stable minerals and heat. The function of a catalyst is to increase the rate of any of these processes, both synthesis and degradation, but it does so selectively in organisms. Many of the organic processes are aided by metal ions as explained above. Catalysts can then lead quite quickly to particular molecular assemblies of great survival value. We write the scheme

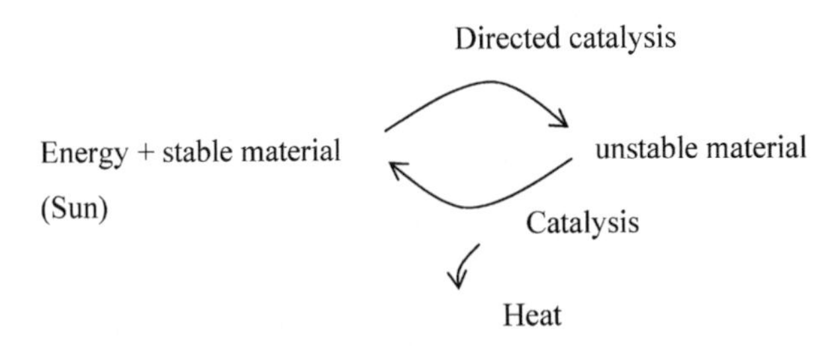

These considerations could not be central to life but for the fact that the capture of energy occurs within a limited space, a cell. We shall find it necessary to divide the unstable material into chemicals organised usefully in cells (life) and waste, energised material, irreversible loss from the cell and dispersed in the environment. The dispersed chemicals, *e.g.* oxygen, react quickly going to equilibrium while the internal cell chemicals react slowly and can be built with larger units. Both of the separated chemicals are involved in further catalysed transformations. Because the new external solutions of oxidised chemicals back-interact with the cell, processing chemicals in them, the two are coupled. These remarks make it clear that an essential feature of life is directed catalysis. We shall ask which elements were used initially and at what subsequent times other elements became involved in the organisms/ environment evolution. The answer shows that the presence of specific elements in the sea, a feature changing in time, is central to evolution.¶ It is the change of the environment therefore that is the basic cause of evolution, preventing any long-term balanced steady state. Now the multiplicity of reactions and the need for material and energy to be distributed to all of them requires balanced exchange of material and energy within cells. Much can be done by organic molecule carriers but the inorganic ions which exchange can also ensure balance by rapid transfer and equal binding. If these ions are often bound in different catalytic molecules so that catalysis is synchronised, they have very different, faster, exchange rates themselves and are then of different control value (Figure 1.10). Most of the rates of on-reactions of inorganic ions are fast as are some of their off-rates compared to organic chemical change so

¶ The importance of directed or selected catalysts is self-evident because a particular set of products is required. Selectivity is achieved through binding of reactants to defined surfaces of folded macromolecules, proteins or RNA, of defined sequences of amino acids or bases. The catalytic metal ions are held in the macromolecule frame. The control over products is clear along the chain of defined surfaces: DNA, RNA, proteins, to smaller molecules.

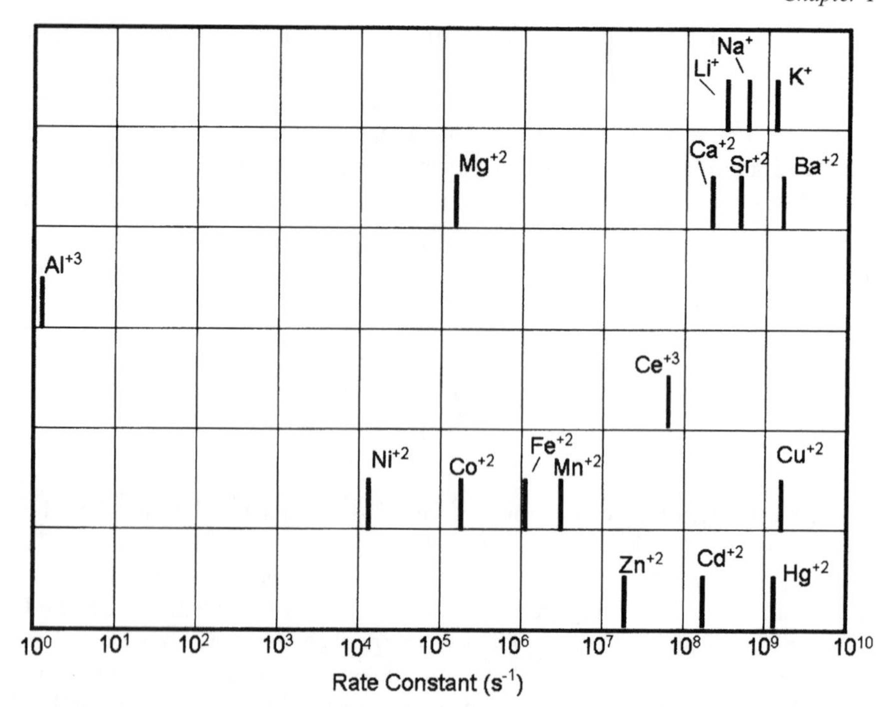

Figure 1.10 Rate constants for water exchange around a variety of cations. Note how much faster the exchange is for Ca^{2+}, Na^{+} and K^{+} than for Mg^{2+}. This difference is of great value in cellular communication. From Ref. 1. Exchange around Zn^{2+}, Cu^{2+} and Fe^{2+} is fast but it is much slower around Ni^{2+}.

that in many cases equilibrium holds. For very strong binding this is not true.[15] Oxidation state changes are also generally fast.

Catalysis does not just apply to chemical transformation of organic molecules but as mentioned before it can be applied to the activation of precipitation. There is an energy barrier to crystal nucleation and thus precipitation of for example calcium carbonates. Particular surfaces can assist the nucleation process (or even stop it) and can be selective in the crystal form chosen, for example aragonite or calcite in shells. The presence of a biomineral is indicative of a particular synthesis of an activating organic polymer, requiring a catalyst. Shape of organisms is then defined by the controlled catalysed synthesis of polymers generally but also by controlled transfer of inorganic ions.

Under the various rate controls in organisms today is that bicoded information, which manages the synthesis of the whole cellular system including the catalysts. However we have to acknowledge that catalysts of a controlled kind existed before DNA or RNA coded molecules existed. It is basic to the origin of life that the solution of elements, organic and inorganic, should be energised and catalysed into synthesis paths *before* there could be coded life. As we shall discuss in Chapter 4 we know that one particular basic organic chemical reaction scheme of proteins, nucleotides, saccharides and lipids lies at the origin of all known life but for these schemes to have arisen there must have been previous systems of high survival strength, that is high kinetic stability. Their chance of survival would have increased if on formation, the molecular products reacted further to generate bigger molecules so as to generate molecules also of high kinetic stability and with selected catalytic capacity. In effect this could give autocatalytic systems for synthesis and could be followed by a slow degradative path, see particularly the ideas of Eigen.[16] The cell could then gain more and more material until it was forced to break up and divide. The best possibility is controlled by timed accurate reproduction which brings us to DNA/RNA codes. We shall explore the use of codes in reproduction in Chapter 7.

Amongst the required catalysts many needed to have been inorganic ions to activate the very inert initial molecules on Earth such as N_2 and CH_4. Later novel catalysts could arise either by the coded synthesis of new proteins, some to bind the existing metal ions or to bind newly introduced metal ions. At least in part this is likely to be by chance exploration of code changes, mutation, but we shall show that change itself is predated by the environment changes and we need to know if they impact on a code. The environmental inorganic changes are themselves initiated by waste from cells. Once waste is formed it has a thermodynamic unavoidable direction toward equilibrium. We must ask whether is there then an unavoidable direction to energised organic chemistry which is governed by long lifetime, due to high thermodynamic stability of molecules, but is dependent on the environment.

If we are to further our understanding of evolution we have to appreciate a further energisation of the original material present when Earth formed. This is the rate of weathering: the physical excitation of seawater to clouds and the consequential movement of material by rain (Figure 1.11). (The formation of raindrops can be catalysed by salt dust, compare biomineralisation.) The resultant weathering with oxidation by oxygen (from energisation of water)

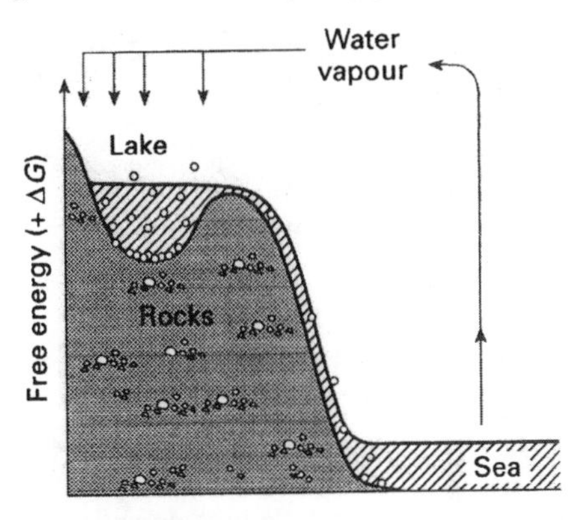

Figure 1.11 A dynamical physical system – water flow causing weathering. Note the energy barrier. Water cycles but material transfer is irreversible giving new opportunities for life later. A lake is a physical intermediate in Fig. 1.7.

increases the amounts of catalytic chemical elements in the sea. Without much further consideration we see that this input of energy without direction is under gravity control and also leads to the irreversible degradation of minerals, giving rise to sediments which provide eventually novel regions, soils, with potential catalysts, for example, for plant growth on land. It is the subsequent role of catalysts to direct chemical and physical change both selecting synthesis and the concentrations of elements and compounds. This is the essence of the evolution of the living process in which weathering becomes central. Notice that much of weathering is not in a cycle unlike the internal chemistry of organisms in steady conditions and perhaps of the chemical content of the atmosphere.

1.7 The Dangers of Catalysis

We have described the importance of directed catalysis to particular products but we need to be aware of risks of more general catalysis because of the instability of organic molecules (Figure 1.12). As an example consider the reaction of oxygen with organic molecules. The eventual products are total degradation to CO_2 and H_2O, which must be avoided, except when energy is required from, say saccharides, while cellular synthesis chemistry proceeds. Moreover, intermediate products, of initially waste oxygen reduction, such as $HO_2^{\bullet}$ (superoxide) and H_2O_2 (hydrogen peroxide) are more reactive than O_2 and must not be free or must be destroyed quickly if life is to persist. Figure 1.12[6] shows that all organic C/H/O molecules are unstable to CH_4 + CO_2 even in the absence of oxygen. Thus selectivity of synthesis has to be coupled with avoidance of destruction and removal of damaging agents, including other hazards such as excesses of metal ions. Cells also had to

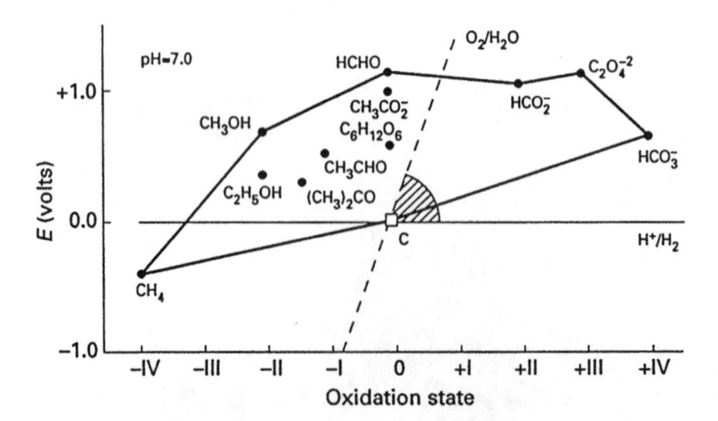

Figure 1.12 The redox potentials of simple organic molecules. Note that all C/H/O combinations give molecules which will disproportionate to CH_4 plus CO_2, they are unstable. They are all unstable to reaction with oxygen too but they are protected from both activities by kinetic barriers.

develop protection against many novel catalytic substances in the environment. However such changes can introduce novelty in evolution.

- Introduction of a novel chemical poison, *e.g.* O_2 or Cu ions, in the environment either by the necessary biosynthetic reactions themselves, oxygen is an example, or by non-biological reactions such as release of Cu from geological deposits of sulfides by oxidation.
- Protection, via destruction or removal, of said chemical poison.
- Use of the chemical in novel reaction in cells, often in catalysis, which becomes essential but now in special compartments.

These steps imply that cells evolve not just to select reagents and metal ions but to manage their concentrations in particular parts of cellular space. To this end they have energised inwardly and outwardly directed pumps. We shall see this progression in the handling of newly introduced inorganic ions, especially of more or less poisonous ones, such as copper, and others which have always been present such as calcium, sodium and chloride ions but are damaging in excess in cells. The development of use of ions, original or novel, and of novel chemicals, such as oxygen in the environment to advance evolution, is then a process of energised, selective handling of transfer or of chemical change before use arose. Concentration control as well as chemical structure is an essential feature of organisms. When we couple these necessities with that of the development of coding for the novel useful activity and reproduction of life the goal of understanding the origin of life seems extremely remote. Notice that a code does not in itself allow understanding of concentration control. Here we shall assume that the initial conditions some 4 billion years ago allowed life to happen and we ask the easier question of how did its evolution come about while it was always at risk by poisons and catalysts created by its own activity?

1.8 Diffusion

This section applies to all parts of chemistry, inorganic and organic, during evolution. A very great difficulty with the description at various times of the conditions in the sea in which gases and ions in solutions are produced in different places is the rate of diffusion. Here diffusion rate includes loss by reaction. Some of the reactants, notably oxygen, are produced near the surface of the sea, by organisms or otherwise due to irradiation, and diffuse both upwards to form an atmosphere and downward toward the solid bottom at very different rates. Only the atmosphere is mixed rapidly. The downward 'diffusion' in the sea is more a matter of rates of reaction and of mixing of water layers to which there are considerable barriers. Environmental studies indicate that oxygen in the atmosphere could well have changed quite rapidly at around 2.5 to 2.0 Ga and again at 0.8 to 0.6 Ga. During this time, even from 2.5 to 0.8 Ga, the oxygen in the sea could have remained non-uniform with low oxygen and a low redox potential in deep water merging gradually to a high oxygen and redox potential in the surface water. The surface would then have supported more aerobic while the deep held more anaerobic organisms for a long period. The extent of the regions probably changed slowly but, it is generally considered[17] that equilibration by diffusion of O_2 was limited until say 0.7 Ga. However even today the presence of large numbers of organisms on the surface can restrict the availability of O_2 and many elements there. Our problem in this book is to describe as simply as possible conditions in the sea during the whole time from about 4.5 Ga to today when we only give rough estimates at the beginning to 2.5 Ga and only reasonably good ones from 0.7 Ga to today. We shall assume for simplicity a slow change in average concentrations and of redox potential in the sea with two somewhat more rapid rises at 2.5 to 2.0 Ga and at 0.8 to 0.5 Ga. In Chapters 2 and 3 the most interesting case of limited diffusion lies in the intermediate period between these two. We shall look for markers of the changes amongst sediments.

Diffusion of gases from the atmosphere to space is more readily calculated with two important exceptions, and leads to a reasonable idea of the rate of loss of H_2, CH_4 and He. The first exception is water which meets a definite cold final condensation trap at some 5,000 metres from the sea. Gravity then ensures its return to the sea. Its flow generates weathering assisted by oxygen and organisms. The second is oxygen which as it diffuses to the outer atmosphere, is strongly exposed to UV light, giving rise to the protective ozone layer.

The opposite 'diffusion' (mixing) from the solid bottom of inorganic ions upwards is bedevilled by the temperature gradients both vertically and horizontally despite some mixing due to eddies, tides and winds, for example. Calculations of differences between regions based on rates of formation and loss of materials as well as of diffusion have been made.[17]

Diffusion of low concentrations of ions or molecules in cells can be very slow, hence rates of transfer of low concentrations to give activity at a distant site requires enhanced transport. In the case of copper ions for which binding reduces the free ion to very low levels (see Figure 1.5), we shall see that carrier proteins which exchange relatively quickly aid transfer to targets which recognise the carrier. This is a very general mode for transfer for many other ions and larger organic chemicals. The carriers for small organic chemical units are frequently coenzymes, small organic molecules which carry energy too, rather than proteins.

A particularly important consideration is the combination of reaction with diffusion. The basic concept was introduced by Turing in 1952.[18] Reaction and diffusion can come into a steady state and we shall use the general notion in the text and several figures but we shall not attempt a mathematical analysis of such feedback systems. It is believed to play a large part in morphogenesis.

1.9 Irreversibility, Chaos and Predictability

Many events have a predictable nature such as the general effect of weathering due to the circulation of water (see Figure 1.11). Although the process is generally irreversible, not at equilibrium, its year-on-year activity is in weather cycles. We experience this ourselves as we live through cold, wet, hot and dry years of summer and winter. On a larger scale the Earth has suffered climatic fluctuations over longer periods of time such as Snowball Earth events when the surface is largely covered in ice and other periods when ice virtually disappears over as long a period as several million years. Other fluctuations in weather over a shorter timescale are caused by volcanoes which can affect Earth's climate for several years. All such events have predictable consequences. Now as Belousev[19] and Lorenz[20] pointed out, although we may write equations which will indicate why these changes occur and even their long-term frequency, timing of them is quite unpredictable. The problem arises because a seemingly very insignificant event can be coupled by feedback to an ever expanding set of consequences. The usual quote is that the flapping of a butterfly's wings in one part of Earth could cause a storm thousands of miles away. These chaotic occurrences can then affect the evolution of life which has an irreversible feedback interaction with the environment. We shall draw attention to periods of large-scale unpredictable extinctions over relatively short time periods. It is expected that they could be at least partially self-correcting by feedback. In this book, however, we are little concerned with such fluctuations though they do affect the history of species, for example the disappearance of dinosaurs. We are concerned with the very long-term periods of evolution which we shall show are due to the overall degradation of energy on Earth as in the Universe, and locally that of the Sun, to low energy heat. In this process life is a rate-increasing set of chemicals as they absorb the light, becoming unstable reduced chemical bodies and then decay while they consequently produce continuous oxidation of the environment. The very nature of the overall process we shall show is a predictable general change of chemistry much though it is open to chaotic fluctuations in both very short and longer periods.

1.10 Summary

This book sets out to describe and analyse the nature of the chemistry on Earth's surface in inorganic atmospheric, aqueous and solid mineral phases and in organisms. Conventional wisdom keeps the two separate as inorganic and organic chemistries. We shall show that they are in fact in a strongly interactive system. In this first chapter we have introduced the major chemical elements of concern which are interactive between the two. They are given in Figure 1.1. The so-called inorganic elements, frequently metals, differ from the non-metals especially those making up the bulk of organism organic chemistry, H, C, N and O, in that we can use reversible equilibria to describe their behaviour in any particular local volume in both the environment and in organisms. The major equilibria of interest are solubility products, complex formation, and oxidation/reduction redox potentials, described in Sections 1.3–1.5. In many of their interactions between the inorganic ions themselves or with organic chemical partners, equilibria are therefore characterised by constants giving numerical value to availability. Such constants apply in organisms as well as in the environment and lead to predictable consequences. The information is of value for different conditions affecting the availability of different elements and is then of immediate use in describing evolution of the whole system, which we shall say follows predictable conditions as availability changes.

The treatment of some inorganic element reactions and the vast majority of those of organic elements, non-metals, and compounds must be very different as they are energised. The organic chemical changes are energetically irreversible and limited by barriers to rates of change such that the major components of life are far from in equilibria. They are in energised states of considerable kinetic stability as are the waste environment chemicals from their reactive organisms. We have indicated the nature of the barriers to the activation of especially non-metal element reactions and of the barriers (see Figure 1.8), which are also important in weathering (see Figure 1.11).

There are strong feedback components between the environment and organisms in this chemistry which produces some predictable and some unpredictable behaviour, against the general trend of predictable increase in oxidative chemistry. Our view is that the unpredictable chaotic activity is not dominant in the overall predictability of evolution and we treat chaos mostly under fluctuations.

In the chapters that follow we shall outline the chemistry of evolution as follows. Earth had a long period in which life was a relatively small contribution to much of the chemistry. In fact there was no life for some 1 billion years. These 'inorganic' changes of the environment have been continuous for all the years of the planet's existence. We describe them in Chapter 2, often referring to 'weathering'. In Chapter 3 we look for the evidence of the chemistry of life of different kinds and stages of development by looking at the firmly dated discoveries of fossils. We know that much of the basis of life lies in 'organic' chemistry of H, C, N and O with some P and S which create the background to these fossils but it is extremely important to note that fossils arose at a given time after much environmental chemical change.

We have to admit that we cannot follow life's evolution directly as organisms have decayed and we have to use our knowledge of today's organisms, of fossils, and of much inferential judgement. We do not wish to use the knowledge of genetics to any great degree in Chapter 4 where we discuss this organic chemistry as simply as possible and elsewhere in the book, because we wish to describe evolution in strictly chemical terms. We consider that coded genetics could only arise after the basic organic chemicals of life had evolved. Chapter 5 returns us to the chemistry of the bulk 'inorganic' elements involved in both the environmental and cellular chemistry, especially that of Na, K, Mg, Ca and Cl. Much of it concerns the stability of cells in the sea, the way information reaches cells from the environment by fast reactions coming to equilibrium in given compartments and communication between cells in later organisms. Chapter 6 introduces the essential functions of trace elements, such as Fe and Cu, which are the major essential catalysts of the difficult reactions in cells of small molecules of H, C, N and O from the environment. There we demonstrate clearly the vital link between the changing inorganic chemistry of the environment, at close to equilibrium, with the evolution of the mixed inorganic/organic chemistry of life. These elements are also directly in the control of cellular processes. In Chapter 7 we bring all the chemistry together in an effort to demonstrate its predictability. We look for these possible links to genetics and the findings of molecular biology. Throughout we keep in mind the 'tree of evolution' proposed by Darwin (Figure 1.2) and we shall put it in the context of today's knowledge of the evolution of chemistry in the total system.

References

1. J. J. R. Fraústo da Silva and R. J. P. Williams, *The Biological Chemistry of the Elements*, Oxford University Press, Oxford, 2001.
2. H. Blatt and R. J. Tracy, *Petrology: Igneous, Sedimentary and Metamorphic*, Freeman, New York, 1994.
3. A. Veis in *Reviews in Mineralogy and Geochemistry, Volume 54, Biomineralization*, ed. P. M. Dove, J. J. De Yoreo and S. Weiner, Mineralogical Society of America and the Geochemical Society, Washington, , DC, 2003, p. 249.
4. M. J. Benton, *The Fossil Record 2*, Chapman and Hall, London, 1993.
5. S. Weiner and P. D. Dove in *Reviews in Mineralogy and Geochemistry, Volume 54, Biomineralization*, ed. P. M. Dove, J. J. De Yoreo and S. Weiner, Mineralogical Society of America and the Geochemical Society, Washington, , DC, 2003, p. 1.
6. R. J. P. Williams and J. J. R. Fraústo da Silva, *The Chemistry of Evolution*, Elsevier, Amsterdam, 2006.
7. C. Darwin, *On the Origin of Species by Means of Natural Selection*, John Murray, London, 1859 and 1872.
8. C. Darwin drew a branching tree-like structure in one of his later notebooks, see ref.7.
9. T. Cavalier-Smith, M. Brasier and T. M. Embley, *Philos. Trans. R. Soc., B*, 2006, **361**, 845.
10. P. W. Atkins, *Physical Chemistry*, 6th edn, Oxford University Press, Oxford, 1998.
11. J. Lovelock, *The Ages of Gaia*, Oxford University Press, Oxford, 1988.
12. C. S. G. Phillips and R. J. P. Williams, *Inorganic Chemistry*, Volumes 1 and 2, Oxford University Press, Oxford, 1966.
13. M. A. Saito, D. M. Sigman and F. M. M. Morel, *Inorg. Chim. Acta*, 2003, **356**, 308.
14. P. A. Cox, *The Elements on Earth*, Oxford University Press, Oxford, 1995.
15. P. Atkins, T. Overton, J. Rourke and M. Weller, *Inorganic Chemistry*, Oxford University Press, Oxford, 2009.
16. M. Eigen, and P. Schuster, *Naturwissenschaften*, 1981, **65**, 7 and 341.
17. D. E. Canfield, *Nature*, 1998, **396**, 450.
18. A. Turing, *Philos. Trans. R. Soc., B*, 1952, **237**, 37.
19. B. Belousev, P. Glandsdorff and I. Prigigone, *Thermodynamic Theory of Structure, Stability and Fluctuations*, Wiley Interscience, London, 1971.
20. E. N. Lorenz, *J. Atmos. Sci.*, 2005, **62**, 1574.

CHAPTER 2

Geological Evolution with Some Biological Intervention

2.1 Introduction

Before reading this chapter we must make the reader aware that the geochemical record has only recently become substantial. We have therefore selected a consensus view of its history and where there is uncertainty we have backed views by strong evidence. We believe the main argument is well grounded: there has been an overall increasing oxidation of the environment from before 3 Ga to at least 0.5 Ga. We also know that fluctuations on a considerable scale are frequent but we have set them aside so that we can consider the overall directional environmental geochemical change which progressed at a variable rate but inexorably. We have done our best to illustrate the position we have taken but we accept there may be one or two mistaken impressions.

Chapter 1 introduced all the principles influencing the chemistry of the environment/organisms system. This chapter will use these principles to analyse first the physical conditions from 4.5 Ga to today (Section 2.2), and then the chemical conditions of Earth and their changes before life formed in 3.5 Ga (Section 2.4 and subsequent sections). In a very brief Section 2.3, between them we outline the modern geochemical use of isotopes in dating and which we show are of the greatest importance for the study of chemical evolution but are little known to most chemists. Returning to Section 2.2 there were many physical changes on Earth, several of which continue to today, and they helped to distribute the chemical elements. One important part of physical change was weathering by the energised flows of air and water around the Earth due to the Sun's radiation. A very different part was the flow of melts, of

Evolution's Destiny: Co-evolving Chemistry of the Environment and Life
R. J. P. Williams and R. E. M. Rickaby

Published by the Royal Society of Chemistry, www.rsc.org

volcanoes and of surface solid plates on the flowing phases beneath them. A separate change, now chemical, was the energisation of chemicals, mainly by the Sun, giving disproportionation into reduced and oxidised compounds, mainly of H_2S and H_2O (see Chapter 4). We treat these three topics in more detail in Section 2.2. We leave much of the connected organic chemistry to one side until Chapter 4, as it is largely confined to organisms but we have to include its simple abiotic beginnings in the sea from CO_2 and simple inorganic sources and the later effect of O_2 released to the environment by organisms. The combination of these energised physical and chemical changes became the major driving force of the evolution of the combined environment/organism system. We shall divide the inorganic geochemical changes into those of major mineral elements and those of the trace elements. It is the dated sedimentary minerals of the bulk elements and their trace elements which provide us with firm direct knowledge of the evolution of the environment (Sections 2.6 to 2.10). Throughout the chapter we can apply the features of the three controlling chemical equilibria reactions: redox potentials of metals, solubility constants of compounds, and the strength of ligand binding, all given in Chapter 1, to understand the evolving chemistry of element availability in the environment during weathering and oxidation. We note that while many inorganic reactions come rapidly into equilibrium in the sea, the sea itself was not always a well-mixed homogeneous phase. Organic chemistry, though often controlled by the influence of these equilibria, is not limited by them (Chapter 4).

2.2 Physical Evolution from the Earliest Times to Today

Physical evolution *per se* is not of great importance in this book. However we must inform readers, especially chemists and biochemists, of it as it is linked to certain chemical changes through the ages.

When Earth first formed it was a hot mass of homogeneous material which remelted and partitioned into layers (Figure 2.1).[1] Even today the interior of a metallic core and a hot lower mantle remain and have evolved only slightly. Interest in this book will therefore be linked to the upper mantle, the crust and the seas because they have evolved very considerably. The initial mineral evolution of Earth's outer layers depended on a sequence of geochemical and petrologic processes, including volcanism and degassing, fractional crystallisation, crystal settling, assimilation reactions, regional and contact metamorphism, plate tectonics, and associated large-scale fluid–rock interactions, much of which we cover only briefly. The upper mantle was hotter than the crust at about 300°K. It underwent partial melting to form the dense crust of the ocean floor and further fractional crystallisation led to the first chemically different low-density continents with their associated near-surface granitoids and pegmatites, hydrothermal ore deposits, evaporites, and zones of surface weathering. We refer to this as land as opposed to the bottom of the sea and

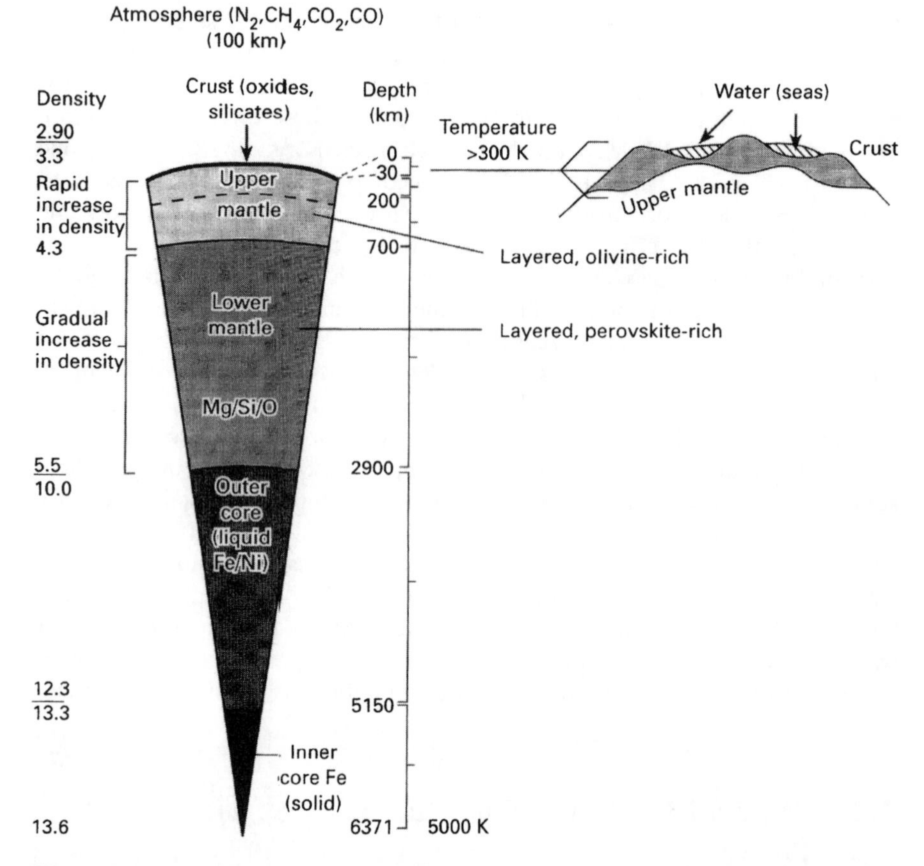

Figure 2.1 The formation of the layers of the Earth. Notice the crust and ocean on the right.

the sea itself. The land floated on the crust of the ocean floor. The minerals of the land, continental zones, were, and still are, rich in silica (60%) and therefore are called granitic while the sea bed was made of basalt, comprising a mixture of minerals with a lower silica content (40%) and some heavier oxides, including olivine, $(MgFe)_2SiO_2$. It is the high olivine content which enriched the early seas in Mg^{2+} and Fe^{2+} (Figure 2.1), both of which are necessary for life.

Boundaries within the ocean crust divided distinct tectonic plates, which effectively floated on the hotter mantle and drifted into collision, either between plates or between the more dense ocean floor and the less dense continents. Following collision the oceanic plate subducted under the lighter continental plate leading to volcanism (Figure 2.2). The further consequences of continental collisions were the raising of mountain chains on the continental zone. Water which percolated in convection paths through the ocean crust reacted with the newly formed hot oceanic crust at spreading centres between

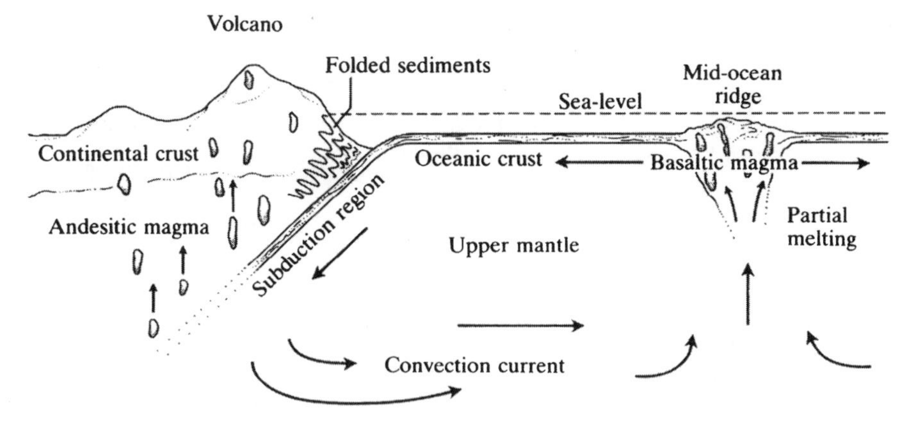

Figure 2.2 The splitting of the crust into the continental and oceanic parts showing subduction and hydrothermal activity.

two ocean plates defined by the mid-ocean ridges, and led to hydrothermal activity and black smokers (Figure 2.2). The smokers ejected reduced chemicals into the sea and both they and other hydrothermal actions gave deposits and supplied chemicals to the seawater. We shall treat these topics of the land and sea floor with spreading under weathering, which also gave rise to sediments noted in several of the following sections. Effectively then there developed three chemically different layers: the continental crust, the basalt overlaying the mantle and the sediment.[2] We shall highlight in this chapter the elements concerning evolution using quantitative isotopic analysis of sediments at different times, pertinent to the central hypothesis developed in this book. Meanwhile the land constantly joined and split though the exact land masses at a given time are hard to reconstruct (see below).

We can consider most of these inorganic physical–chemical events as the inevitable effects of cooling towards thermodynamic equilibrium endpoints but with some kinetic limitations, which continue until today (Figure 2.3). The input from the Sun is different, moving material away from equilibrium on Earth. Weathering is energised to some degree by this source, the Sun.

In the earliest period of Earth's existence, from 4.6 to some 3.5 Ga, there was less land (continental crust) than today. Life in the sea began by some 3.5 Ga but on land it seems to have developed seriously much later, only perhaps after 0.50 Ga, and earlier it must have been a place where life's chemistry could hardly survive. Much mineral sediment deposition and its decay of early life therefore occurred in the sea and hence considerable interest in this book concerns the sea. Not only must we observe which chemicals were likely to form as Earth cooled and seas developed but we must note that the solubility of minerals in the sea and the stability of some of them are both temperature-dependent, usually much less so than the stability of organic molecules. It is important therefore to understand the reason for the approximate stability of

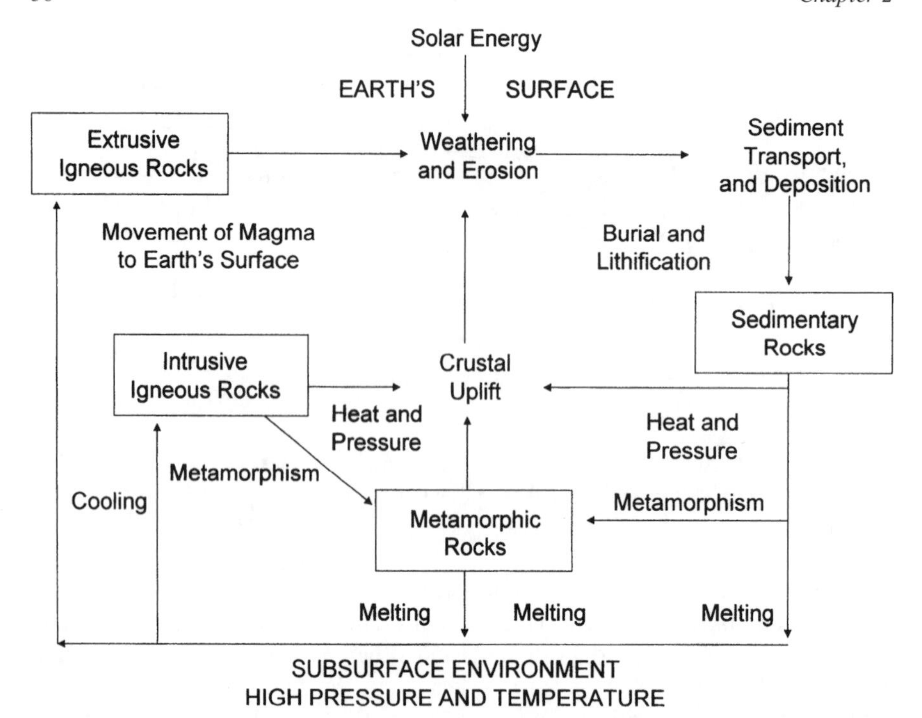

Figure 2.3 A general scheme of the movements of the top levels of the Earth due to their energisation by the energy input, the temperature gradient and gravity. Much of the system is going slowly to equilibrium but note the sun's input.

the temperature of the surface, established some 4.0 Ga, before looking at Earth's chemistry changes.

It is considered that Earth's temperature cooled until water, much of which was imported by collisions with comets, condensed to a very large degree, and then fluctuated by up to some 50°C from very cold to very hot conditions but always in the range of say −10°C to +90°C. Certainly Earth did not boil or freeze entirely from 3.5 Ga to today but smaller variations did have some chemical effects. Two factors mitigated against the gradual, further cooling of the surface, which would have prevented reasonable reactivity of organic molecules. The first was heat given to it from the huge hot reservoir of Earth's centre through the mantle together with the fluctuating volcanic activity at the edges of the plate-like structures of Earth.[1] The second one was the Sun's radiation which inevitably energised the surface water and atmosphere. The amount of energy retained from the Sun was dependent on the concentration of greenhouse gases, such as CO_2 and CH_4, and on the total adsorption and reflectance of different areas of the Earth's surface. Fortunately the Sun increased somewhat in energy output in the period 4.5 to 3.5 Ga, which compensated for the loss of CO_2, due to weathering and calcium carbonate formation and for the loss of CH_4 from the atmosphere to space. Overall the

Earth's surface became of a relatively fixed temperature to today. A rough and ready 'thermostat' was maintained by the compensation of greater heat loss from the surface by the extra adsorption of energy from the Sun and its capture within the atmosphere. We stress that this thermostat makes for a fixed set of chemical equilibrium constants (Sections 1.2 to 1.4). However, as noted, there were considerable fluctuations, due for example to periods of intense volcanic activity and those which gave rise to Ice Ages (see Table 3.4), which affected the absorption and reflection of the Sun's energy differently and to which we turn later. Snowball Earth conditions occurred between 2.4 and 2.1 Ga and around 750, 710, 645 and 570 Ma,[2] when considerable areas of Earth were covered by ice and snow.

Just as there have been very cold periods, as illustrated by the Snowball Earth conditions with average temperatures below 0°C, so there have been very warm periods mostly due to periodically increased greenhouse gases. Leaving aside the earliest times when the surface cooled from above 100°C to around the average of 15–20°C there have been periods when the average temperature was well above 20°C. At such times there would have been little ice anywhere and the sea levels would have risen considerably, even by tens of metres. A few degrees change in average temperature has a huge effect on the sea levels and some biological organisms which today causes increasing alarm about global warming. Judging by such levels the sea was warmer than it is today, for example between 500–360 and 250–50 Ma. In the very long term physical–chemical consideration of the evolution of the ecosystem such changes are but fluctuations which have to be taken together with other natural disturbances such as are brought about by tectonic shifts with volcanoes and earthquakes (see Section 1.9). They can have relatively small effects on overall long-term chemical and biological evolution.

During the period from 4.0 Ga the continents have increased in area due to further partial melting and crystallisation of the ocean crust and the original continents alternately started to break up and coalesce. Super-continents formed now and then when all the continents merged. Their more recent history can be mapped. They are known as Nuna from 2000 to 1800 Ma, Rodinia from 1100 to 750 Ma,[2] and Gondwanaland from about 600 to 540 Ma (Figure 2.4). Gondwanaland broke up and then in about 300 Ma Pangaea formed but later it too split so that from 300 to 150 Ma there was considerable volcanic and tectonic activity. We draw special attention to the later periods from 1000 to 200 Ma when the rise of many organisms, including most mineralised examples, took place, and which is of special interest in this book (see Table 1.2). The greatest volcanic activity we may assume was at a very early period before 3.5 Ga. We must be aware that geochemical findings in a given region of land today could have been formed in a very different part of the globe at the particular time of their formation.

Apart from large-scale movements of solid mineral and melts an additional cause of physical change, mentioned earlier, was due to the energisation and flow of air and water from the sea, one form of weathering. These flows caused

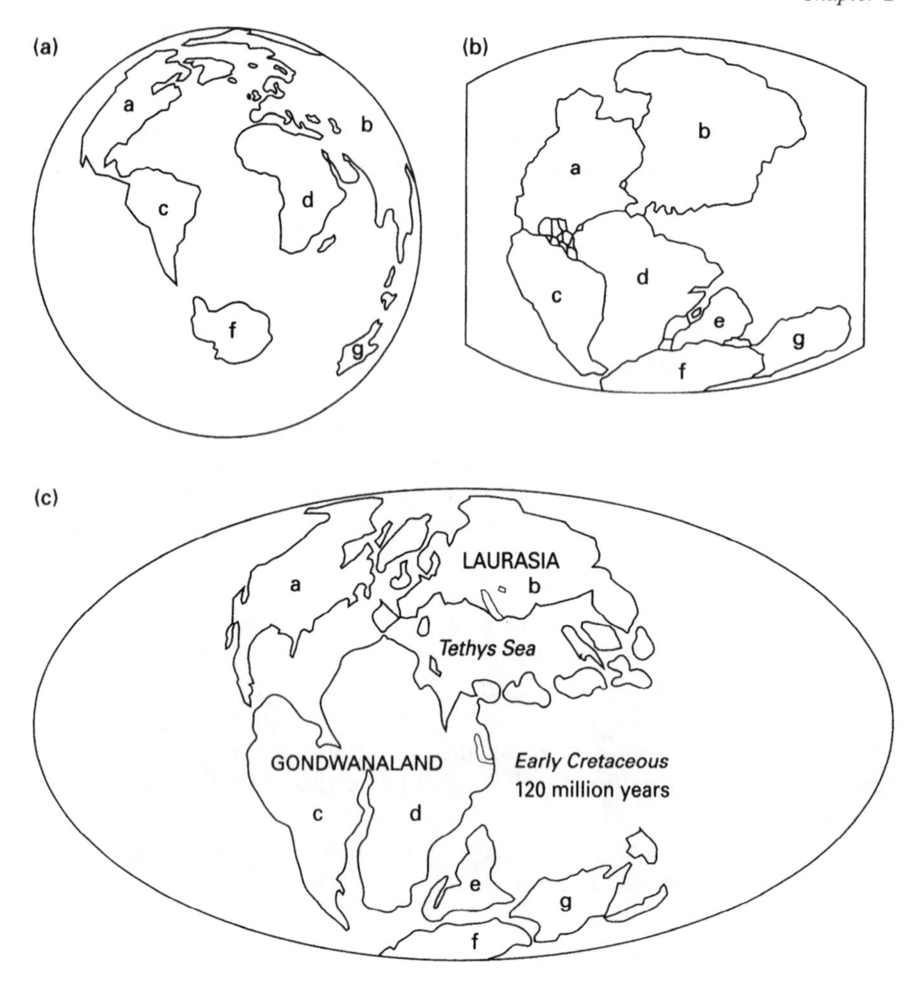

Figure 2.4 The movement of the continents from a single supercontinent (b) to (c) and then (a) as it is approximately today.

erosion, especially of the land masses, and movement of land toward the sea, giving sediments and changes to the chemistry of the sea. Sediments accumulated around edges of land masses and later some became soil which is of great chemical potential for life. We have already described carbonate formation and a further deposit was that of silica and silicates. It is therefore important to introduce the chemicals in the flows between reservoirs not only due to those carried in air and water but to solutions from the rocks. We shall describe the original inorganic chemicals of Earth's surface in Section 2.4, separately from the later intermediate conditions with oxygen increasingly present until about 0.5 Ga. We consider that much of the physical beginnings of evolution are downhill energetically and predictable in outline.

In Sections 2.4 and 2.5 we describe too the chemical rather than the physical effects of energisation though they are often linked and how they led to the beginnings of organic chemistry in anaerobic conditions before 3.5 Ga and hence to the initial period of anaerobic life in cells in 3.5 to 3.0 Ga. Sections 2.6 to 2.11 give the evidence for the major changes of Earth environmental chemistry as oxidation increased. After about 0.5 Ga it is considered that there was little further gross oxygen increase or chemical change, so we discuss how modern conditions arose with fluctuations including the story of weathering from 0.54 Ga in Section 2.12.

Before we describe the way the chemical element combinations formed the physical phases we must be aware that the elements themselves (see the Periodic Table in Figure 1.1) are mixtures of chemically very similar atoms of different atomic weight called isotopes (Section 2.3). Although the isotopes and their changes do not affect chemistry to any degree, they are excellent markers of chemical change. Knowledge of changes in isotope composition of an element allows us to date many of the important changes in the element's chemistry. Changes in isotopes of heavy as well as light elements will again and again be referred to in this book.

2.3 The Value of Isotope Studies: Indicators of Chemical Changes and Geochemical Dates

An important property of isotopes (atoms with nuclei of different mass) is that they are not all stable and their slow radioactive decays, especially those of the heaviest elements, to stable atoms, radiogenic isotopes, provide a means by which we can date the history of the time of formation of chemical compounds from the abundance of isotopes in decomposition products, for example of carbon.[3] We return to particular cases in Section 2.7.

By contrast to these radioactive isotopes, there also exist stable isotopes of elements.[3] Stable isotope fractionation occurs in chemical reactions and physical phase changes and measurement of the resultant isotopic fractionation recorded in geological materials provides invaluable insight into past climates, environmental chemistry and life processes.

Initially, gas-source mass spectrometers were developed which allowed only resolution of the distribution of the isotopes of the 'light elements', namely carbon, oxygen and sulfur, which we use in several sections. They have provided extremely valuable data on chemical change at selected times. It has only been in the last 30 years, with the development of high-resolution plasma source mass spectrometry that many heavy element isotope fractionations have been revealed which act as aids to dating and understanding chemical change. Examples we use later are the isotopes of molybdenum, of the rare earths, for example neodymium, and of rhenium, especially to follow the evolving increase of oxidising conditions after 3.5 Ga (Section 2.6).

2.4 The Early Chemical Development of the Environment before 3.0 Ga

We have described the major physical phase separation of the gaseous atmosphere, the liquid aqueous sea and the solid ocean floor, before and after land appeared. In addition we need to know their chemical character.[4] The main control over the initial chemical composition of the surface was the abundance of the elements (see Figure 1.9). As a consequence of their combinations the early atmosphere was reducing, containing no or very little oxygen, but much N_2, CO and CO_2 (much greater than today) and some methane, ammonia, hydrogen and hydrogen sulfide. Light gases such as H_2, NH_3 and CH_4 were largely lost from the atmosphere to be replaced later by N_2, CO_2 and oxygen (dependent on life) but note that there is still much buried methane (and oil and coal later). Much of the free CO_2 may have been lost as carbonate before 3.5 Ga but it continued to fall almost to 0.54 Ga with a resultant decrease in bicarbonate in the sea. We summarise the major chemical changes of the atmosphere with time in Figure 2.5 before organisms formed.

Most of the minerals were overwhelmingly formed from sulfur and oxygen, including silicate, bound with the heavy metal elements from Earth's origin. Details of the probable nature of the major minerals are given in the references.[1,2] We shall be mainly interested in metal oxides, silicates, carbonates and sulfides, some from the original mantle but some formed subsequently in sediments[5] and their equilibration with the sea. The hydrothermal vents (see Figure 2.2) were the early source of much Fe^{2+} and Mg^{2+} with H_2S and $Si(OH)_4$ leaching from the ocean crust to the oceans, and were partly responsible for sediments and the sea's contents. The minerals are interactive due to the motion of the tectonic plates which caused the surface minerals to be constantly chemically transformed by subduction, hydrothermal alteration, and volcanoes (see Figures 2.2 and 2.3).

The chemistry of the sea in these times was also added to by the processes of weathering giving movement of elements from the land and then sedimentation from the sea controlled by solubility products and oxidation potential rise as oxygen rose (Figure 2.5). Hence the ions in the sea depended on the equilibrium properties of the mixture of elements from the periods of Figure 1.1 described in Sections 1.2 to 1.4. Weathering is a term used here very broadly to describe environmentally energised change in the sea, mainly of abiotic, inorganic materials. From 4.5 Ga to today, many changes of the sea were due to such physical–chemical processes in the environment (see Figure 2.3).

Certain minerals are common and more soluble than others and in particular those of Na^+, K^+, Mg^{2+} and Ca^{2+} dissolved to a greater degree than those of Al^{3+} and ions of heavier elements, so that the sea quickly became rich in them. These heavier elements dissolved to give only traces in the sea before 3.0 Ga. While the weathering and other changes, such as oxidation, are generally in one direction, there was considerable chemical cycling during the long periods of time we are considering. A good example of such cycling in the

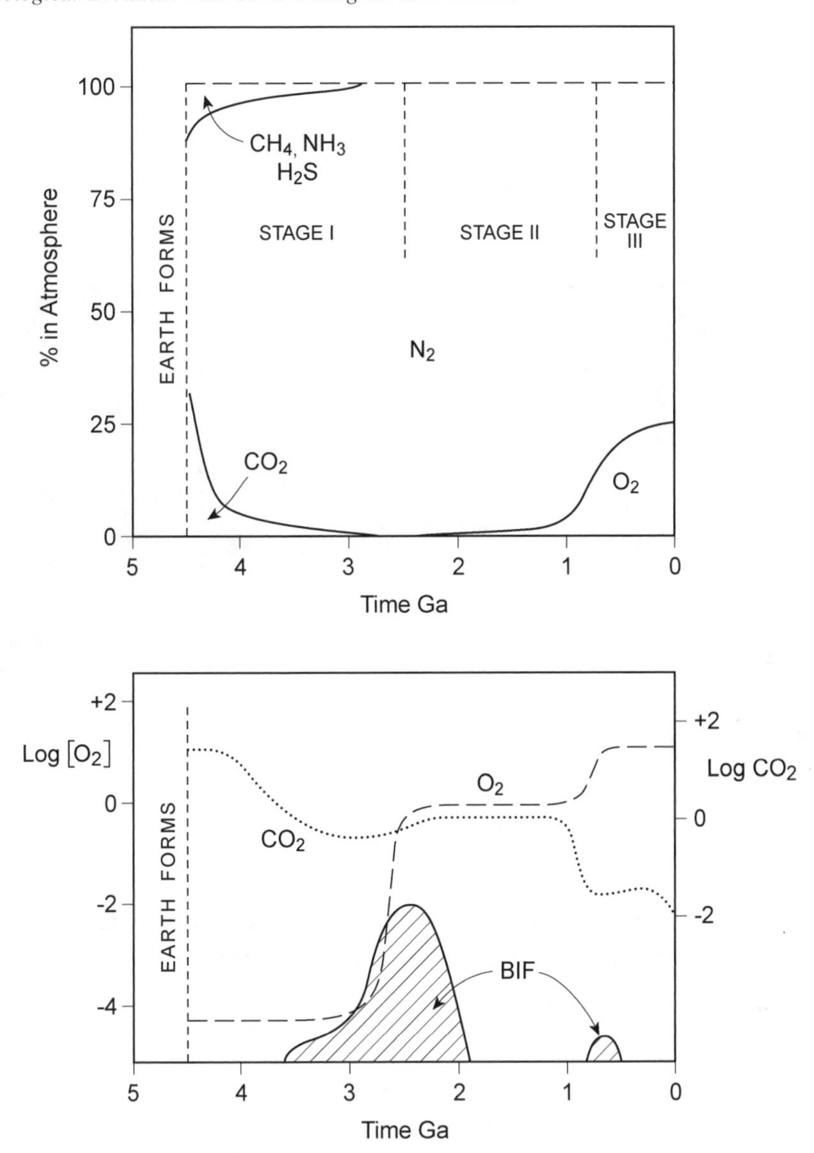

Figure 2.5 The changes of the atmosphere (above) and the appearance of Banded Iron Formations (BIF) against the early scale of CO_2 loss and the late rise of O_2 (bottom) against time. Stage I to II separation is the first rise of oxygen and stage II to III its second rise.

dominant controls of some elements is the long-term largely abiotic cycling of carbon (Figure 2.6). Once biota were established in the sea after 3.5 Ga they added to the purely inorganic processes to some degree. The relative importance of these two processes, weathering and organism activity, has fluctuated through time, leading to oscillating CO_2 concentrations in the

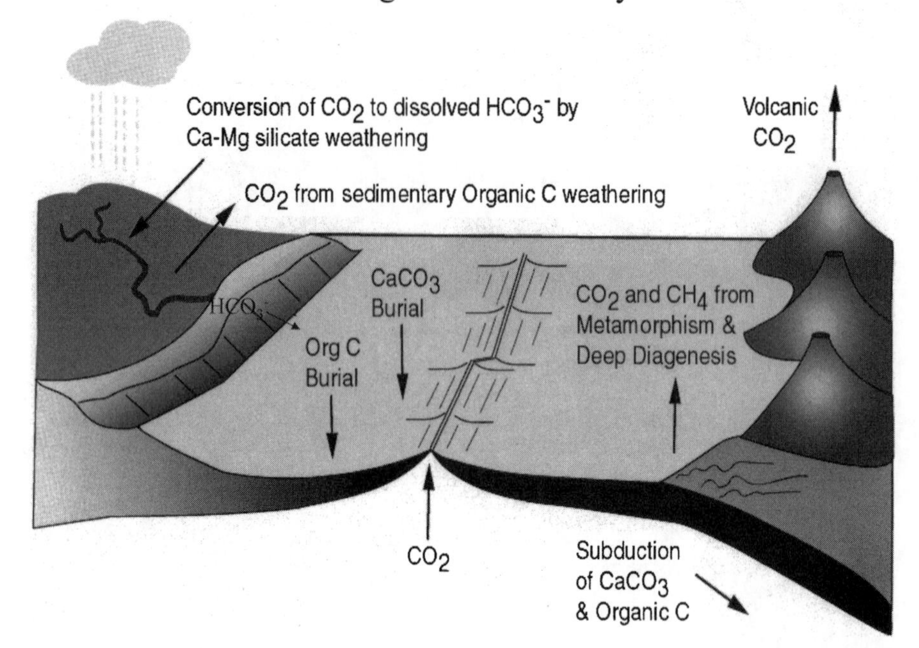

Figure 2.6 The long-term carbon cycle, see section 2.4.

atmosphere, and also changes in the relative composition of some major ions in the ocean (see Sections 2.8 and 2.11).

2.5 Energy Capture and Surface Geochemical Changes: The Beginning of Organic Chemistry and Oxygen in the Atmosphere

Subsequent to Earth's formation there were several chemical changes in the direction of stabilisation. One direction was the loss of very volatile gases such as hydrogen and methane and another was the weathering due to wind and rain mentioned in Section 2.1 and which we reintroduce in Section 2.12.

While the energy for the initial organic chemistry could have arisen from the Sun there are other possible sources. As we mentioned earlier the Earth has a cold surface broken by hot vents, injecting unstable (energised) inorganic compounds to it. Of great interest are not the sporadic volcanoes we still see on land but the more persistent activity of these hot vents, often at edges of tectonic plates, called generally black smokers. Some such vents are known to produce acidic water relative to the pH of the sea and some novel compounds. Provided these gradients of heat or chemicals can be captured across a vesicle membrane, then, given that the vent contains metal ions with carbon oxides and is also strongly reducing, it could have initiated organic chemistry by a simple route. (A membrane is not absolutely required because trapped solutions could carry out the reactions using localised gradients of acid and base reactions. We shall not speculate further here but return to the subject in Chapter 4.)

The primary reaction of any organic from inorganic chemistry must be the reaction of the oxides of carbon with hydrogen to give formaldehyde. The reaction of CO itself does not require energy but needs catalysts:

$$CO_2 + H_2 \rightarrow (HCHO + H_2O) \quad (2.1)$$

$$CO + H_2 \rightarrow HCHO \quad (2.2)$$

$$CO + H_2 \rightarrow (CH_2)_n + H_2O \quad (2.3)$$

Before we continue the development of the way in which organic chemistry can be achieved we wish to draw attention to a more general route than by H_2 to reduction of carbon. While the above reactions are possible we describe next that known later in all cells.

We treat this subject here as abiotic. It is very well possible that some sulfur, and later oxygen arose by the irradiation of the sea by light in the presence of iron. The photochemistry of H_2S could have become important at first after H_2 had been lost:

$$2H_2S + h\nu \rightarrow 2H_2 + (S)_n \quad (2.4)$$

However any oxygen produced by the reaction:

$$2H_2O + h\nu \rightarrow 2H_2 + O_2 \quad (2.5)$$

would have slowly removed H_2S. It is the production of hydrogen which it is necessary to stress, because it is essential for the beginning of organic chemistry. The sulfur was quite possibly deposited as such while the oxygen reacted further with any available reduced material in the environment. Subsequent maintained separation of H_2 from $(S)_n$ or O_2 in compounds by reactions of hydrogen with carbon, particularly later inside cells, could well have been started from formaldehyde (equations 2.1 and 2.2) and more reduced carbon compounds from reaction 2.3.

Loss of environmental oxygen, redox buffering, at first would have been in the reactions with sulfur and ferrous iron:

$$2H_2S + O_2 \rightarrow 2H_2O + S_2 \quad (2.6)$$

$$S_2 + 3O_2 + 2H_2O \rightarrow 2H_2SO_4 \quad (2.7)$$

$$H_2O + Fe^{2+} + O_2 \rightarrow Fe^{3+}(OH^-)_3 + H^+ \tag{2.8}$$

The iron hydroxide/oxide was seen in banded iron formations (BIFs),[5] (Figure 2.5) some quite probably before there were organisms.

The slow 2.1 to 2.3 reactions of CO (CO_2) + H_2, are known to require various trace element catalysts in model compounds (and later in cells) such as Fe, Mo(W) and Ni sulfide (selenide) complexes (later in proteins) (see Section 2.14). Surfaces of minerals are another possible source of catalyst. There was the possibility of the production of reduced carbon compounds, which could react further. These are the basic reactions that are the beginnings of organic chemistry. We treat them here as abiotic as they can be catalysed by purely inorganic compounds and belong in this chapter on geochemistry. Table 2.1 includes some of the early metal catalysed reactions with later additions. In Chapter 4 we shall detail how energy from the Sun (or elsewhere) can be used to drive more complex organic chemistry in organised, controlled ways, also employing compounds of phosphorus and sulfur as intermediates and using several trace elements as catalyst centres. Additionally the volcanic activity and hydrothermal vents helped the supply of reduced compounds and still help to maintain reduced conditions and a supply of sulfide locally but, as mentioned before, its input would have also given rise to chemical changes including pH in the oceans.

A distinctly different energisation from compound formation is the production of gradients of ions or chemicals[6,7] which must also have predated the appearance of life. Again a vesicle is needed. This will also be described in Chapter 4.

Table 2.1 Major Elements in Catalysis in Cells

Element	*Catalysed Reaction*
Iron	Redox in the cytoplasm and membranes/bioenergetics Electron transfer in bioenergetic apparatus Hydrogenases
Manganese	Oxygen production (uniquely required) Glycosylation in vesicles (later)
Copper	Oxidation outside cytoplasm (later) Absent from earliest cells
Molybdenum (Iron, Vanadium)	Nitrogen fixation O-atom transfer in early reduction later in oxidation
Nickel	Hydrogenases, urease required early
Zinc	Hydrolysis of especially peptide bonds. Essential in digestion and in multicellular growth
Magnesium	General transfer of phosphate and phosphate metabolism. Essential in energy transduction and condensation

N.B. The later appearance of porphyrin compounds of iron, cobalt and nickel are not included.

We stress that although all the steps of chemistry we have described have become associated with life (see Chapter 4), it must have been the case that before life, considered as a reproductive activity in cells, could begin, the basic steps of it had to be a consequence of the abiotic energisation of simple environmental chemicals. The steps had to be qualitatively inevitable, they also had to give compounds of considerable kinetic stability, but we do not know the probability of them coming together in trapped complex systems to give rise to life.

Now we wish to put together some of these chemical abiotic changes with the physical changes of Section 2.2 in two periods of energisation because the oxidation potential of the environment took place most strongly in two stages, around 2.5 to 1.8 Ga and around 0.75 to 0.5 Ga. Remember that the chemical and physical changes are interactive in the total developing evolution of the environment/organism system.

2.6 The Environment after 3.0 Ga: Revolution in Redox Chemistry before 0.54 Ga

In previous sections we have described, in a somewhat speculative fashion, the only times chemical conditions were largely in a reducing atmosphere. For 1 billion years the environment changed very little but the reduction of carbon oxides led to the beginning of organic chemistry and rejection of oxygen. The first anaerobic organisms had little effect on the environment and it and organisms changed little. We now come to the most intriguing development of the chemistry of living organisms and the environment as oxygen production increased in two considerable steps from 3.0 to 0.5 Ga. After 0.50 Ga evolutionary change was not so chemically interesting as oxygen levels were roughly constant much though it is the period of greatest interest to biologists as a huge variety of organisms appeared and it is the easiest period for genetic studies. Before beginning we note that some of the chemistry of the period 3.0 to 0.5 Ga is still open to considerable debate and we have chosen the most certain of the main changes on the basis of today's evidence. The evidence concerning the chemistry of the environment comes mainly from isotope studies of elements in sediments following their rising oxidation states (Section 2.3). It is the rising oxidation conditions of elements which we believe to have been the main driving force of the evolution of organisms (Chapter 4).

We shall limit description to those elements which are found in both the environment and in organisms or can be used as time markers of chemical oxidation change and evolution of the environment/organisms system. Uses of elements in organisms will be analysed in detail only in Chapters 4, 5 and 6. There is some uncertainty in our quantitative knowledge of the trace elements in the environment at these times, *i.e.* before 0.54 Ga, the Cambrian Explosion. The elements present in the sea in this period are the bulk metal elements Na, K, Mg, Ca, Al (Fe and Mn), the non-metal elements H, C, N, O, Si, P, S and Cl, and the trace elements. We shall not dwell on the ions of Na, K, Ca, Mg, Al, Si, P and Cl here as we believe they changed but slightly from 3.5 Ga to today in minerals or in the sea (Table 2.2). The trace elements of concern did change, some not directly involved in organisms but most of them are essential for it. The sources of information of both bulk minerals and of elements in the sea are described next.

We do not need to record the progression of formation of the underlying core of rocks beneath the surface minerals, but vents, smokers and volcanoes did add elements to the environment and were oxidised with elements from other sources such as weathering. Our knowledge of the distribution of the elements and their changes with time in the environment comes from sediments from all sources and we shall take them as reflecting conditions in the sea. The major sediments are silicates and oxides, including a variety of shales and clays, and carbonates and some sulfides. Variation of trace element ion concentrations in the sea subsequently can be inferred relative to amounts of aluminium in sediments as a recorder of sedimentary detrital input to the sea because aluminium is insensitive to ocean chemical redox change. The inference is supported by calculation of the probable chemical equilibria below. The chemical processes which elements have undergone can be estimated from their concentration, oxidation state and isotope composition in particular sediments of given times. There are also some changes of bulk sediment, whether it is mainly oxidative or sulfidic (*e.g.* pyritic) and can be readily determined. Unfortunately, it is considered that the chemical conditions of the sea during some of these times are somewhat confusing as there is poor mixing,[8–10] and the general view of overall chemistry is biased by the tendency of the sedimentary record to reflect shallower water depth deposits. We shall ignore these problems until Section 2.10 onwards.

We begin the discussion by following the evidence for the oxidation from 3.0 Ga to 0.5 Ga using the revealing studies of changes in the isotopes of

Table 2.2 Logarithm of the Metal Ion Content of the Sea*

Earliest Times > 3 Ga (ln)	*Middle Period 2.5 to 1.0 Ga (ln)*	*Latest Period 1.0 Ga to Today (ln)*
Na^+ –1	–1	–1
K^+ –3	–3	–3
Mg^{2+} –2	–2	–2
Ca^{2+} –3	–3	–3
Mn^{2+} –8	–9	–9
Fe^{2+} –5	–7	–9 to –10 (Fe^{3+})
Co^{2+} –8	–9	–10
Ni^{2+} –9	–9**	–8
Cu^+, Cu^{2+} < –20	–15 (Cu^+) to –25**	–9 (Cu^{2+})
Zn^{2+} < –15	–12 to –17*	–8
Mo^{n-} very low (–10)	–10	–7

*The numbers are estimates largely based on Williams and Fraústo da Silva (2001)[15] and Saito, Sigman and Morel (2003).[17] The values ** are in dispute. The value for Nickel is taken from ref. 34

sulfur.[9] We then turn to some changes in minerals, firstly of iron oxides/silicates together with their soluble ionic species and then to the relative changes in the pair of minerals, uranite and thorium oxide. Because we shall then have confidence in the outline oxidation state changes we shall consider a quantitative analysis of these changes. We follow this calculated chemistry with that of changes of trace elements, especially in aluminium silicate sediments. We shall conclude that there is a proven increase of the oxidation of the environment. In doing so we shall observe a shorter period more carefully from about 2.5 to 1.0 Ga. The sulfur chemistry and that of iron dominated the buffering of the environment in this period (Figure 2.7). We shall show that it underwent but a small change in redox potential while there were two more rapid changes in redox potential from 3.0 to 2.5 Ga. Once we have a secure knowledge of the environment in these times we shall be able to examine the evolution of cellular life (Chapter 4). The revolution in the chemistry from sulfur being the main element of important trace element binding to oxyanion binding, controls the changing release and availability of trace elements in the sea due to oxidation.[15,17,18] It is important to note the contrasting solubility product of sulfides and oxides (hydroxides) (see Figure 1.4). The changes and the dates for some of these most important reactions in evolution can be found both from the study of the isotopes of S (see Section 2.7) and deposits of iron oxides.

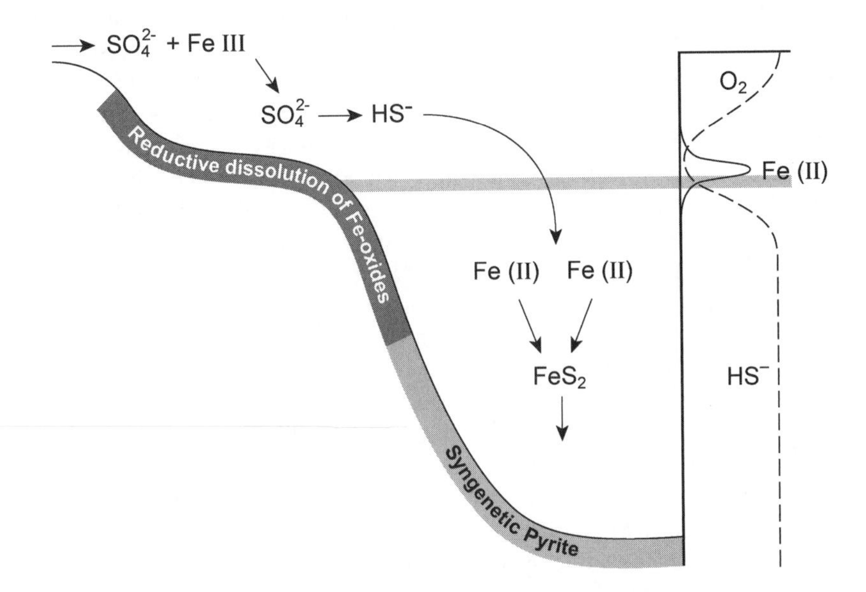

Figure 2.7 Schematic to show the iron and sulfur cycling associated with euxinic conditions at depth, and oxic conditions on the shelf. The columns depict idealised Fe speciation between an upper and a lower layer corresponding to an oxic-to-euxinic redox transition. Modified from reference 8.

2.7 Sulfur Isotope Fractionation from 3.5 to 0.5 Ga; Dominance of Iron/Sulfur Buffering

The transformation of Earth from a dominantly oxygen-poor world with some SO_2 present from 3.0 Ga to one with an atmosphere that became and remained oxygenated and with no SO_2 soon after 0.75 Ga is most profoundly evidenced by the disappearance of mass-independent sulfur isotope fractionation ($\Delta^{33}S$).[8–10] The $\Delta^{33}S$ in Precambrian sulfide and sulfate minerals suggests that photochemical reactions played a role in this mass-independent fractionation prior to 2.45 Ga but these reactions were significantly damped as oxygen rose in the atmosphere after this time.

The oxidation of sulfur to sulfate, from 2.5 to about 1.00 Ga, can also be estimated from the mass-dependent fractionation of the isotopes of sulfur. Negative values of these data indicate that there would have been a low redox potential of below −0.1 V (probably until 2.5 Ga) before sulfate appeared in any quantity. (We shall need to consider the major influence of organisms on the SO_4^{2-}/S^{2-}.) We must also note the buffered redox chemistry of sulfur with iron in this period (see Figure 2.7; Section 2.8).

Beyond the loss of mass-independent fractionations at ~ 2.4 Ga, there is a substantial broadening of values of the $\Delta^{34}S$ of pyrite, FeS_2, away from the mantle value. This pattern has also been attributed to increasing sulfate concentration in the ocean, which would yield greater fractionation during bacterial sulfate reduction. Enhanced supply of sulfate to the ocean is a natural consequence of increasing atmospheric oxygenation and weathering and would accompany an enhanced supply of metals from previously insoluble sulfide.[10]

Despite such an increase in supply, there is growing consensus that sulfate concentrations in the mid-Proterozoic, ~1.5 Ga, were still only a minor fraction of the 28 mM of the ocean today, primarily inferenced from the frequently isotopically heavy pyrites, FeS_2 found in sediments of this age, and the rapid rates of S isotope variability inferred for seawater sulfate. This scenario would result from a large supply of sulfate to the ocean from continents, but its transformation to H_2S by organisms, which used it as an energy supply, in the persistently anoxic bottom waters. This resulted in copious pyrite formation and burial which strips the sulfur out of the water column (and contemporaneously some metal ions key for life such as Mo and Fe). The sulfate only rose to present levels around 0.75 to 0.5 Ga. Thus we have a picture of a sulfide/sulfate cycle in which the sulfide also equilibrated with pyrite (FeS_2) and Fe^{3+} over a period of about 1.5 Ga. The appearance and later loss of this sulfur was key to the changes in iron and other elements (see Section 2.8). It ran out as oxygen production increased, which gradually changed all available sulfide to sulfate. Organisms evolved more and more able to use oxygen directly as an energy source, rather than sulfate, as oxygen became more available. We stress the striking changes in oxidation potential due to the times of oxygen input into the environment which took place in three stages as shown by the sulfur isotope studies. There were two rapid rises at about 3.0 and around 0.75

Ga with a slow change in between. In Chapter 4 we shall see equally striking changes in organism chemistry in these same stages.

2.8 Evolving Mineral Outputs from the Ocean: Further Evidence for Redox Chemistry to 0.5 Ga

2.8.1 Banded Iron Formations and the State of Iron in Solution

In this section we summarise the major chemistry of the environmental redox change of iron. We have already referred to its changes in several previous sections but we have yet to bring all the information together. Importantly we describe its concentration in the sea as Fe^{2+} and Fe^{3+} and we shall have to repeat some earlier remarks about these concentrations in iron oxide sediments.

Although iron is a trace element not exceeding 10^{-7} M Fe^{2+} in the reducing conditions of the early sea and falling to below 10^{-10} M Fe^{3+} today, there has always been a huge reservoir of reduced iron as olivine (with sulfur) from smokers (see Figure 2.2). This Fe^{2+} iron supplied the sea slowly but it was increasingly rapidly removed by oxidation when oxygen increased.[11,12] The loss of Fe^{2+} concentration was reflected in the quick rise at 2.5 Ga of redox potential from very early periods of Earth's existence extending from 3.5 to about 2.5 Ga, as shown by Banded Iron Formation (BIF), Figure 2.5. The soluble iron was important in assisting the strong buffering between about 2.5 and 1.0 Ga after oxygen became involved in the sulfide/sulfate cycle (see previous section). This period of iron formation of mainly pyrite is sometimes termed 'the boring billion years' of evolution. Sulfide (HS^-) was virtually completely oxidised by 0.75 Ga in all but the outpourings from the black smokers. The residual iron in solution then became oxidised and the last banded iron deposits appeared around this time (see Figure 2.5). An interesting feature of the deposition of iron oxides mentioned above is that the earliest deposits contain Fe_3O_4 with Fe_2O_3 (that is partial oxidation) while later the deposit is of Fe_2O_3 alone. This is a strong indication of the increasing oxidation of the environment. The reservoir of olivine is still present but now very slowly dissolving. Iron has become a limiting nutrient for life which, while entering from ocean vents for example, is rapidly captured by organisms and by particles as ferric oxide which settle eventually dispersed in sediment. We turn to the other mineral changes which support the strongly held view of the long-term oxidation of iron in the sea from all Fe^{2+} to all Fe^{3+}.

2.8.2 Uranium and Thorium Minerals

The modes of accumulation and even the compositions of uraninite, as well as the multiple oxidation states of U (4+, 5+ and 6+) are sensitive indicators of global redox conditions. In contrast, the behaviour of thorium, which has only a single oxidation state (4+) and that has a very low constant solubility in the absence of aqueous F-complexing, cannot reflect changing redox conditions.[13] Geochemical concentration of Th relative to U at high temperatures is therefore limited to special magmatic-related environments, where U^{4+} is preferentially removed by chloride or carbonate complexes, and at low temperature by mineral surface reactions.

The evolution of uranium minerals and ore deposits on Earth parallel the four distinct phases in the evolution of the chemical environment. Between 4.5 and 3.5 Ga, both U and Th were concentrated from their initial uniform trace distribution to highly enriched magmatic fluids, from which they were incorporated as minor elements into zircon. Subsequent fluid enrichment led to the first uranium minerals, predominantly uraninite (UO_2) and coffinite ($USiO_4$). From 3.5 to 2.2 Ga the formation of detrital Th-bearing uraninite took place in Witwatersrand-type quartz–pebble conglomerates deposited in a relatively anoxic surface environment. Shortly after the time of the first oxygen rise, at 3.0 Ga, the uraninite was deposited from saline and oxidising hydrothermal solutions which transported uranium in the (6+) oxidation state as $(UO_2)^{2+}$ complexes. At this time, most uranyl minerals would have been able to form through weathering processes in the O_2-bearing near-surface environment. The rise of the land plants, ~400 Ma, led to a fourth phase of uranium mineral deposition as low-temperature, oxygenated, U-rich near-surface waters precipitated uraninite and coffinite at reduction fronts in organic-rich continental sediments. We consider that the rise is well documented qualitatively by these geological studies but we shall question that the rise follows a quantitative analysis of equilibrium redox potentials.

2.9 Quantitative Analysis of Oxidation Conditions

We began the description of the advent of oxygen by reference to the qualitative deductions which can be made from the chemistry of sulfur, iron and some minerals in the environment in the above sections. The changes are dominated by the large amount of iron and sulfur in the environment. Here we shall use quantitatively the oxidation/reduction potentials of these two elements (Section 1.5) to guide us to the timing of the changes of other elements in the environment (see Figure 1.6).[15–17] The Fe^{3+}/Fe^{2+} standard redox potential at pH = 7 is +0.1 V, which is close to that of the SO_4^{2-}/S^{2-} couple at +0.0 V (see Figure 1.6). The actual redox potential of Fe^{3+}/Fe^{2+} is lowered to less than +0.0 V by the greater binding of Fe^{3+} than Fe^{2+} by various anions in the sea. As a result we can calculate that much of the ferrous iron would be quickly lost as ferric precipitates before sulfide could be oxidised to sulfate. We shall take it that this is the first part of a continuous process of oxidation of the two bulk reductants, iron and sulfur, in the available environment. If we assume that there was continuous further addition of oxygen to the environment, then the oxidation potential at fixed pH of the sea must follow a somewhat distorted redox titration curve while the bulk of the iron and sulfide was oxidised. After this stage the residual iron would have been oxidised and deposited, seen as the last small BIF at 0.75 Ga (Figure 2.5) while the remaining sulfide became sulfate. We have not needed to introduce any biological complications in devising this basis of the simple inorganic redox titration curve (Figure 2.8). In the presence of organisms at the time of this oxidation the picture given of the oxidation has been outlined in Figure 2.7. Here organisms, sulfur bacteria in particular, catalyse the oxidation. This will distort the simple redox curve of Figure 2.8 but does not damage the outlined derivation of it. The distortion will arise through the changing rates of reactions and in some additional temporary side reactions. Keeping these caveats in mind we can take the redox curve as requiring time when the 'boring billion years' is no more than the time for the greater part of any redox titration centred on the redox potential of about 0.0 V to take place. The redox potential could well have started from about −0.4 to about +0.1 V, then it would rise to be buffered again as oxygen reacted slowly with the organic material in the sea and would finish below a potential of O_2/H_2O. At the same time cells would have come under threat. The whole process was inevitable given that organisms released oxygen. It also means that since inorganic oxidations proceed rapidly as the redox potential rises there would be corresponding selective rises of the states of the trace elements so important for

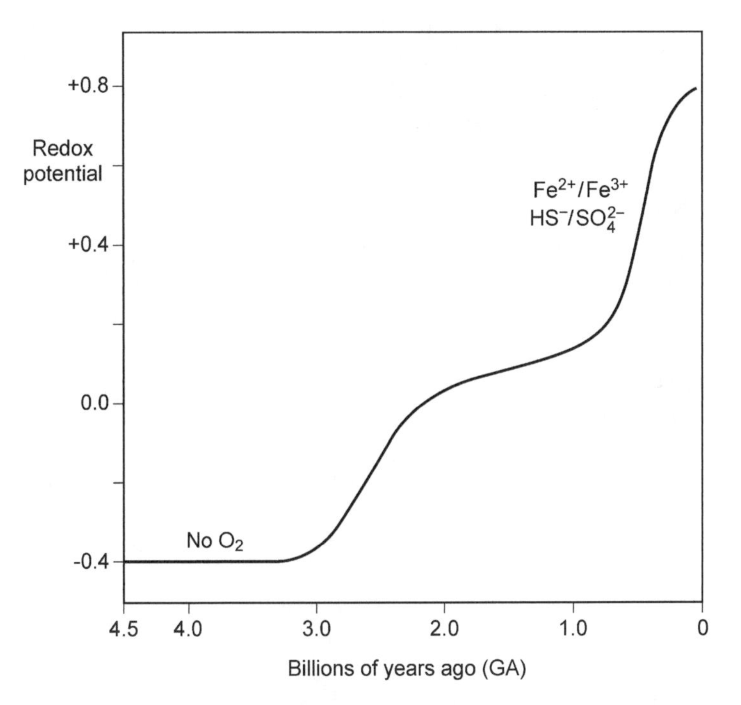

Figure 2.8 The variation of the average redox potential of the sea with time deliberately drawn as a redox potential change with a fixed buffer content of reduced Fe^{2+} and HS^-. Time is that of oxygen release by organisms. Except for the release the process is purely geochemical. The time axis could well have been affected by catalysis by cells of some sulfur reactions, see Fig. 2.7.

life. We give experimental evidence for this unavoidable environmental change below. In particular the rise of the calculated redox potential would have oxidised elements of the metals released from sulfide, Mo, Co, Ni, Cu, Zn, Cd, in order of redox potential (see Figure 1.5), so that the last to be released would be those with the most insoluble sulfides, Cu, Zn and Cd. In Section 2.11 we shall describe these and other changes in more detail to confirm this description. In passing notice that life could get access to these elements as new catalysts and so increase activity. We shall show why we believe the cooperativity of these changes initiated the Cambrian Explosion. Note that many sources of high redox potential became more available for redox reactions but it is the rises of unicellular eukaryotes around 1.5 Ga and multicellular eukaryotes around 0.75 to 0.54 Ga which are the most striking events. Clearly they are linked to the inevitable chemistry of the elements Fe and S with oxygen.

(In describing the redox conditions in this way, treating the environment as a uniform whole we are avoiding the problem of poor mixing.)

2.10 Geochemical Changes of Trace Elements

The first trace element markers we shall consider are those that generally could not be affected by biological activity, the rare earths (lanthanides), and elements such as Re and Os and their isotopes.[12,14–20] We make an exception of molybdenum which is of biological importance, but which is an excellent marker of the change of the environment.[19,20]

2.10.1 Rare Earth Probes of the Environment

The rare earth elements (REEs) are valuable probes of environmental conditions through two features of their chemistry.[21–26] Firstly they are a series of 15 elements of similar but graded ion size M^{3+}, when the presence of the concentrations of them in sediments can be used to estimate the general Lewis acid/base conditions in the seas over time, for example the concentration of complexing anions such as carbonate. Secondly two of them have variable oxidation states and they sense changes in the oxidation/reduction potential of their environment (Table 2.3). The chemistry behind the first use of rare earths is that the decrease in ion size along the series is known to be reflected in the decreasing solubility products of their hydroxides. Let us assume that the source of the lanthanides is from a high-temperature melt, then, provided solubility in the cool sea is the only controlling factor, and provided that they were in equilibrium with oxide and silicate sediments, the gradient along the series would be expected to give a considerable increase in deposits on oxide surfaces by the end of the series. The sediment of any BIF is such an oxide deposit. The major deposits in shales are silicates and we expect a smaller gradient of solubility than for hydroxides. These decreases in solubility of lanthanides along the series in the sea would be offset by the increase of complexing by carbonate by an increasing factor of 10^2 from the beginning to

Table 2.3 Relative Amounts of Two Lanthanides in Banded Iron Formations and the Modern Sea*[5,26-28]

Date (Ga)	*Eu*	*Ce*
3.8	Very High (x2)	Equal
2.8	High (x1.5)	Equal
2.3	High (x1.5)	Equal
0.75	Equal (x1.0)	Low/equal
Sea (Today)	Equal (x1.0)	Low

*Data are amounts relative to the composition of North America Shale × 1.0 normalised to an average for all other lanthanides

the end of the series. In all cases the variations of the natural universal abundances of the elements in the series can be normalised against the abundances in the chondrite meteorites. There are no solution constraints in meteorites. In fact while some measurements on sediments show the expected direction of change others do not, so we can draw no firm conclusions.[21–26] The examination of the total content of all REEs in sediments (Figure 2.9), is more easily studied than the gradients and shows that from 3.8 Ga to today it increases by about 10^4, which closely follows the estimates of decrease of

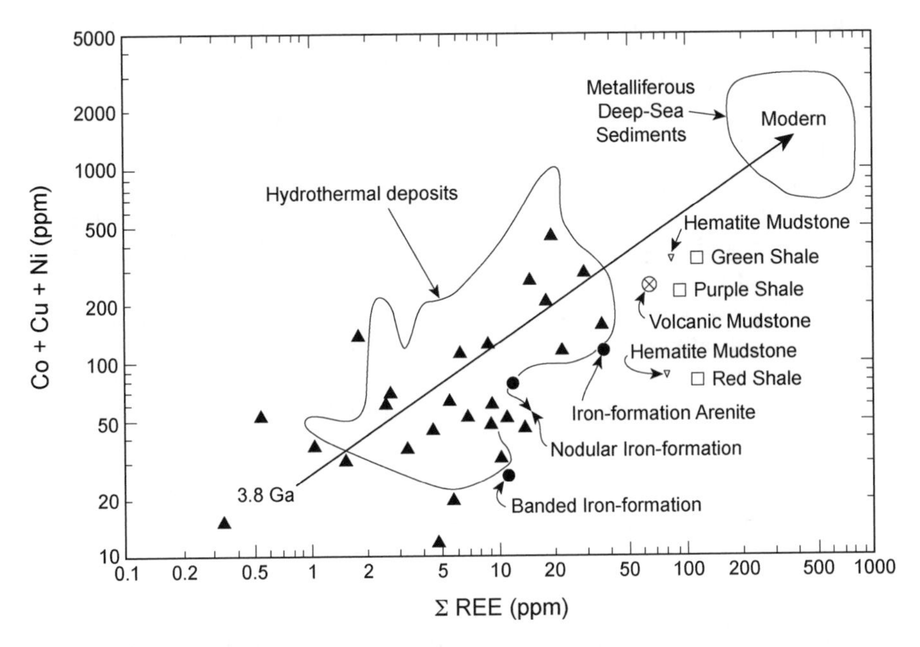

Figure 2.9 The changing sum of elements, Co, Cu and Ni, all probably derived from oxidation of sulfides compared with the sum of all the rare earth elements in sediments (which likely indicates the degree of free REE as opposed to that complexed with the carbonate ion). The straight arrow is an indication of time from 3.5 Ga to today. From Klein Ref. 12.

carbon dioxide in the atmosphere and hence of carbonate in the sea during the same time. Assuming a fixed supply of lanthanum ions in the sea the rare earths in sediments are then good markers of an evolutionary fall over time of the concentration of the CO_2 in the atmosphere at equilibrium with the sea which is a major buffer of acid/base conditions. Further evidence of lanthanide changes from early times has been obtained from the study of neodymium isotopes.[26] The general trend is exactly that discussed in Sections 2.3 and 2.5.

The change of oxidation states of the rare earth ions can also be followed as their solubility decreases from M^{2+} to M^{4+} states and is then related to general environmental redox changes.[31] In oxidising conditions we expect cerium to be in the oxidised state Ce^{4+} and to be increased in scavenging by sediments, while in very reducing conditions europium is expected to be low in sediments as its reduced state Eu^{2+} is more soluble than its M^{3+} state. Cerium is found to be decreased in today's sea but not in ancient sediments before about 1.0 Ga, indicative of the already noted change from low reducing conditions at 3.6 Ga to more oxidising conditions by 1.0 Ga.

A major source of the lanthanides in the sea is probably from the hydrothermal vents.[23] This means that the elements are fractionated from the melt state into the solution state in strong reducing conditions. Europium in the melt is in the greatest concentration as it is readily solubilised as Eu^{2+} there. Entering an oxidising sea it is transformed to Eu^{3+} and is then scavenged in sediments more strongly than other lanthanides than from a more oxidising initial condition. The states of all lanthanides have been frequently analysed in sediments of different times. It is clear that early reducing conditions change to oxidising conditions in the environment at the expected times.

A further check on scavenging for the presence of carbonate is provided by the environment chemistry of yttrium.[27] This element is in the second row of transition metals and as such has a higher affinity for hydroxide and silicates but little change in carbonate complexing compared with a rare earth ion of the same charge and size. As expected it is frequently found in higher concentrations in later sediments. Putting all evidence together we consider that the changes along the series of these elements and of their oxidation states are very useful markers of both carbonate and oxygen concentrations, roughly buffered close to equilibrium over the whole period from 3.8 Ga to today and that they confirm earlier calculations.

2.10.2 Trace Transition Metals in the Sea

All the elements in the second and third transition series have been analysed in sediments. They are the most important metal ions in our analysis concerning evolution.[12,17,18,27–30] Except for those containing molybdenum they do not provide much quantitative information about the environment but the changing Re/Os ratio[19] confirms the general increase in oxidising conditions with time (Sections 2.3 and 2.5). The chemistry of molybdenum is the most understood.[8,19,20,31] It has a very insoluble sulfide, MoS_2, slightly soluble salts

$M^{2+}MoS_4^{2-}$, but soluble salts of MoO_4^{2+} ions. Hence it is readily precipitated in reducing sulfide media and soluble in oxidising media. It is very readily available in the oxidising sea today and has been valuable in cells, possibly from around 2.5 Ga, before which time it is probable that tungsten was used in its place. Tungsten is less easily reduced and does not readily form such an insoluble sulfide so that it was more available than molybdenum in the very early sea. We observe that molybdenum in sediments increases greatly and steadily from 3.8 Ga to today by more than a thousand-fold and it and its isotopes are very sensitive probes of the redox evolution of the sea (Figure 2.10). Moreover these changes parallel our calculation of the change from reducing to oxidising conditions[31,32] deduced in the previous section and fit in with the view of the quantitative changes of iron and sulfide in the sea expressed in an equilibrium redox curve with oxygen (compare Figures 2.8 and 2.10). A useful indication of the way in which selection of elements into minerals is given by data on basic magma.[33]

We can turn now to the elements of the first row of transition metals from titanium to zinc which are so useful in cells. Their chemistry is that of elements of decreasing ionisation state from Ti^{4+} to Cu^{2+} (Cu^{+}) with the elements Mn to Zn all having a frequently observed M^{2+} state but different oxidation

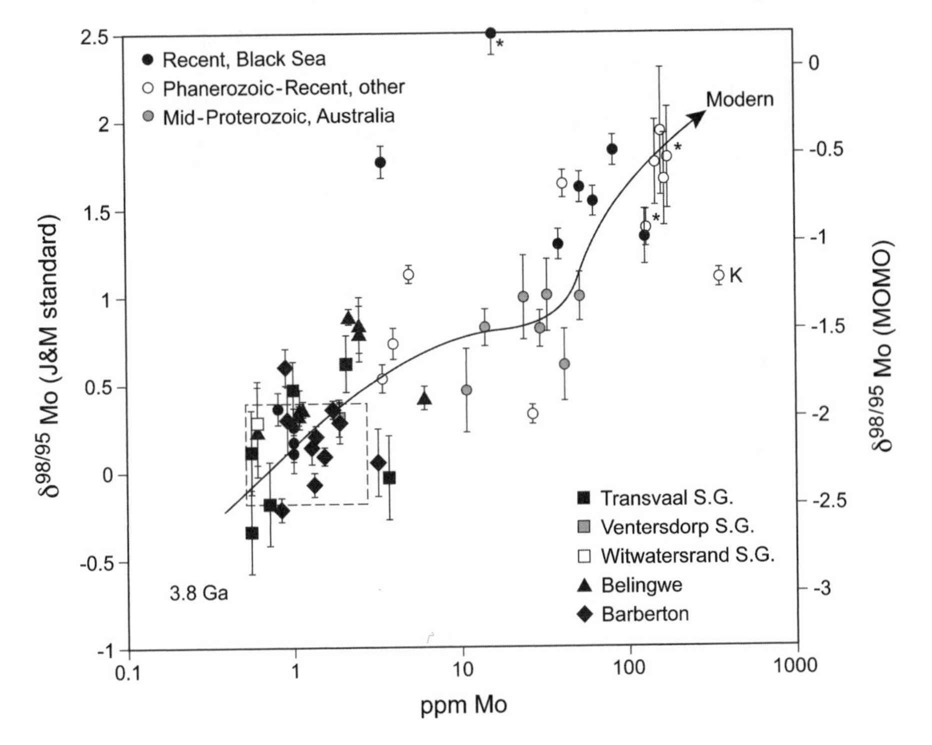

Figure 2.10 The changes in molybdenum isotope ratio and the concentration in shales. The straight arrow is an indication of time from 3.8 Ga to today. From Siebert Ref. 19.

conditions give rise to other states for some. We have seen in Chapter 1 that the six elements, Mn to Zn, form a series of increasing binding strength, the Irving–Williams series (see Figures 1.3 and 1.4), such that they occur in the mantle as oxides in the case of Mn^{2+}, oxides and sulfides for Fe^{2+} (and Fe^{3+}) and predominantly as sulfides for the last four. Like molybdenum we expect and observe that these sulfides are found as impurities in early sedimentary minerals but as silicates in sediments after 1.0 Ga.

There have been two detailed examinations of BIF, giving us by inference the required analytical data of the transition metal content of the sea. The first by Klein and Beukes[12] showed that in certain locations, Rapitan, where the deposition is more recent, around 0.75 Ga, the concentration of the trace metal ions Zn and Cu increased over that in early deposits from 3.8 to 2.7 Ga (Table 2.4). The early deposits were formed in reducing conditions (see Section 2.8). While the second investigation[34,35] did not observe these changes, Rapitan was not studied, but it did point to the high values of nickel around 3.5 Ga which were not seen later perhaps because silica in the sea increased. Now Klein and Beukes also pointed to the close parallel between the sum of the concentrations of the three trace elements Co, Ni and Cu and that of the lanthanide sum already discussed (see Figure 2.9). The fall of lanthanides was attributed to the fall in carbonate but these three metal ions and that of zinc, which also increased, do not have carbonates of sufficient complexing strength to affect their presence at any time from 3.8 Ga to today. Their increase must be due to release from sulfide by its oxidation. They, like molybdenum and zinc, have insoluble sulfides and all such metals increased in solubility and in silicate sediments almost continuously from 3.8 Ga to 0.50 Ga due to the oxygen rise causing continuous sulfide oxidation. Hence we can be very confident that the increasing amount of Co, Ni, Cu, Zn and Mo reflected in sediments of two kinds, BIF and many shales (see Figure 2.8), and the disappearance of sulfide from the sea are due to increased oxygen (see Figure 2.7). The switch of iron

Table 2.4 Trace Elements in Banded Iron Formations (ppm)[12]

		BIF			*BIF*	*BIF*	
		Kuruman				*Rapitan*	
Location	*Isua Basalt*	*(a)*	*(b)*	*Ouplaas*	*Sokoman Hemoolite*	*granular*	*nodular*
Date (Ga)	3.0	2.4		2.4	1.9	0.8	
V	33.2	<150		<150	—	—	
Cr	31.6	3.23	2.75	3.6	3	25	18
Co	11.0	0.59	0.37	0.49	8.08	4.27	1.3
Ni	58.2	29.5	29.5	46	13	14	9
Cu	27.5	<15	<15	<15	25	93	67
Zn	83.7	25.5	25	<20	44	53	41
Si	55	52					
Fe^{3+}	11	1.7				46	
Fe^{2+}		31				1.4	

from a soluble Fe^{2+} with sulfide to an insoluble Fe^{3+} oxide is in accord with these findings. Manganese in the sea is also oxidised as it forms insoluble oxides, M_2O_3, in more recent times from the earlier Mn^{2+}.

The impression is strengthened further by calculations of solubility products of the sulfides from Mn^{2+} to Zn^{2+} as sulfide is reduced and the reduction in sulfide is estimated from isotope data (Section 2.7) and from the non-linear oxygen rise of the atmosphere with time.[17,18,36] While others have chosen to use the stepwise oxygen in the atmosphere as a direct guide to the contents of the sea, we have used a more smoothly changing description of the sea's oxidation potential,[15] as in any redox titration (see Figure 2.8). The firm establishment of the general continuous change to oxidising from reducing conditions is not in doubt from the total evidence and the associated changes in trace elements are of the utmost consequence for life. A further test of the likely analytical content of the early sulfide sea is provided by the contents of euxinic seas such as the Black Sea.[8,36] It is the comparison with the evolution of organisms, notably Zn and Cu contents, which demonstrates undeniably the coupling of the environment changes and those of life as we shall show in later chapters. We shall also note that the increasing uses of selenium and iodide follow the same pattern of oxidation close to 0.5 Ga.

In Chapter 1 it was shown that nickel differed from cobalt, copper and zinc in that the difference in the solubility products of its oxide and sulfide was considerably smaller than those of the other elements. It is also observed that nickel is common in the $(MgFe)_2SiO_4$ mineral, olivine. The reasons for the relatively high affinity of nickel have been explored in the mineral silicate of Skärgaard Intrusion.[33] (It was shown that Ni^{2+} has a higher affinity for octahedral sites in oxides and silicates but not in sulfide lattices than most other metals of the transition metal series.) As mentioned above this could explain the difference in behaviours of nickel relative to other metals in the transition metal series in evolution.

2.11 The Non-Uniform Sea

In the last paragraphs the sea has been taken to be homogeneous but it is not thought to be so nowadays.[8,10,16,36] In the periods after 2.5 Ga the sea could not be uniform in composition because oxygen was produced by sunlight at the surface (or by organisms there) but much reducing solid mineral material lay at the base of the sea (see Figure 2.7). The surface of the sea was also warmed by the Sun and the steep gradient in temperature acted as a barrier between it and the denser cold bottom water. Additionally the top was and will always be more oxidised and the bottom more reduced, by reduced material from the hydrothermal black smokers and sinking organic waste from organisms. It is believed that mixing of top-level oxygenated water with bottom water was poor before the continents broke up after 1.0 Ga and only fully developed later. The general statements in previous paragraphs concerning the average redox potential need correcting at times as it is necessary to take into account this late mixing of the upper and lower zones. Even after an approximate, evenly mixed, state in the ocean and atmosphere was realised from, say, 0.6 to 0.4 Ga, there has remained and still remain immense reserves of reduced minerals below the surface which are not a part of any steady state, including some buried carbon and huge quantities of sulfide minerals of all kinds. This will continue to affect gradients of trace element concentrations to which we refer below and they can greatly influence any final steady state due to their biological influence. It is also likely that the sea as a whole will be more reducing than land water which is removed from sources of sulfides except from rare volcanic action. An important consequential observation is that extremely different chemical activity could exist in different places coincidentally – on the surface of the sea (photochemistry) with release of oxygen and at the bottom of the sea in niches added to by decay of debris from above, and in non-marine environments such as lakes or lagoons. Clearly shallow water is the most difficult to be given an oxidation potential.

In passing remember that in Chapter 1 we have drawn attention to the separation of chemicals across boundaries, especially with reference to organisms and in the solid environment. We return to this theme in Chapter 4. In such cases we can treat local concentrations assuming separate equilibria are present in the two or more volumes identified by boundaries, membranes. The situation is not the same in the sea where separate regions merge gradually over large volumes. In some respects an assumed average of the whole sea is of greatest value but in other circumstances we have to treat very local volumes, for example in hydrothermal vents or black smokers. There are also very still deep small regions of the sea, such as the Black Sea and locally in the Mediterranean, which are very reductive. They are useful as possible models for the very early anaerobic sea. We have now completed the main conclusions as to the chemical oxidative changes.

2.12 Summary of Weathering from 3.5 to 0.75 Ga

The basic settling of Earth's physical–chemical condition at 20°C from 3.5 to 0.75 Ga[37,38] as described in Section 2.2 was

1. Movements of tectonic plates.
2. Volcanic and hydrothermal vents.
3. Circulation of water, movement of water and ice.

Steps 1 and 2 release energy while Step 3 requires energisation. Together they led to the content of the sea in a reduced state with particular concentrations of Na, K, Mg, Fe salts of silicate and carbonate at pH = 7.5 and some sediments of iron oxides, mixed silicates and calcium carbonate by 3.5 Ga.

4. Photochemical or otherwise disproportionation of H_2S and H_2O to give starting organic compounds and sulfur and oxygen. The processes and

those of some other elements resulted in greater amounts of sedimentary iron oxide and the solubilisation of certain sulfides.

5. In some unknown way living organisms appeared which accelerated Step 4 giving rise to novel organic and inorganic compounds later in cells.
6. The non-linear increase in oxygen and oxidising agents in the environment.
7. The increase in organisms mainly anaerobes at first, then increasingly aerobes.

We have used these steps of the dated sediments during this period to give us the composition of the sea from 3.5 to 0.75 Ga, assuming that inorganic reactions attain equilibrium. Moreover we shall use the continuation of these steps later to aid our insight into the evolution of organisms.

After 0.75 Ga with the rise of oxygen, organisms expanded greatly in kinds and numbers. Oxygen and aerobes now became a more important part of weathering processes.[37,38] Throughout this section, as in the whole of this chapter, we shall keep reference to organisms and organic chemistry to the minimum required to give a reasonable understanding of the environmental change. In discussing the period after 0.75 Ga and after the Cambrian Explosion of living forms at about 0.54 Ga, we must keep in mind that all these physical–chemical steps continued to form a background to these oxygen-dependent changes. By 0.54 Ga the sea became close to the sea of today in chemical element content. It took about 4 billion years from Earth's formation of the planet to reach this physical–chemical stage. Until about 0.75 to 0.54 Ga there is only continuous change and certainly no possibility of a steady state. Change was slower in the billion-year period from 2.0 to 1.0 Ga due to the Fe^{2+}/Fe^{3+} and HS^-/SO_4^{2-} buffering.

Now especially after 0.54 Ga all these flows acted as considerably increasing traps for large quantities of organic detritus from organisms. Unless quickly consumed by organisms as food, detritus became part of a store of organic compounds in the mineral earth. In the next section, we consider a possible set of reactions, involving carbon compound deposits, much favoured by geochemists, in an explanation of the major effects of weathering in this period. The major feature was oxidation and reduction in bulk element cycles of C, O, S and Fe on the Earth's surface and including the buried stores in sediments. The system is written as cyclic, kinetically reversible in total amounts of material in a cycle, but as stated above, the biotic and some abiotic steps involving weathering and biological oxidation/reduction are energy requiring and, overall, they are then irreversible. The pathways of given chemicals in the forward and reverse steps of the equations may well be very different. The scheme allows estimates of the oxygen and carbon dioxide content of the atmosphere from 0.75 Ga. We stress that we now enter a period where interaction of the organic chemistry of life and that of the inorganic chemistry, both internal and external, environmental, is inseparable in evolution and both changed little in general outline. We shall see that the system has however remained one of considerable environmental fluctuations around a chemical mean. The fluctuations did not greatly alter the chemistry of organisms overall but quite probably they changed the evolution of particular organisms.

2.12.1 Weathering and Chemical Conditions from 0.75 Ga

The proposal that is accepted by many Earth scientists is that the levels of O_2 and CO_2 in the period after 0.75 to 0.54 Ga, when life began to play a more considerable role (see Figure 2.6), can be followed by using a computerised model covering the rates of three equations, six rate constants:[38,39]

$$CO_2 + (Ca, Mg)SiO_3 \rightleftharpoons (Ca, Mg)CO_3 + SiO_2(\text{weathering}) \quad (2.9)$$

$$CO_2 + H_2O \rightleftharpoons CH_2O + O_2(\text{biological}) \quad (2.10)$$

$$15O_2 + 4FeS_2 + 8H_2O \rightleftharpoons 2Fe_2O_3 + 8SO_4^{2-} + 16H^+(\text{weathering}) \quad (2.11)$$

The first equation is an expression of the reaction of silicates with CO_2, termed chemical weathering. It is reversed by subduction and metamorphism. The second equation gives the burial rate of sedimentary organic matter from biological sources and the reverse oxidation of this buried material. (The reverse reaction does not include that of organism metabolism, see below.) The third equation gives the weathering of pyrite oxidised by the atmosphere to give iron oxide and sulfate in the sea and the reversal of the reaction by bacteria. Equation (2.10) shows energisation by the Sun and not each step is in equilibrium. All the steps are assigned rates based on experimental information. When equations are integrated in million-year periods they give a graph of CO_2 and O_2 levels against time based on an initial condition at 542 Ma. The resultant calculated history of O_2 is given in Figure 2.11 and shows that oxygen partial pressure averages around 20%, *i.e.* PAL (present atmospheric level). This figure is generally accepted and is not very different from previous descriptions. The most notable feature is that the O_2 pressure during this period can vary no more than two-fold. The rise between 358 and 300 Ma is due to the burial of reduced carbon. This is called the Carboniferous Period and its products are man's fuels. The CO_2 variation is shown separately in Figure 2.12 where the present CO_2 level is taken as a reference for the ratio $R(CO_2)$. The important generally agreed features are the drop before 542 Ma and again between 358 and 300 Ma and the fluctuations within the slow fall to today. The scale of these changes is much larger than that we relate to global warming today, but not different in kind, and they would have been associated with temperature changes. The CO_2 concentration from the time Earth formed had fallen by at least 100-fold by 3.5 Ga and fell almost another 100-fold before the Cambrian Period (see Figures 2.5 and 2.12).

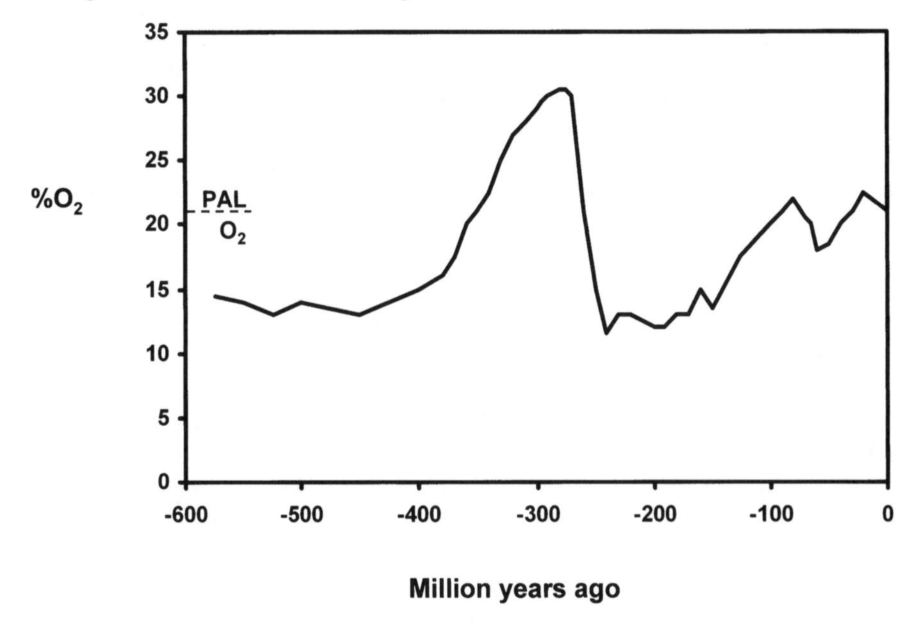

Figure 2.11 The changes in percentage of oxygen in the atmosphere in the last 600 million years. Compare Fig. 2.5. From Berner Ref. 39.

We can give some quantitative estimates of the above carbon cycling. The quantities of carbon on the surface of Earth as given by Berner[39] are

Carbon in rocks	$5{,}000 \times 10^{18}$ moles of carbonate carbon
Organic carbon in sediments	$1{,}300 \times 10^{18}$ moles
Biomass carbon today	0.05×10^{18} moles

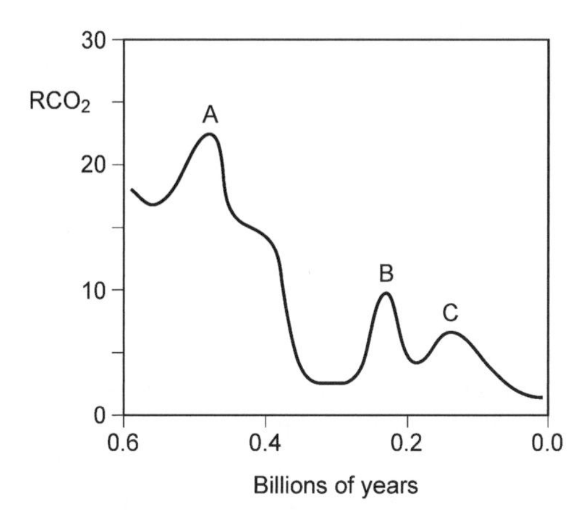

Figure 2.12 The changes in CO_2 relative to today's value taken as 1.0 in the last 600 million years. Compare Fig. 2.5. From Berner Ref. 39.

By far the greatest amount of carbon is locked in minerals and sediments and not in organisms. The surprising figure is the organic carbon in sediments. The half-life of reactions (2.9) to (2.11) is about 0.01 million years (10^4 years), giving a turnover of carbon greater than 10^{17} moles per year, which exceeds or equals that of living matter today but, considering that biomass in earlier periods was less, while carbonate was similar, it was the non-living carbon that dominated at earlier times. It is the opinion of Berner[39] and others that the dominant influence over the O_2/CO_2 conditions at any given time is due to the inorganic carbonate and the organic debris. Biology maintains the debris in sediments. Clearly, however, although certain amounts of debris became trapped in coal, oil and gas over long periods, almost permanently, from around 358 Ma (see Section 3.9), they are insufficient to affect the long-term changes. (We shall return to the O_2/CO_2 ratio in the course of evolution and its relationship to plant life in Section 4.18).

A final point concerning equations (2.9) to (2.11): the first of the two reactions of (2.9) are catalysed continuously after the appearance of land plants and both the reactions of (2.10) and the back-reaction of (2.12) are under some biologically controlled catalysis. In these reactions the pathways are complex and many different catalysts are required in addition to the bulk elements in the steps. Here therefore it is not bulk non-metal element cycles alone which are important, but also those of trace elements. These elements are from the environment and organisms are dependent largely on their availability. Apart from magnesium, also involved directly in equation (9), as is iron in equation (11), they include catalytic amounts of V, Mn, Fe, Co, Ni, Cu, Zn and Mo (see Table 2.2). We have looked carefully at the changes of their concentration with time both before, when they were considerable, and after 542 Ma, when they were small, as they are critical to evolution and may well be far from steady states. We must also be aware that addition of other poisonous heavy metals can adversely affect the rate of step (10).

The generality of this approach to modelling atmospheric evolution was affected by many strong fluctuations. For example, in the Permian Period following 300 Ma there was considerable sea-floor spreading with a rise in sea levels. At the same time there was new formation of much dolomite. Together they account for some changes in composition of the seawater in this period. Both the concentrations of Mg^{2+} and SO_4^{2-} were also high in the immediate Precambrian and remain so today with a lower Ca^{2+} concentration in the sea,[4,15,40] (Section 5.6). Interest also lies in the variations of the above trace elements (see Section 2.11). Here we note that these element changes occurred in the same periods as the intense volcanic activity associated with separation of continents and connected with the spreading of sea floors. These periods, like the Precambrian/Cambrian boundary, are associated with extinctions of many organisms (as are the Snowball Earth periods) (see Section 3.3), but life persisted through these variations of times of considerable loss. (The main loss was of eukaryotes in each case.) These data and the very use of equations (2.9) to (2.11) show the very strong coupling of organism and geological chemistry in an ecosystem making it impossible for us to look upon evolution of organisms other than as strongly and systematically coupled to the environment. We stress that the environmental changes overall are unidirectional both in the effect of weathering and in the increase in the oxidation potential though fluctuations occur. They are coupled and inevitable. The first is mainly of inorganic chemistry of minerals while the second is mainly that of the sea but the two are intertwined, especially through cellular catalysis and control by trace metal ions. To them we have to add the growing effect of organisms in both. They are continuous from the earliest times though the basic chemistry changed only relatively slightly in the last 542 million years, but this does not exclude variation in types of organism which suffered extinction or evolution. We shall consider in the next chapter that organisms also developed complexity in a systematic manner in concert with smaller changes of the environment but again with fluctuations. Only on a very short time scale is there a possible Gaia-like steady state (see Chapter 7).

Apart from extreme cold periods and warm volcanic periods of extinction, there are other periods of less warmth and glaciation (Section 2.2). The longer periods correspond to the fall in CO_2 (cold) and its rise (warm), the strongest of which appear in Figure 2.12. One such glaciation period was in the Carboniferous from around 0.358 Ga, already noted, while the warm period between 0.25 and 0.05 Ga is due to increased volcanic activity. In Chapter 3 we shall note the changes in organisms in these periods. The general fall in CO_2 after 0.542 Ga is due to the rapidly increasing carbon trapped in living organisms and a general cooling. Smaller glacial–interglacial (ice sheet fluctuation) cycles are common and have little effect on overall evolution.

We can use a particular stage in the evolution of organism chemistry and the changes in the minerals, namely the coal of the Carboniferous Period, to illustrate the intimacy of the development of chemicals of organisms and those of the geological formations, after BIF was complete.

2.12.2 Changes in Major Non-Redox Mineral Elements in the Sea from 0.54 Ga

The main mineral elements in the sea are Na^+, Cl^-, Mg^{2+}, HS^-/SO_4^{2-}, K^+ and Ca^{2+}, with some Si and P (Figure 2.13; Table 2.2). They became important in the evolution of life and most must have been important even in the origin of life but they show relatively small changes in concentration in the sea over time, except for HS^-/SO_4^{2-} as described in Section 2.7, Chapter 5 and in Figure 2.14. They only have single valence states and are not susceptible to redox changes of the environment. They do change with weather variation. Included in these small changes are also recent decreases in silica content.

Mg^{2+} and Ca^{2+} are of interest as they show fluctuations, not generally unidirectional long-lasting changes, due to continental weathering (glaciation and high temperature), precipitation, subduction, hydrothermal activity and volcanic and continental break-up. Silica varied somewhat in the ocean[41] and

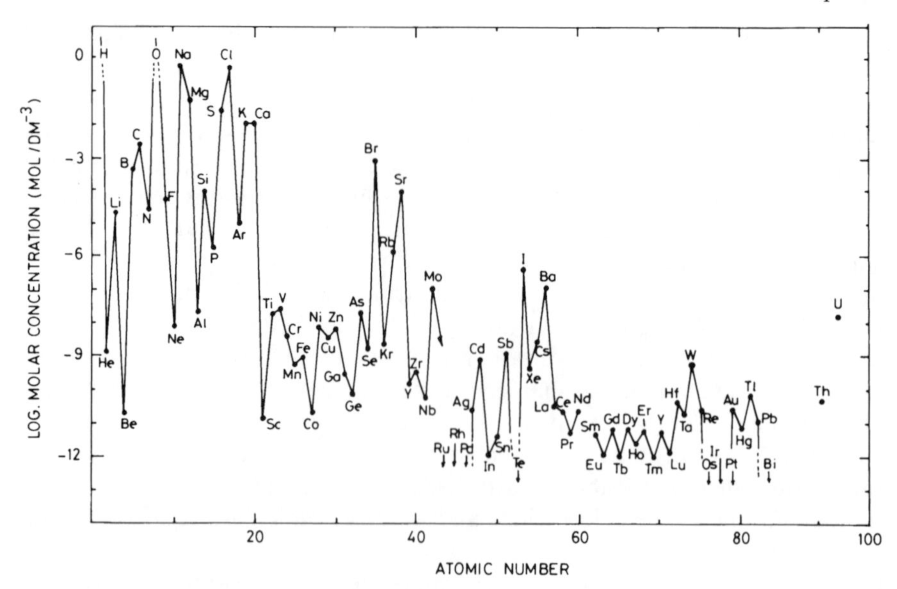

Figure 2.13 The concentrations of the elements in the sea today. Note the peaks for the available elements.

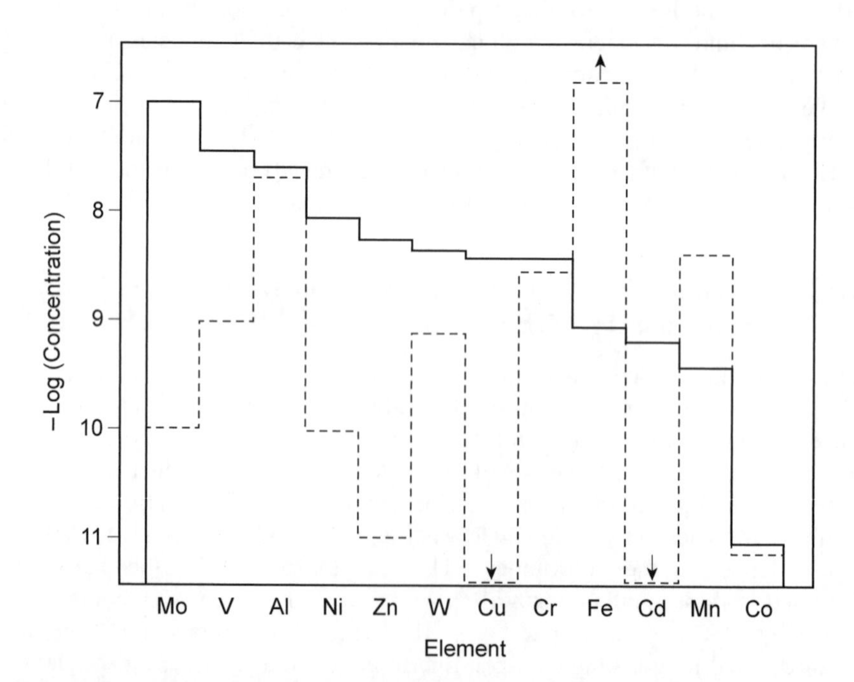

Figure 2.14 Concentrations of some elements in the sea before 3.0 Ga, broken line, and today, full line.[4,15–17]

in relatively recent times. Around 300 Ma, the oceans were saturated with silica due to volcanic activity and thereafter there is a biological sink which was reduced as silica declined. Na^+ and K^+ concentrations are not affected. When we turn to trace elements we have noted that silicates, like oxides, are preferentially formed with small higher tetravalent and trivalent cations but somewhat less strongly with small divalent ions such as Ni^{2+} and even less so with larger ions, such as Mn^{2+}. All larger ions may be more favoured by association with carbonate and sulfate, *e.g.* Sr^{2+}. We turn to the biological significance of these interactions in Chapters 3 and 5.

As an indication of the small variations of the major salt concentrations in the sea the values of inorganic deposits from inclusions of the sea have been studied. They show that while the magnesium concentration varied slightly from 50 mM the calcium concentration fluctuated from below 20 mM to above 50 mM. These variations are strongly correlated with the appearance of aragonite when the Mg/Ca ratio is greater than 2.0 and Ca^{2+} is less than 30 mM, but of calcite when Mg/Ca is below 2.0 and calcium is above 50 mM. Now the geological periods of the appearance of the two allotropes are that calcite occurs from the Cambrian, 542 Ma, to the Carboniferous, 358 Ma, then aragonite from the Permian, 300 Ma, to the end of the Jurassic,150 Ma, after which time calcite became most common almost to today, 25 Ma ago, when aragonite again dominates (see Section 5.6). The dates around 550–500 Ma and 300–150 Ma are close to the beginnings of two major phases of biomineralisation in the sea,[42–44] and may follow from concentration changes due to volcanic action. Certain periods of new biomineralisation may be linked to these periods (Chapters 3 and 5).

The reader may be puzzled as to why we have introduced the times of the appearance of different carbonates. As they form sediments both from organisms and from direct precipitation they provide definite dated information about evolving conditions. In fact the change of calcite to aragonite, and its reverse, occur in exactly the same time periods for both environment and some organisms. We can follow these and other changes by isotope studies. It is very likely that the changes in conditions of the sea change the nature of biomineralisation which is not then due to gene change.

2.12.3 Carbon Isotopes

A further major source of information for the above quantitative interpretation of weathering is the distribution of isotopes of carbon, hydrogen and oxygen. Sediments contain carbon compounds in organic compounds from the chemistry of reduction of CO_2 and CO or in deposited carbonate from depleted CO_2. The isotope distribution, $\delta^{13}C$ relative to a standard, the PeeDee Belemnite in this case, in carbonates is considered as reflecting the $\delta^{13}C$ of the ocean, itself an indicator of the degree of reduction at a particular time and the burial of reduced carbon (organic) relative to that of $CaCO_3$ (deemed oxidised).[44,45] Changes in the chemical redox activity of the oceans whether they arise from inorganic or life's processes are then reflected in carbon isotope/time data. A very large bank of information has been collected from sites all over today's continents (Figure 2.15). They show that there have been major changes in $\delta^{13}C$ from what is referred to as a typical value of +2 to +4‰ during time relative to excursions to large negative values, *e.g.* −6‰ have been found particularly and generally around 635–632, 580 ± 1, 555–550 Ma (and to a lesser degree around 642) as well as later and strongly at about 358–300 Ma. It is generally considered that the very large negative change in ^{13}C isotopes is through the decreased burial of organic (reduced) carbon due to extinctions of organisms. Also high-resolution perturbations are being identified and attributed to massive injection from degradation of methane hydrates. Curiously it is after the major isotopic changes before 542 and 358–300 Ma that biominerals appear in new organisms (see Tables 3.2 and 3.3). These times are also those of major geological events such as glaciation, 630 and 580 Ma, or volcanic activity starting at 300 Ma. Maybe they gave rise to high calcium and silica in the sea.[41] The isotope data for very early periods, perhaps at all times before 1.0 Ga have been open to much debate and we prefer to leave them aside until experts agree. The dating of these periods has been described in Table 1.3 and is reliably given by radioactive degradation reactions of heavy element nuclei, *e.g.* of uranium (Section 2.3).

2.12.4 Oxygen Isotopes

The $\delta^{18}O$, or relative abundance of ^{18}O to ^{16}O compared to a reference standard ($\delta^{18}O$), of well-preserved shell material is thought to provide some

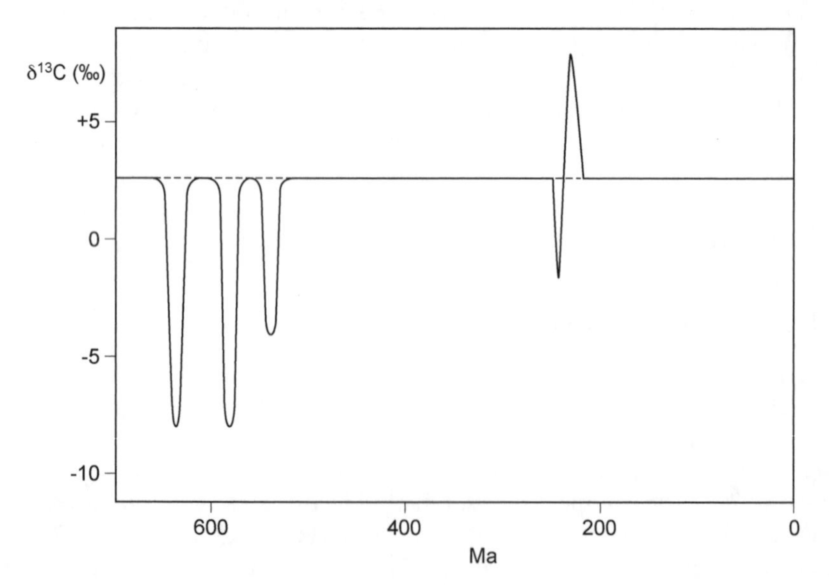

Figure 2.15 A very schematic view of the major changes in carbon isotope fractionation with time to illustrate coincidence with extinctions or major climate changes followed by rapid recovery.

record of global climate over the last 500–600 million years.[3,46] There are two components captured by the calcium carbonate isotopic signature of $\delta^{18}O$. First, there is a temperature-dependent equilibrium fractionation of oxygen isotopes between calcium carbonate (generally precipitated by a biological organism) and the seawater composition of $\delta^{18}O$. The colder the temperature, the greater this fractionation, and the isotopically heavier the carbonate. This provides the basis for the wide application of $\delta^{18}O$ in fossils as a paleothermometer. However, the isotopic composition of the carbonates will also be determined by the isotopic composition of seawater from which the carbonate is growing. On shorter timescales (tens of thousands of years) the dominant control on seawater $\delta^{18}O$ is the expansion and retreat of ice sheets which lock away preferentially the light O isotope due to the Rayleigh fractionation of isotopes during precipitation from atmospheric vapour. Hence as the world enters an Ice Age, climate cools and ice sheets grow which both contribute to an isotopically heavier carbonate (more positive $\delta^{18}O$ signature) compared to the warmer periods. These large climatic changes outweigh the imprint of physiology on the $\delta^{18}O$ of the biominerals or so-called 'vital effects'. Over long timescales (order of 100 Myrs), the geological water cycle can play a role in determining the isotopic composition of the ocean and it is necessary to correct for an evolving baseline to derive climatic information.

The data show that there was a cold period around 450 Ma and close to today while warmest periods were some 250–50 Ma, and between 500 and 360 Ma with a possible end Ordovician glaciation. The oxygen isotope data with that of carbon are extremely useful in following change in the coupled environment/organism system much as were the isotopes of many heavy elements in following a wider range of chemical change before 0.54 Ga.

2.13 Summary of Geological 'Inorganic' Chemistry Evolution

Before we introduce the evolution of biological organic chemistry in Chapter 4 we have shown that it must have had imposed upon it the limiting global conditions from the continuously evolving inorganic chemistry of Earth's atmosphere, mineral surface and sea from 4.6 Ga as described in Sections 2.2 and 2.3. No matter how much of this inorganic chemistry came to be due to products from organisms after 3.5 Ga, the non-biologically inorganic chemistry changes have continued until today. The basic inorganic chemistry evolution was and remains in large part irreversible. It was initially dependent on element abundance, and the cooling from the state at the time of the Earth's origin, leading to the separation of compounds in the atmosphere, sea, and solid minerals in a reduced condition and close to at a fixed temperature by 3.5 Ga which has lasted until today. At the low temperature at this time, chemical equilibrium had been approached between the surface components in all the three phases for the rapidly reacting inorganic ions, but not for any organic chemicals. Throughout this account we have then been able to use quantitative data on solubility products, complex formation constants and oxidation reduction potential to understand if not to predict inorganic chemical events leading to the time 3.5 Ga when there is evidence for, but also to see how it guided organic cellular chemistry from 3.5 Ga to today.

In the earliest period inorganic chemicals are excellent markers of the major changes of chemicals due to weathering including volcanic actions, upwellings from ocean vents. During this time from 4.5 to 3.5 Ga the atmosphere changed considerably as it lost H_2 and CH_4 to space and CO_2 to sediments of carbonates. Much of the chemistry of environmental weathering is initially energised by the Sun or temperature gradients but it goes to equilibrium in the sea. We have seen here and shall see more extensively in Chapter 4 that between 4.6 and 3.5 Ga energisation of Earth's surface also produced unstable non-equilibrated organic chemicals and then life, and as a subsequent consequence, oxidising agents through the oxygen released to the environment. We then considered how the slow change from a reducing to an oxidising environment due to oxygen, induced new inorganic chemistry on Earth's surface first in the period 3.0 to 0.75 Ga. For a long time, even until, say 0.54 Ga, the time of the Cambrian Explosion, the dominant effect was of continued weathering, including oxidation of minerals of the elements which entered the sea. We considered that equilibrium in the sea, and between it and the underlying mineral surface, notably sediments, was still attained at any given time in these more oxidising conditions. Under conditions of environmental equilibrium, we treated the resultant changing availability of the trace elements in preparation for the discussion of the interaction of them with organic chemicals in cells (Chapter 4). We were able to confirm the changes from dating of isotope changes and of the impurity contents of sediments at various times. In this account we described the equilibria in the sea as if there were a gross average over all depths. We do know that this is not the case, as the surface was often more oxidised than the deep, most probably from 2.5 to 0.75 Ga. Such inhomogeneity is very hard to detail, so we used the changing average redox potential to describe the way ions in the sea were slowly modified. In particular, the most dominant chemical changes were that of the large amount of Fe^{2+} much of which is seen in BIF, with the eventual rise of sulfate by oxidation of sulfide to the high concentration in the sea today recording changes in sulfur chemistry. During this period of oxygen release by a relatively large number of cyanobacteria, that is from 2.5 to 0.75 Ga, these two oxidative reactions buffered the sea between an equilibrium redox potential of -0.1 to $+0.1$ V. This period includes the 'boring' billion years from approximately 2.0 to 1.0 Ga when change was slowest. This slow change is typical of the central part of a redox titration (see Figure 2.8). We concluded that much as organisms could have catalysed the oxidation of these, the two dominant and originally reduced elements, the underlying evolution is due to simple geochemistry once there was oxygen. A number of different trace element redox potentials, including the rare earths, were examined to confirm the analysis of the changing redox conditions and of CO_2 levels. We noted that the steep initial variation of solubility of the initial sulfides across the series of divalent ions Mn > Fe > Co > Ni > Cu > Zn in reducing conditions, was replaced by the shallower variation of oxides (hydroxides) in oxidising conditions, shown in Figures 1.4 and 1.5. While in reducing conditions Mn and Fe dominated, in oxidising conditions, Cu and Zn and other metal ions of Mo and Se were shown to have increased considerably in the sea (Figure 2.16), and they became outstanding essential catalysts particularly in later cells.

Quite clearly as equilibrium conditions of each element were controlled by the Fe/S redox curve (see Figure 2.8), they arise in turn in the environment and hence these inorganic chemical equilibria go on to influence inevitably the types of cells which can evolve in turn. Meanwhile the constant bulk inorganic ions of the sea remained the effective components of osmotic and electrical charge balance (Chapter 4). The necessary energisation of their gradients must have been formed in the precursors to cells before there was reproduction. Both sets of ions became involved in controls of organic chemistry later. Soon after 0.75 Ga the Cambrian Explosion occurred, probably due to the second rapid rise of oxygen accompanied by increases in availability of many of the metal elements from minerals.

Meanwhile, the inorganic chemistry of the non-metals changed in a quite different way to somewhat inaccessible organic products of oxidation, for example from CO to carbonate and from NH_3 to N_2 and nitrate. They could not

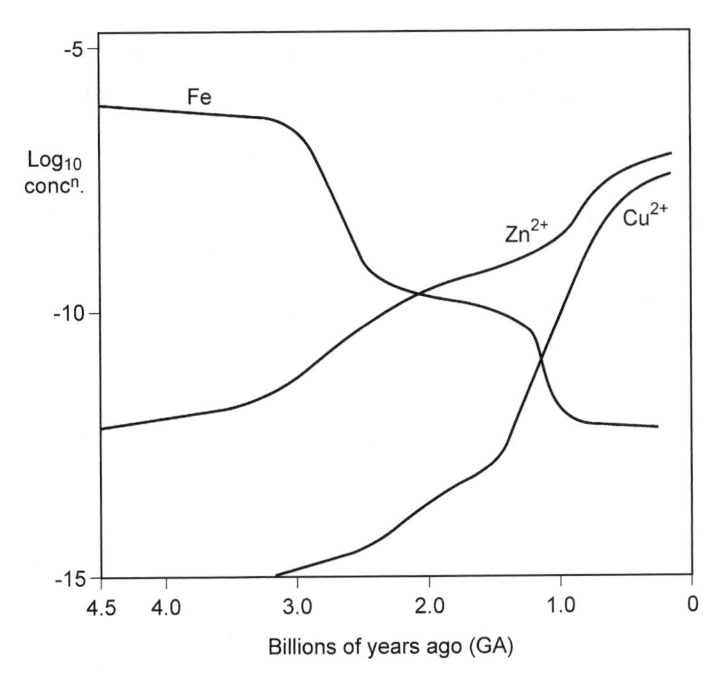

Figure 2.16 A more detailed examination of the probable concentrations of three elements in the sea over time. The estimates are based on the change in redox potential in Fig. 2.8 and 2.10. See also references, 5, 17, 31.

equilibrate quickly in compounds much though they were changed in order of their redox potentials (see Figure 1.6). This led to great difficulty in their reduction to produce organic molecules a process which needed the newly available elements as reduction catalysts.[12] As a final remark we stress that, in the changing mixed system of organic and inorganic chemicals, non-metals and metals, toward oxidising conditions, the very fact that the inorganic elements reacted rapidly and so changed first, meant that they would lead development of any new organic chemistry. It is the rejection of oxidising agents especially O_2 by organisms themselves which caused the change in the sea but at the same time, increased weathering and introduced more ions in the sea. They also reacted rapidly toward equilibrium. The evolution of the use of all these inorganic ions in organic chemistry and cell development was then predictable. After 0.5 to 0.4 Ga the evolution of basic chemistry was largely complete, as shown by the approximate constancy of O_2 and CO_2 in the atmosphere (see Figures 2.11 and 2.12), but we must look closely at the causes of later extinctions and recovery from them. To support the evidence for the relationship of the development of organic chemistry with that of inorganic chemistry we look next at the historical record of organisms seen as fossils as direct evidence for chemistry evolution in Chapter 3. Only toward the end of the book shall we consider how genetic instructions evolved, and are interactive with these chemical changes of the environment. The question arises: was evolution from 4.5 to 0.5 Ga random or directed in predictable outline?

2.14 A Note: The Relationship between Metal Structures in Organisms, Minerals and Chemical Models

In passing we note that the study of certain metal salts and clusters in abiotic laboratory chemistry and the ways in which they are synthesised in laboratories is often of assistance to our understanding of the ways in which the metal ions are held by organisms.[15] The obvious relationship between the building blocks in biological minerals and both geological minerals and synthetic crystals is one side of their overall structure in organisms, *e.g.* in shells, Section 2.6. However, these high symmetry units in such shells as those of cockles, mussels ($CaCO_3$) and Acantharia ($SrSO_4$) bear little relationship to the final shape of the biomineral structure or shell often of little or no symmetry. There is evidence too of the relationship between biological silica structures which are based on hydrated SiO_2 and silica in synthesised compounds and they are quite difficult to model.

Of greater interest are the parallels between some metal centres in enzymes and laboratory synthesised molecular complex ion structures. A simple example is the Fe_4S_4 cubic complex common in electron transfer proteins and readily synthesised in the laboratory. Less easily explained is the finding of Fe, Ni, clusters of complexes of CO and CN^- in hydrogenases which are so common in synthesised complexes (Chapter 6). These compounds are not always of high thermodynamic stability. The principles of selection in the binding of metal ions to the organic side-chains of proteins in marked contrast have often been shown to arise from considerations of the thermodynamics of binding constants as seen in model complexes in the laboratory, for example following the Irving–Williams series. Once again it is the details of both lengths and angles which are distinctively different in low symmetry sites in enzymes leading to them being given the special name of entatic states. These states are particularly important in catalysis by transition metalloenzymes. The description of all these metal sites and the parallels or lack of parallels with minerals and models has been summarised in several books on biological inorganic chemistry. The view is sometimes expressed there that many of the parallels simply arose from the incorporation of abiotic fragments of mineral/abiotic/metal complexes into proteins. Recently this view has been extended to the description of the $CaMn_4$ cluster site of oxygen synthesis in photosystem II. It has been shown that minerals and models with composition close to this can be activated at high concentration and adsorption to release oxygen from water.[47]

References

1. J. Press and R. Siever, *Understanding Earth*, W. H. Freeman, New York, 1994.
2. R. M. Hazen, D. Papineau, W. Blecker, R. T. Downs, J. M. Ferry, T. J. McCoy, D. A. Sverjensky and H. Yang, *Am. Mineral.*, 2008, **93**, 1693.
3. R. E. Zeebe and D. Wolf-Gladrow, *CO_2 in Seawater: Equilibrium, Kinetics, Isotopes*, Elsevier Oceanography Series, Volume 65, Elsevier, Oxford, 2001.
4. P. A. Cox, *The Elements on Earth*, Oxford University Press, Oxford, 1995.
5. C. Klein, *Am. Mineral.*, 2005, **90**, 1473.
6. R. J. P. Williams, *J. Theoret. Biol.*, 1961, **1**, 1.
7. P. Mitchell, *Nature*, 1961, **191**, 144.
8. T. W. Lyons, A. D. Anbar, S. Severman, C. Scott and B. C. Benjamin, *Annu. Rev. Earth Planet. Sci.*, 2009, **37**, 507.
9. J. Farquhar, H. Bao and M. Thiemens, *Science*, 2000, **289**, 756.
10. C. Scott, T. W. Lyons, A. Bekker, Y. Y. Shen, S. H. Poulton, X. Shu and A. D. Anbar, *Nature*, 2008, **452**, 456.
11. H. Ohmoto, Y. Watanabe, K. E.Yamaguchi, M. Naraoka, T. Kakagawa, K. Hayashi and Y. Kato, *Geol. Soc. Am., Mem.*, 2005, **198**, 291.
12. C. Klein and N. J. Beukes, *Econ. Geol.*, 1993, **88**, 542.
13. R. M. Hazen, R. C. Ewing, D. A. Sverjensky, *Am. Mineral.*, 2009, **94**, 1293.
14. H. Elderfield, H. D. Holland and K. K. Turekian (eds), *The Oceans and Marine Geochemistry*, Elsevier, Oxford, 2004.
15. R. J. P. Williams and J. J. R. Fraústo da Silva, *The Chemistry of Evolution*, Elsevier, Amsterdam, 2006.
16. D. E. Canfield, S. W. Poulton and G. M. Narbonne, *Science*, 2007, **315**, 92.
17. M. A. Saito, D. M. Sigman and F. M. M. Morel, *Inorg. Chim. Acta*, 2003, **356**, 308.
18. C. L. Dupont, A. Butcher, R. E. Vales, P. E. Bourne and G. Caetano-Andies, *Proc. Natl. Acad. Sci. USA*, 2010, **107**, 10567.

19. C. Siebert, J. D. Kramers, T. Meisel, P. Morel and T. F. Nagler, *Geochim. Cosmochim. Acta*, 2005, **69**, 1787.
20. R. Bolhar, B. S. Kamber, S. Moorbath, C. M. Fede and M. J. Whitehouse, *Earth Planet Sci. Lett.*, 2004, **222**, 43.
21. C. Bonnot-Gurtois, *Mar. Geol.*, 1981, **39**, 1.
22. H. Elderfield and M. J. Greaves, *Nature*, 1982, **296**, 214.
23. H. Elderfield, *Philos. Trans. R. Soc., A*, 1988, **325**, 105.
24. C. R. German, G. P. Klinkhammer, J. M. Edmond, A. Mitva and H. Elderfield, *Nature*, 1990, **345**, 526.
25. J. B. Corless, M. Lyle and J. Dymand, *Earth Planet Sci. Lett.*, 1978, **40**, 12.
26. B. W. Aledanda, M. Ban and P. Anderson, *Earth Planet Sci. Lett.*, 2009, **283**, 144.
27. J. D. Vine and E. B. Touretelot, *Econ. Geol.*, 1970, **65**, 253.
28. B. Lehmann, T. F. Nägler, H. D. Holland, M. Wille, J. Mao, J. Pan, D. Ma and P. Dulski, *Geology*, 2007, **35**, 403.
29. H. D. Holland, *Econ. Geol.*, 1979, **74**, 1676.
30. A. D. Anbar, *Science*, 2008, **322**, 1481.
31. K. M. Meyer and L. R. Kump, *Ann. Rev. Earth Planet Sci.*, 2008, **36**, 251.
32. W. Yang and H. D. Holland, *Am. J. Sci.*, 2003, **303**, 187.
33. R. J. P. Williams, *Nature*, 1959, **184**, 44.
34. K. O. Konhauser, E. Pecoits, S. V. Lalonde, D. Papineau, E. N. Barley, N. T. Arnott, K. Zahnle and B. S. Kamber, *Nature*, 2009, **458**, 750.
35. K. O. Konhauser, 2009, pers. comm.
36. G. L. Arnold, A. D. Anbar, J. Barling and T. W. Lyons, *Science*, 2004, **304**, 87.
37. W. Bland and D. Rolls, *Weathering: An Introduction to Scientific Principles*, Hodder Arnold, London, 1998.
38. R. A. Berner, *Geochim. Cosmochim. Acta*, 2006, **70**, 5653.
39. R. A. Berner, *The Phanerozoic Carbon Cycle: CO_2 and O_2*, Oxford University Press, Oxford, 2004.
40. S. M. Stanley, *Paleogeogr., Palaeoclimatol., Palaeoecol.*, 2006, **232**, 214.
41. M. Maldonado, M. C. Carmona, J. Briz and A. Gruzado, *Nature*, 1999, **401**, 785.
42. P. M. Dove, J. J. De Yoreo and S. Weiner (eds), *Reviews in Mineralogy and Geochemistry, Volume 54, Biomineralization*, Mineralogical Society of America and the Geochemical Society, Washington, , DC, 2003.
43. A. Sigel, H. Sigel and R. K. O. Sigel, *Biomineralisation: From Nature to Applications*, John Wiley and Sons, Chichester, 2005.
44. J. L. Payne, D. J. Lehrman, J. Wei, M. J. Orchard, D. P. Schrag and A. H. Knoll, *Science*, 2004, **305**, 506.
45. J. A. Karhu and H. D. Holland, *Geology*, 1996, **24**, 867.
46. K. Wallman, *Geochim. Cosmochim. Acta*, 2001, **65**, 2469.
47. M. M. Najafpour, T. Ehrenberg, M. Wiecher and P. Kurz, *Angew. Chem., Int. Ed.*, 2010, **49**, 2233.

CHAPTER 3

Organism Development from the Fossil Record and the Chemistry of the Nature of Biominerals

3.1 Introduction

[In order to give the physical scientist reader an appreciation of the part fossils have played in all evolutionary studies we give in Sections 3.2 and 3.3 descriptions of them and their evolution. The rest of the chapter is devoted to their chemistry].

In this chapter we begin to link the evolution of the very different organic chemistry of organisms to the evolution of the geological inorganic chemistry using fossil evidence. The most observable link is via the biominerals which are largely inorganic but in part organic structures bound to or within organisms. They are hard structures and have a great importance for our knowledge of the evolution of organisms as they are preserved directly while soft structures only appear in the fossil record as imprints in sedimentary rocks. The earliest soft structures have been dated, around 3.0 Ga long before hard structures arose after 0.75 Ga. Both give us also knowledge of the development of external shapes of organisms and some knowledge of their internal structure. We stress immediately that while we have an interest in fossils *per se* they also tell us when a certain type of biological structure evolved. They then help us to see the timing of biological evolution in parallel with the geochemical changes using inorganic elements. Our greatest interest is then in the kind of chemistry which evolved from making soft organic structures to hard mineral chemistry protecting cells. Biominerals tell us about the history of precipitation and the

Evolution's Destiny: Co-evolving Chemistry of the Environment and Life
R. J. P. Williams and R. E. M. Rickaby

Published by the Royal Society of Chemistry, www.rsc.org

biochemistry of the elements in biological solid structures. The finding of imprints indicates forms of earlier life with only soft organic structures. Their structure and biochemistry are assessed from parallels with the shape and chemical content of today's organisms. Their importance resides in that the fossil record[1] is the only direct history of organisms, dated by the known ages of volcanic rocks to the sediments in which they were buried, or from carbon and other isotope data, based on chemostatigraphy where that is available (see Table 1.3). A somewhat more difficult material to assess are so-called organic molecular fossils, inert chemicals found in sediments of different periods and probably related to organism chemicals (Section 3.9). A quite separate indirect record of life in some sedimentary rock formations is that of the inorganic chemical products probably from reaction of chemicals from organisms with components of the sea of the same period. We have already referred to BIFs as one example mentioned in Chapter 2. They were formed, at least in part, by the reaction of oxygen, a waste product of very early organisms to today, with ferrous iron, much of it from the sea. All other descriptions of the evolution of organisms are inferred from knowledge of organisms using extrapolation in time by either a comparison of related organisms of today or of chemicals such as DNA and RNA of today's organism. We also make use of novel small chemicals which appeared in particular groups of organisms at given times, usually calibrated by fossils and their datings. An account of this inferred organic chemistry of evolution, and direct involvement of inorganic elements in its catalysts and controls, is given in Chapter 4. Once we have firmly in place a fossil history of organisms themselves in this chapter and organism's organic chemistry, with passing reference to the use of inorganic chemicals in their synthesis we shall turn in detail to the way evolution of the environmental inorganic chemistry described in Chapter 2, influenced this and all other biological organic chemistry.

We begin this chapter then with a general description of the kinds of fossils and their datings (Section 3.2). We follow it with a description and times when fossils are found only with considerable difficulty. In both sections we note coincidental features of the physical and chemical environment such as the coming of oxygen or of unusual temperature with the appearance of particular kinds of fossil. Other types of evidence, which can be obtained directly from the study of fossils are times of appearance of different kinds of biominerals (Section 3.4). The chemistry of the organisms of fossils both inorganic (Section 3.5), and organic (Section 3.6), is very likely to be related to today's chemistry of similar organisms. We describe chemical fossils, including complex molecule findings in Section 3.9, and fossil fuels in Section 3.10. We draw together all the information in the section into our conclusions in Section 3.12.

Before we commence this description we must stress that the fossil record, the only direct evidence of the history of life, is not perfect. It is biased by the strength of preservation of particular life forms. Preservation refers to kinetic trapping as in Figure 1.8. In the sea or in anaerobic conditions preservation is expected to be greater than on land, where fossils are nearly always exposed to oxidation and predators. Soft-bodied organisms are not expected to be preserved as well as hard-bodied organisms, which includes those with hard crosslinked plastic constructs with the biomineralised life forms. Confidence in the quantitative history of life is only possible when direct fossil evidence is put together with as many other forms of information, much gained by inference, as we can find and as we shall stress in the last chapter.

3.2 The Fossil Record

Life is thought to have begun some 3.5×10^9 years ago, a billion years after Earth formed, so we can start from that time to inspect the fossil record using probable imprints of soft structures of organisms, the record of any chemical traces of life in sediments such as unusual organic molecules, or particularly by biased local concentrations of trace inorganic elements, and isotope composition variations of all elements which could indicate cellular activity.[1] There is

Figure 3.1 Four examples of fossil imprints from 0.75 to 0.55 Ga modified from ref. 3.

in fact very little direct information until there are biomineral fossils, which are relatively recent, the earliest from around 0.75 Ga[2,3] and which are one such readily recognisable biological concentration of elements (Figure 3.1). They also often have structural features, helping us to draw conclusions about cell organisation.

In the earliest period of life between 3.5 and 2.5 Ga the fossil evidence is no more than clusters of vague imprints resembling the shapes of today's bacteria.[1] There may have been only two kinds of small soft-bodied single-cell prokaryotes, Eubacteria and Archaea, including by 3.0 to 2.5 Ga, more than one kind of photosynthesising organism, *e.g.* cyanobacteria with different pigments. None of these prokaryotes had other than simple shapes, judging mostly by today's examples, and are known and classified by chemical differences observed in today's related organisms for the most part (see Section 4.2). They are grouped together before 2.5 Ga as anaerobes. Judging by today's organisms they also had some internal physical organisation (Section 4.5). Starting at a date of around 2.0 to 1.0 Ga there is some suggestive fossil evidence of larger organisms. Most if not all of these novel organisms are aerobes. From 3.5 Ga to even the end of this time, 1.0 Ga, there is still no evidence of any biomineralisation except stromatolites which are prokaryotes with accretions of sedimentary material. However some of today's cells, thought to have originated by 2.5 to 2.0 Ga, have inclusions which can include storage of minerals such as sulfur and perhaps phosphorites (apatite?) which are found in Eubacteria, for example the giant sulfur species, Sulfulobus.[4] Some of these bacteria may also have had a thin phosphorite $Ca_3(PO_4)_2$ covering reminiscent of an eggshell in character. Similar inclusions of mineral salts are found even in the cells of modern eukaryotes, for example in snails.[5] Most phosphorite mineral deposits of this early period seem to have resulted, however, from external inorganic precipitation from saturated solutions and are not known to have been selectively induced by cell surface interaction. The next inferred related chemistry and their possible imprint fossil evidence has led to the strong belief that, at least three main lines of the first eukaryotic photosynthesising algae, green, red and brown, arose around 2.0–1.5 Ga. (We can confirm the date of the earliest red/brown algae to around these times from genetic data; see Figure 3.4 and Chapter 4.) The three are related to major life forms today, mostly red and brown algae (plants) in the sea and green plants on land. There are also possible imprint fossil indications of previously existing non-photosynthesising eukaryotic (unicellular?) organisms in groups related to modern fungi and animals. However there is little firm evidence of any kind for biominerals until after 0.75 Ga when there were some mineralised multicellular organisms. We shall describe the biochemical differences between Eubacteria, Archaea and eukaryotes from today's knowledge in Chapter 4.

We cannot place great confidence in this fossil history of the nature of life before 0.75 Ga. The fossil record from 0.75 to 0.54 Ga, the main Ediacaran Period,[2,3] is better as there are persistent indications of imprint fossils of organisms of a 'designed' shape, which are considered to be both of large unicellular eukaryotes, and even some of multicellular origin, together with prokaryotes.[2] Some do have a complex structure indicating that it is likely that they are truly multicellular with differentiated cells (see Figure 3.1), and some are possible images of primitive sponge- and 'jellyfish'-like organisms. A difficult case is the description of the organism *cloudina*. There is also a slow increase in the small numbers of apparent eukaryotes seen as mineralised fossils (Table 3.1). In modern eukaryotes the interior of the cells is organised in compartments but their arrangement is not necessarily fixed and classification of eukaryotes in this early period is more associated with peculiarity of overall fossil shape than of compartments or of chemistry.

The absence of any considerable fossil record of any kind just before the Cambrian, 0.54 Ga, shows that there was an intense extinction of life then,[2] in which many of the above eukaryote Ediacaran species, soft and hard bodied, disappeared, but it is believed that through this period phytoplankton (and some jellyfish and sponges?) survived. They are very adaptable. We have already noted that there were extreme fluctuations in the environment at about this time (see Section 2.1). There then followed from 0.54 Ga a very rapid increase in fossil numbers and variety of fossils, mainly multicellular eukaryotes, many with outer shells of calcium carbonate, silica and calcium phosphate in several lineages (Figure 3.2; Table 3.2). The period is therefore known as the Cambrian Explosion.[6] These are the first definite forms of extensively mineralised organisms with the possible exception of some acritarchs, a name given to recognisable early fossils of no recognised family. We shall look for an explanation of the Cambrian Explosion later.

Some small organic compounds of the early period have a very long life, *e.g.* the hydrocarbons of saturated H/C compounds, and have been extracted from or associated with the fossils and their surrounds (see Section 3.9). We shall see in Chapter 4 that there are now additional ways of examining the history of organic compounds, particularly of biopolymers, in the period from around 1.0 Ga to today. Particular metal elements are essential for catalysis of certain chemical reactions leading to these organic compounds in shells, for example copper. We know that their availability evolved in or around this period (Section 2.11). It is through deduced knowledge of the change of particularly metal elements in organisms, the metallome,[7] associated with life forms known today and seen as fossils that we see a link to inorganic elements in sediments. They are especially connected to catalysts: and through them we shall be able to make some of the best connections between the development of the biological organic chemistry of fossils since we have lost much of the detailed

Table 3.1 Early Groups of Organisms with Biominerals

Porifera	*Cnidaria*	*Cloudina (Namapoikia) ?*
Silica/Carbonate Sponges, Diatoms	Carbonate Corals	Carbonate (Extinct)

These early organisms may be unicellular or colonies of them

Figure 3.2 Three examples of mineralised structures. Calcium carbonate of the multicellular cockle shell, calcium carbonate of the foraminiferae and silica of the crown of thorns radiolaria, all formed by 0.5 Ga, from ref. 7.

Table 3.2 Multi-cellular Mineralisation

Organism	*Mineral*	*Date (Ma)*
Sponges	SiO_2	Pre Cambrian? (Ediacaran)
Sclerosponges	SiO_2, $CaCO_3$	Early Cambrian ~540
Molluscs	$CaCO_3$	Cambrian 542-488
Vertebrates	$Ca_2(OH)PO_4$	Ordovician 488-443
Cloudina etc.	$CaCO_3$	Pre Cambrian* Ediacaran

*Multicellular nature uncertain, extinct

knowledge of organic chemical changes. If trace inorganic chemicals and/or their isotopes associated selectively with fossils could be identified clearly we would even be able to search for evidence of life on other planets with confidence for the elements and their isotopes are strongly selected quantitatively (Chapters 5 and 6). At the same time it is important to follow the division of space into compartments in cells by organic membranes and into differentiated cells seen amongst fossils of multicellular eukaryotes. Most importantly we observe a vast variety of shapes of fossils, internal and external, in the Cambrian Period related to ever-increasing varieties of species (Figures 3.2 and 3.3). From this time biodiversity and numbers of organisms increased very rapidly. Interestingly from about 0.54 Ga onwards the vast majority of evolutionary lines of future species, especially their basic chemistry, are defined even to today by these fossils.

Now in this period, 0.54 to 0.50 Ga, biomineralisation of certain non-photosynthetic unicellular eukaryote cells, Radiolaria, also appear in the fossil record. Photosynthesising cells only became mineralised later, however, and there are peculiarities in the appearance of biominerals in algae (plants). In the sea the first $CaCO_3$ minerals in them are the coccoliths, at about 0.25 Ga. The dinoflagellates arose at about the same time followed by the diatoms (Table 3.3) evolved somewhat later. On land, where multicellular plants thrive, the horsetails mineralise earlier, 0.4 Ga, but some of the grasses much later, 0.1 Ga. These organisms all contain silica biominerals. We need to understand the early mineralisation of animals and the later appearance of it in plants. In general, why did mineralisation happen at a particular time?

There are two very important further developments. Slightly later than silica and carbonate shells, around 0.40 Ga the calcium phosphate skeleton, bone, developed in animals. This is an extraordinary change of location of a biomineral but there are other examples of similar internal frameworks. One later example is that of the $SrSO_4$ of Acantharia (see Figure 3.6), where the pattern of the skeleton is based symmetrically on the centre of a single cell.[8] The geometry of the whole pattern, unlike that of bone, is rigid and closely related to the symmetry of the crystals of this compound. They probably evolved at about 0.2 Ga but structures of $SrSO_4$ are not stable in the sea as fossils so an exact dating is impossible. There are also later examples of internal isolated small crystals of calcium carbonate in the structures of land

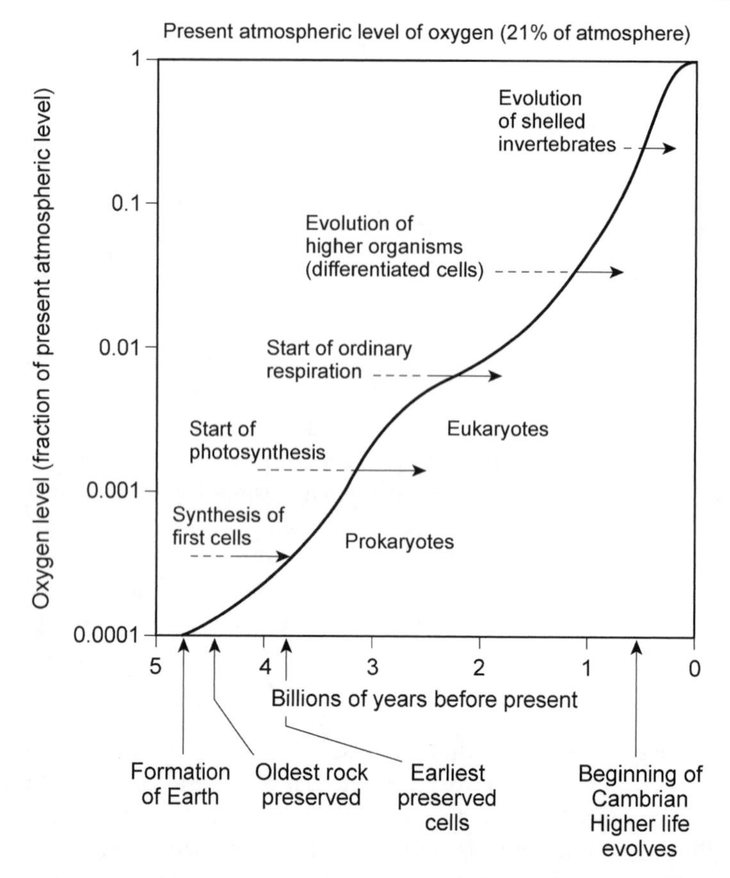

Figure 3.3 The very late arrival of biomineralisation shown against the background of the evolution of the environment linked to that of organisms, and especially that of multicellular organisms and the second steep rise in oxygen, see also Fig. 2.5.

Table 3.3 Single-Cell Eukaryote Mineralisation

Cells	*First Mineralised Fossils (Ma)*	*Mineral*	*Shape*
All kinds of Algae	Pre Cambrian/Cambrian	$CaCO_3$	No Shape
Radiolaria (Animal)	Early Cambrian (540)	SiO_2	Shape
Foraminiferae (Animal)	Early Cambrian (540-500)	$CaCO_3$	Shape
Corals (Mixed)	Ordovician (490-440)	$CaCO_3$	No Shape
Coccolithophores (Plant)	Triassic (250-200)	$CaCO_3$	Shape
Dinoflagellates (Plant)	Triassic (250-200)	SiO_2, $CaCO_3$	Shape
Diatoms (Plant)	Cretaceous (150-110)	SiO_2	Shape
Acantharia (Animal)	Cretaceous (150-110)	$SrSO_4$	Shape

N.B. Mineralisation before 540 mya is not certainly unicellular animals, *e.g.* in the ear, and of larger otoliths in fish, both connected to balance sensing. Some plants have internal $BaSO_4$ crystals, *e.g.* Desmids and Loxedes, presumably as gravity sensors. The properties of these minerals are very different from those of bone and they are difficult to find in fossils. In Chapter 5 we shall discuss the remarkable nature of bone. A variety of other biominerals related to known fossils, are described in many books on the topic[9–11] but they do not yield much information about evolution.

Although much of this description of evolution from fossil evidence is somewhat tentative there are the generally agreed evolutionary trends. The two major events or rather periods of dramatic change are associated with the proposed rise of oxygen partial pressure suggested as taking place in two sloping steps (Figures 3.3 and Section 2.9), the first precedes the somewhat tentatively placed evolution of single-cell eukaryotes and the second coincides with the appearance of multicellularity and many biomineralised organisms (Section 2.4). The two steps are apparently common to chemical changes between many groups of organisms as are the kinds of biomineral in the second period. We shall follow shortly what changes in organic chemicals associated with today's biominerals took place mainly at the above times much though more gradual change would have been taking place all through the period 2.5 to 0.55 Ga. They are plastic polymers, proteins and saccharides. In this way we shall link biomineralisation to the organic molecule changes upon which it clearly depends for nucleation and growth. The coincidental rise in multicellular organisms and of biomineralisation in many groups of eukaryotes is a clear indication of a change of organic structures on the outer surfaces of cells or within internal vesicles around 0.6 Ga. We ask also if the second notable feature of later biomineralisation of plants arose close to the times of 'violent' changes in the environment.

A further very important fossil source of information now concerning plant evolution is in the deposits of reduced organic compounds, oils with gases such as methane, and of coal, a mineral from biological waste (debris) to be described in Section 3.10. These deposits, which are to be likened to mineral fossils, and contain many fossil imprints formed at about 0.4 to 0.3 Ga, the Carboniferous Period, from the degradation of large plant biomass (trees?) and the inability of animals and plants to oxidise them quickly.[12] However, words of warning are again necessary. The synthesis of much earlier organic material leading to very early carbon as graphite (or diamonds) has to be treated as a not very probable marker of life's chemistry. There are associated, purely inorganic, processes which can also give rise to graphite. However the findings of cyclic hydrocarbons with coal is taken as strong evidence for certain kinds of life.

Summarising the fossil evidence, the variety of novel biological minerals and special plastic polymers in their composites and separately, hard plastics as in many 'shells' and incorporating lignin, all appeared strongly in the fossil record only between 0.55 and 0.40 Ga within many groups of species. This is the period of the expansion of a variety of multicellular organisms. We know the earlier date correlates closely just after or with the second rise in oxygen (see Figures 2.4 and 3.3), and its effects on the chemistry of all the elements, but we need to find a causal chemical relationship between these chemical changes and biomineralisation if we are to attribute them all to events which are not just coincidental chance phenomena. The biomineralisation at later times from around 0.2 Ga must have a different explanation. All the developments of organisms must be coded in DNA and we need to find a further connection between the code and the organisms and then environment evolution. The clear links between them all must be in protein chemistry (see Chapter 7), lost from the fossil record.

The progression of the use of inorganic (environmental) elements of animal organisms from the Cambrian to today is also of complexity, and diversity of species, and of greatly increased total biomass, mostly of organic chemicals. The inorganic chemistry seen in fossils changes too but not dramatically until man evolves. Man however is able to exploit to the full the chemistry of all the elements[13,14] in the environment both of the sea and of the land, geological deposits, and represents a quite new chemotype. He is able to work all the minerals of Earth external to himself to obtain new and synthesise long-life chemicals such as plastics to build what will appear in or as fossils later. While the use of energy, from the Sun, increased rapidly before the rise of man, man has required many new sources of energy input and created novel waste materials (see Chapter 7). This will lead to a huge range of future artefacts (fossils) as is obvious already from archaeological exploration. These activities are not easily placed in a Darwinian scheme of chemical evolution, and we must therefore not expect to place all chemical developments with genetic change. Genetic change is closely understood in the case of man, but his activities which will appear in fossils will not be open to correlation with genetic observations.

3.3 Extinctions

As we have stressed in Sections 2.1 and 2.2 the physical changes on Earth's surface were not simply continuous in one direction, hence we cannot expect biological evolution to be a simple sequence as it is dependent on the environment. We may not need to worry about fluctuations before 1.0 Ga when early single cells dominated and as we know phytoplankton for example appear to have not been much affected by whatever environmental stress they met. Rapid mutation also helps them to adapt. (Note the 'trees' of evolution do not easily show the losses – some branches terminate and others thin greatly in fact.) We have already indicated however that in the period between 1.0 and 0.5 Ga particularly, the Earth was not smoothly changing but suffered Snowball Earth periods as well as the break-up of land masses with volcanoes (Section 2.1, Table 3.4), and now and again bombardment by meteorite impacts, as in earlier times. As we noted, in extreme cases this caused occasional massive loss of living species, especially eukaryotes, as shown especially by the greatly reduced number of fossils, for example around

Table 3.4 Mass Extinctions

DATE (Ga)	*Class*	*Strength*
2.3 – 2.1	First Snowball Earth Periods	?
0.71, 0.645, 0.57	Second Snowball Earth Periods	?
0.65 – 0.55	Pre-Cambrian Extinction	Very Strong
0.41-0.35	Devonian Extinction Glaciation	Weak
0.36 – 0.30	Glaciation, Carboniferous	Weak
0.30 – 0.25	Permian Extinction (Volcanic)	Very strong
0.25-0.15	Triassic/Jurassic Extinction (Volcanic)	Very Strong

N.B. Minor extinctions have occurred more recently.

0.54 Ga, called periods of extinction.[15] They are often followed by exceedingly rapid recovery with many new species for example in the Cambrian. The rate of evolution is observed to increase immediately following this extinction perhaps because competition is reduced but we need to look carefully at element availability too. Now a rather different change in conditions occurred especially in the Permian, 0.3 Ga, and at the beginning of the Triassic/Jurassic Period, 0.20 Ga, when intense tectonic and accompanying volcanic action decimated the population of organisms, especially of eukaryotes noted again by their absence in the fossil records. These mass extinctions are of great interest as they, like the extinction of before 0.54 Ga, were followed by a surge in mineralisation in the sea. We shall return to the problems of the chemistry associated with these relatively recent extinctions. To appreciate the link between all the novel mineralisation events and recovery from extinctions we must outline the nature of the types of cell that form biominerals and the chemistry of minerals themselves examining the availability of chemicals and their changes during and after extinction periods. This will lead us to ask detailed questions about organic chemistry as it is novel organic chemicals which form the framework and catalyse nucleation of biominerals.[10] At the same time mineralisation in the sea does require an approach to super-saturation of a chemical, hence a quite high concentration of certain ions, *e.g.* Ca^{2+}, or of a small molecule, *e.g.* $Si(OH)_4$, is of great consequence. Exception to this generalisation follows if an organism accumulates the required elements. As we have stressed in Chapter 1, saturation, the solubility limit, or small degrees of super-saturation are sufficient to cause precipitation, but it requires nucleation.

We have now described in outline the evolution of fossils from direct evidence. Their value is that they show when a certain change in chemistry made it possible to synthesise organic covers with hard plastics, then cells with biominerals both of animals and plants in several periods. This leads us to the second part of this chapter which is to look closely at the nature of the inorganic and organic chemistry and in particular what environmental changes made it possible for these biological changes to occur. We do so in Sections 3.4 to 3.6, indicating from fossils and knowledge of the chemistry of present-day shells of organisms. We can then turn to the shapes of the shells asking what controlled them – physical stresses or genetically coded information. The first chemistry to note is that of the biominerals.

3.4 Types of Biominerals

We have indicated that biominerals are a very important part of the fossil record of evolution. Following this general description of the record we shall describe next the types of biominerals and their record, as they are indicative of the changes of organisation of organic chemicals in different organisms as well as reflecting the environment at a given time.[9–11] The particular points of interest are the chemistry of the elements concerned and that of the supporting matrices as well as the shapes of fossils. We shall exclude from this discussion of biologically mediated biominerals the accidental collecting of preformed geological particles of any kind into a mat by cells. They are well known in the stromatolites where cyanobacteria formed mounds of their cells with silicate and carbonate particle debris. They form a mound possibly so that these photosynthesising cells can gain from lying as close to the surface as they can, and in sunlight. These stromatolites predate biomineralisation, going back to at least 3.0 Ga. We also exclude calcification of cyanobacteria and other organisms at periods of general marine precipitation, that is when the sea is heavily saturated with calcium salts, and also of precipitation from concentrated solution in inclusion bodies as mentioned in organisms such as giant sulfur bacteria. These precipitates have no particular shape and may be colloidal. We shall further leave to one side dated deposited minerals due to oxidation of the environment by organisms but not directly interacting with them, including those in Table 1.1, and already described in Section 2.8. We describe here therefore only those which were part of a living organism by 'intentional' or controlled precipitation and the first formation of which can be dated by fossils.

Using our present-day knowledge we can see that there are several different styles of biomineralisation which have evolved at different times.[9–11] At first we do not refer to the (inherited) shape of the mineral deposited. (The references to organisms in this section are kept to commonly observed species and to as little of their organic biochemical structure as possible.) The modes of biomineral formation are:

1. Single cell external nucleation and growth of minerals, sometimes within cellular colonies.
2. Single cell internal, in vesicle, nucleation and growth of minerals.
3. An intermediate case of single cell internal nucleation and limited growth followed by ejection and extended external growth.
4. Single cell internal nucleation and growth of complete sections follow by export to the outside.
5. Multicell nucleation and growth on the exterior of an organism following total soft-body shape.

6. Multicell growth internal to the whole organism and hence between cells.

We list next known mineralising organisms of these kinds in two sections, A and B, not in order of their evolution, see also Table 3.3 for single cell and Table 3.2 for multicellular eukaryotes.[9–11] Figure 3.4 illustrates their dates.

A. Photosynthesising organisms (related to red and green algae): later plant related

1. Coccoliths, dinoflagellates, diatoms, all derived from red algae, based on single eukaryote cells, following one of the above patterns (1) to (4).

Not so many single-cell green algae are mineralised. There are a few multicell plants related to red algae, *e.g.* some seaweeds, but they are not mineralised. This mineralisation arose after 300 Ma.

2. Many higher plants on land including horsetails and grasses following the sixth of the above examples of mineralisations. All these plants have chloroplasts, and are related to green multicell algae. Most however are not mineralised, including green seaweeds. Mineralised grasses are relatively very recent (50 Ma).

B. Animal or animal-related (non-photosynthesising)

3. Radiolaria, Acantharia,[8] single cell and in colonies, following one of headings (1) to (4). They may well have arisen close to 0.5 Ga but the history of Acantharia is very unsure.
4. Foraminiferae[16] related to 3, single-cell but multicompartment, which are confusingly like pattern (5) (see Figure 3.2). They probably arose around 0.5 Ga.
5. Shells, generally of molluscs, snails, *etc.* (invertebrates, multicell) dating from the Cambrian Period in essence following the fifth pattern. Some, such as worms and insects, are not mineralised, but have chitin, plastic 'shells' (Figure 3.5).
6. Many classes of higher animals, fish, cat, dog, mammal families (vertebrates, multicell) following the sixth pattern. The earliest vertebrates arose around 0.4 Ga.

C. Mixed systems of A and B

7. Coral (often colonies of single cells) – often made from red algae (plants) associated with single-cell non-photosynthesising (animal) eukaryotes, following pattern 1.

Note fungi do not give rise to minerals, neither do broad-leaf plants, and plant roots generally have small mineral content, if any.

The chemical nature of the minerals in these organisms is also listed in Tables 3.2 and 3.3. Examination of the above list indicates that biomineral formation is not just found in one or two types of organism but is very general.

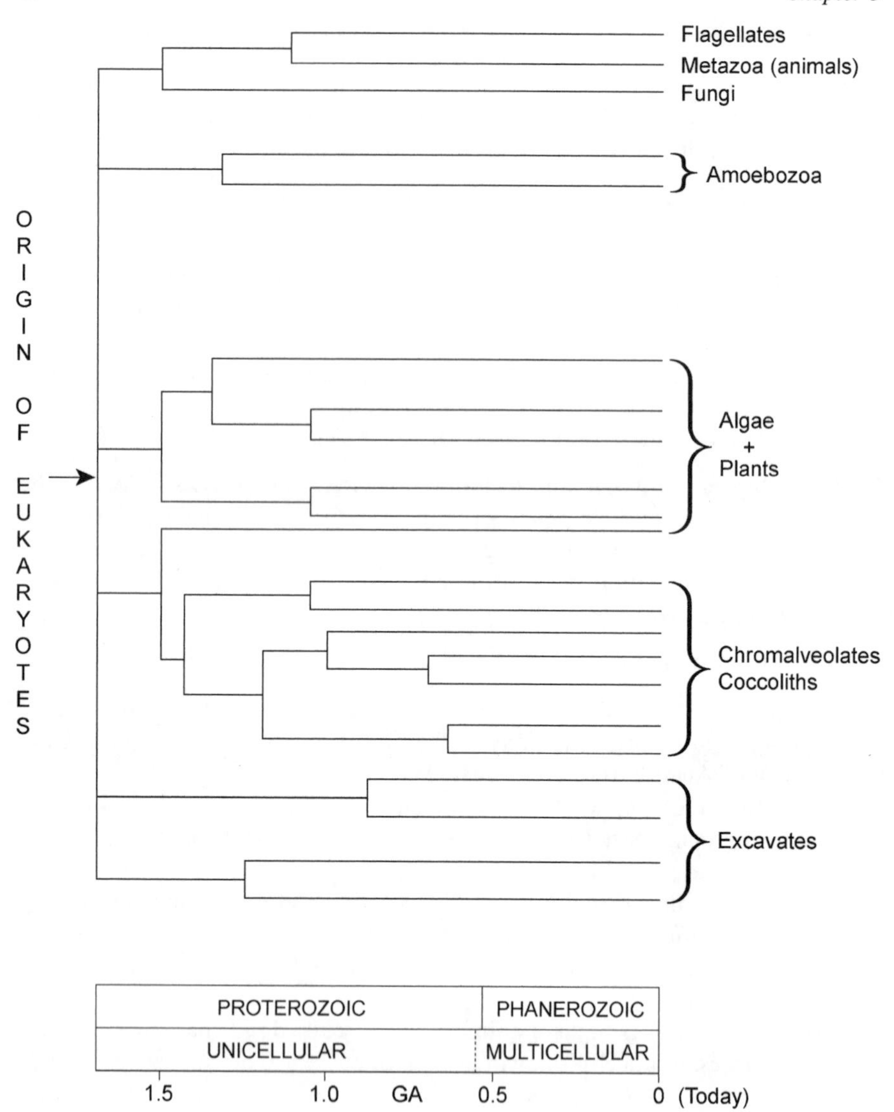

Figure 3.4 A tree of eukaryotes from fossil and gene data. The divergences and their times are described in Ref. 20. The organisms are a simplified classification.

Immediately it is clear that many animals developed calcium carbonate and calcium phosphate mineral structures while a few have silica structures. On the other hand the plant minerals are mainly silica with a very few such as coccoliths having calcium carbonate though several have calcium oxalate crystals. However the mode of synthesis of calcium carbonate in animals is different from that of the coccolith, which makes the carbonate structure piecemeal in vesicles. We shall examine the handling of the elements involved including those of calcium and silicon and of the anions carbonate and

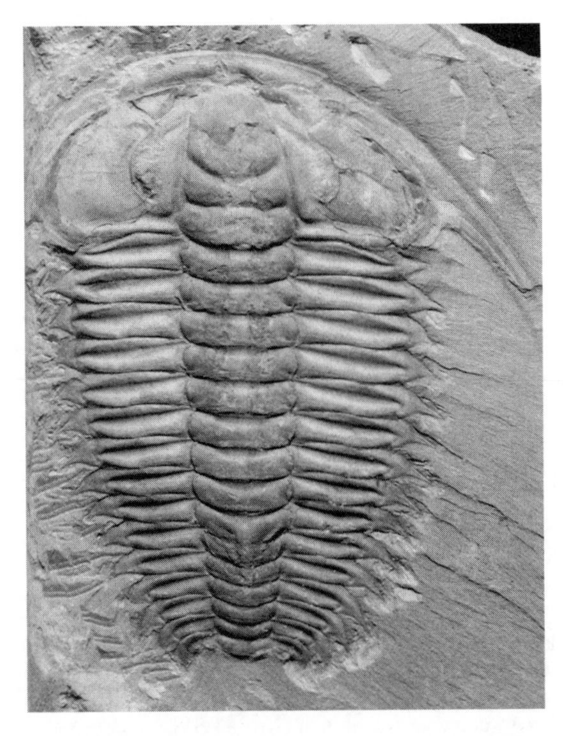

Figure 3.5 The framework of the trilobite which has a largely plastic coat often impregnated with calcium carbonate.

phosphate in the next section but mainly in Chapter 5. It is the case that the extracellular fluids of plants are more acid making carbonates and phosphates much more soluble. The concentration of salts in the circulating fluid of plants is far from that of sea while that of animals is much more closely related to it. This is extremely important for bones (and nerves) of animals. One important point in this section is that while there is great diversity of minerals in many different groups of organisms generally speaking they all arose over two or three relatively short periods. It is very important for us to know the times of development of the different types of mineralisation and in which organisms they evolved. We turn, in Section 3.6, to a more detailed examination of the biominerals when we shall show more clearly that they are not simple inorganic minerals. In order to concentrate, nucleate and incorporate these units into biominerals a special set of proteins had to evolve.

3.5 The Chemistry of Biominerals: The Handling of Inorganic Elements

The inorganic elements in the biominerals are most frequently calcium salts and silicon hydroxy oxides.[2–11] A peculiarity of this silicon is that it is not contaminated by metal ions. These two elements, Ca and Si, had little or no

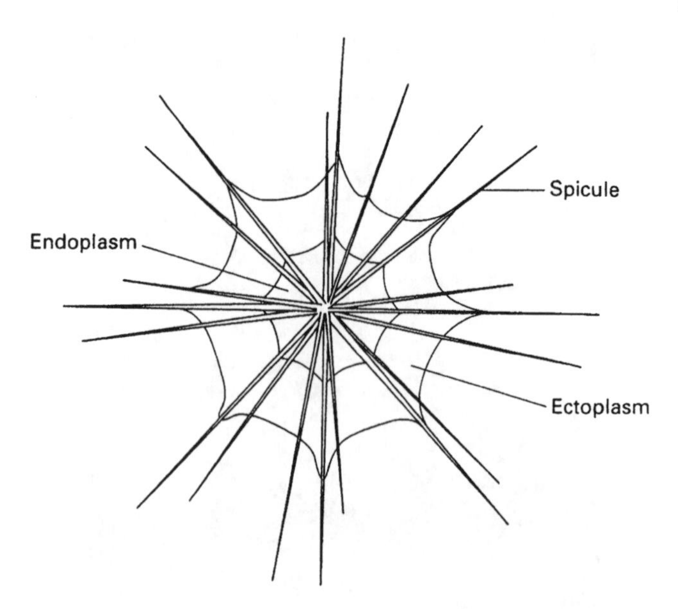

Figure 3.6 The structure of the acantharia's crystalline spikes. The disposition of the spikes is that of 4-fold symmetry based on the orthorhombic shape of $SrSO_4$ crystals. The final growth outside the cell membrane is under genetic control and there are many different species. However locally 4-fold symmetry is maintained. From ref 8.

functional use in the cytoplasm of cells. Because they were quite common, close to 10^{-3} M, in the sea, they had to be rejected from the interior of the original cells where calcium, especially, is poisonous above about 10^{-6} M. The mechanism of rejection of calcium by pumps is well known and will be discussed together with the handling of silicon in Chapter 5. Here we stress that in evolution it is quite common for elements rejected from the interior of cells to become utilised outside cells. In the cases of calcium and silicon, their rejection increases the concentration in the fluids immediately outside the cell's cytoplasm so that the solubility limit of the salts of calcium, for example, can be exceeded in the presence of particular anions. In the case of silicon, it is the solubility of rejected $Si(OH)_4$ which is exceeded. Through the mechanism of rejection then biominerals can be formed in extracellular spaces, even in environments where the surrounding aqueous phase is far from saturation, for example in fresh water. There are also clear-cut examples of formation of the biominerals inside organisms, in vesicles, not only of the two types of compounds mentioned but of salts of strontium, iron, barium and zinc ions, all of which are pumped out of the interior fluids of cells with particular anions.

We have not described the availability of selected anions of the biominerals. They are, for calcium, mainly carbonate, commonly available, and phosphate which has to be raised from 10^{-5} M in the sea. In extracellular fluids of animals it is raised to 10^{-3} M. In the case of strontium sulfate, it is the strontium which is raised in the vesicles as sulfate is very available from the sea.

Notice that throughout this description, the equilibrium solubility products are controlling, aiding formation and restricting dissolution. The formation of the minerals and their retention in selected shapes required an organic matrix which with ion pumps, was critical to the evolution of biominerals because ambient solubility products are rarely sufficiently exceeded. We return to the discussion of solubility and pumps in Chapter 5. There is a further need for inorganic elements, different from those in the biominerals but in the catalysts for the organic matrix synthesis which supports the inorganic material. It is the timing of their fossil appearance which gives evidence for the catalysts's evolution.

3.6 The Chemistry of Biominerals: Organic Components, Composites

We need next to appreciate the complex nature of these biominerals. Although they are given the names of simple inorganic compounds such as silica or calcium carbonate they are usually made of small crystalline or amorphous inorganic units held together in an orderly way by organic polymers, either proteins or polysaccharides. (Deposits in inclusions of prokaryotes are not made in this way nor is the $SrSO_4$ of Acantharia. They are pure inorganic materials isolated in vesicles.) The composite structures of the biominerals under discussion have great strength against fracture, unlike many single crystals.[9–11] In some sense they have properties in common with plastics. Many organisms build shells from these composites, *e.g.* shellfish, but there are huge families, which build 'shells' from purely or nearly pure organic 'plastics'. Some of these plastics have remained as fossils. The most well known are those of the arthropods, insects and animals such as crabs, which make their outer coats from chitin with crosslinking. Again trees, a later arrival, have outer structures of cellulose, chitin and – of special note – lignin. The presence of lignin gave strength to outer plant structures, helping them to grow and protecting them from outside attack (see Section 3.10). The trilobites (Figure 3.5) are an example of animal 'shells' of chitin but often include some calcium. Dating their appearance indicates that the hard plastic 'shells' evolved at about the same time as the composites. There is every grade of intermediate between highly mineralised and overwhelmingly plastic 'shells', but we shall treat their organic chemistry together. Chitin was part of 'shells' in the near Precambrian and the Cambrian Period, but we know that its N-acetylglucosamine polymer was even a component of Eubacterial cell walls, while cellulose is just as early.[17] It is the crosslinking that evolved later and made for a hard plastic coat or 'shell'. Crosslinked lignin seems to have evolved around 0.4 Ga close to the time after plants invaded the land and some also developed siliceous frameworks then. The animal shells, mostly of aragonite, calcite or apatite, sometimes called phosphorite, and of silica were held together by essential proteins which also control their shape including especially crosslinked collagen. In Chapter 4 we shall examine its evolution. The essential point is that all these mineralised structures owe themselves to the prior synthesis of an organic matrix acting as a nucleating agent and a growth control. The deposition of the composites usually follows from the outer arrangement of organic molecules and of cells. Before the time of the Cambrian Period or approaching it organisms were of softer structure.

We pause for a moment with our general description of fossils and so-called biominerals, and turn to the special case of the development of the bone made from apatite and quite different polymers, internal proteins, often strongly phosphorylated, and saccharides mostly attached to proteins in glycoproteins.[18] We can only give the most brief introduction to these skeleton structures with reference to a huge literature. The evolution of the composite bony skeleton appears to follow from a multicellular soft structure, the notochord, a collection of cells arranged in a linear arrangement connecting the head to the digestive region of animals first observed close to 0.5 to 0.4 Ga. This is approximately the line of the spine and spinal cord, now part of the brain, in modern animals. The notochord is seen first in the sea squirt but evolves into a bony structure in fish by 0.4 Ga. The bone structure is internal to the organism and is intimately associated with its growth. It can be said to be a 'living' structure, unlike an outer shell, because even its earliest parts can be modified during growth. Built in sections, repeated bone composites in joint/bone combinations, it becomes flexibly jointed unlike shells or even teeth. The very important point is that the building of a shaped internal skeleton, just like that of a shell, is due to the ability to organise cells, but now, in particular elastic arrangements. It is a consequence of shaped connections between cells, extracellular organic structure, and of the development of controlled extracellular fluids within a relatively impermeable outer covering of the whole organism, the skin, mainly of crosslinked keratin. In passing we have not referred to teeth. In teeth there is less organic material and while many animals have teeth made from calcium phosphate, some have teeth of iron oxides.

To explain the evolution of the 'biominerals' we see that we need knowledge of the origin of novel ways of handling certain inorganic elements, and of all trace elements needed in the catalysis of oxidative crosslinking of organic polymers for the matrix due to the variable oxidation states (see Figure 1.6; Table 6.1). The evolution of some novel proteins and other polymers could then be used to maintain ways of organising the inorganic materials to create fixed overall shapes or patterns. As mentioned earlier some further information about the early organic chemistry of oxidation is to be had from the nature of certain cyclic hydrocarbon organic compounds formed in ancient sediments, and more oxidised recently, and we shall include these compounds with the details of cellular organic chemistry (see Chapter 4), but they are legitimately part of the fossil record. However most knowledge of it is by inference from the composition of organisms alive today (Chapter 4). Finally the quantitative analysis of isotopes of elements such as C, O and S from ancient sources also is a 'fossil' record but again we consider the data are better placed in Chapters 2 and 4, with the presumed chemistry of their origin.

Before we turn to a discussion of the evolution of different inorganic and organic compounds required in the formation of the fossils of various kinds, we note in more detail the different growth patterns of the materials of the fossils.

3.7 Shapes of Organisms and Biominerals and Genetics

There are two interesting features for this book in the study of shape. The first is the question as to what controls shape. The possibilities are random searching for the best shape using genetic variation while the second is that shape reflects mathematical form necessary for growth. Within the second possibility is that shape is dictated by crystal habit of the chemical forming the biomineral. The two could operate together in that underlying features follow from physical control by chemical principles while variety is due to genetics.

The chemistry of any cell has to be contained and protected. The simplest construction is a sphere or an ellipsoid of a single cell with a wall, as seen in many bacteria. Their growth patterns can be controlled by unusual environmental chemicals to obtain filamentous forms. An adjustable shape, as is seen in simple but much larger unicellular eukaryotes, amoeba, is advantageous in that it can allow change of volume to match the physical environment and can engage in trapping by engulfment. Alternatively, but later, the soft body can be enclosed in a skin or shell after growth as is often seen in fossils. A multicellular organism's growth with protection and access to food and waste rejection is ideally managed in a mineralised tube or cone. D'Arcy Thomson[19] showed that these tubes and simple and twisted cones were the simplest of physical patterns and only the cone angle and twist of the cone needed genetic control. The cone or tube itself is a physical necessity of an organism's size during growth, which is able to protect and accommodate increasing size. D'Arcy Thomson showed how various monovalves, *e.g.* limpets and snails, and bivalves, *e.g.* mussels, conform to one underlying pattern. A tube with separate in and out paths remains a strong feature of many forms of elementary and advanced life. As the complexity of organisms grew the organisms needed to have shape linked to other functions and genetic control became more dominant. Plants needed vertical growth to gain access to light while increasing permanent stem thickness but animals needed mobility to scavenge rapidly as well as for protection. A fixed exterior did not meet the whole of animal requirements optimally and mineralisation became associated with the growth of internal jointed structures. In these cases of internal mineral growth the inorganic deposition now follows internal organic shape and physical forces. In the special case of skeletal bone, growth requires remodelling of the mineral phase shape continuously with growth and here apatite has a special value, it is piezoelectric. In this case, although it is the organic matrix which decides the general shape and is genetically linked, the exact shape is dependent on external physical forces and genetics, as is well shown by deliberate human manipulation of bone shape in a few cases. Despite this genetic link D'Arcy Thomson showed that the problem of carrying a body's weight is solved as expected from stress/strain diagrams. He likened the backbone structure of the two-legged dinosaurs to a single span of the girders of the Forth Bridge. We conclude that many features of growth and form of biominerals but not genetics conform to physical–chemical principles. All features of biomineralisation are clearly linked to organic structure synthesis and we shall seek an explanation of why this changed at a particular time.

The geometry of crystals of $SrSO_4$ in Acantharia spicules indicates again physical as well as genetic control and is species-specific (see Figure 3.6).[7] One of the greatest difficulties in the discussion of shape is how it can develop in an organism. It requires genetic instruction but it also needs messages between cells. These are often referred to as morphogenic fields, which we mentioned in Chapter 1 and were described by Turing. These fields are an additional physical–chemical part of evolution. Our next concern rests with the changing mangement of the chemistry of the shapes of organisms with time and the parallel changes in mainly the organic chemistry leading from ill-defined to the well-defined shapes observed in biomineralisation and in growth generally.

3.8 Induced and Controlled Biomineralisation and Genetics

Our description of fossils so far was given largely with little detail of the management of the chemistry of their formation. Generally the biomineralisation by cells has been classified into two kinds of underlying organisation: biologically induced and biologically controlled. Under control the mineral takes on a preferred, inherited, shape. Now the shape can be based on organised small crystals or amorphous precipitates. The growth of crystals of a defined allotropic crystal symmetry requires a pattern of binding sites of any nucleating (catalytic) surface which is not required for amorphous solids. This introduces an extra demand for controlled kinetics of crystal growth usually in addition to the control of the total shape of an organism.[9–11]

The first definite biologically induced mineralisation seems to have occurred before 0.54 Ga in the form of external organic surfaces aiding nucleation of $CaCO_3$, $Ca_2(OH)PO_4$, or $SiO(OH)_2$ in such organisms as *Cloudina*, *Brachiopodia*, and *Namapoikia*, but the nature of their structures is ill defined, as are the earliest coral minerals, dating from about the same time. The structures do not require a well-defined organic matrix of crystallisation but random points of nucleation. (It may be that some algae, green and red, induced such mineral, $CaCO_3$, growth around 0.75 Ga but this is very unsure.) The first defined and controlled mineralisation of silica and carbonates was in sponges dated around 0.60 to 0.54 Ga and certainly not much before. They are multicellular organisms (see Figures 3.2 and 3.3). It is defined crystal shape and points of deposition even of amorphous material that are genetically controlled and which govern this type of mineralisation. The genetic control of precipitation rests in the synthesis of a matrix. The organic matrix is laid down first and presumably it limits growth in three dimensions, and it is this structure which is mineralised. It is undoubtedly then this organic matrix which has to be understood including its evolution. The growth is different in different cases. That of $CaCO_3$ usually yields one of two allotropic controlled forms of crystalline carbonate, calcite or aragonite, an indication of particular surface catalysts or solution conditions, and laid down in the outermost 'skin' of the organism. $Si(OH)_4$ gives amorphous small often spherical particles, which are laid down in bead-like patterns following what are often layers or tube-like organic structures. Calcium hydroxyphosphate, apatite, is microcrystalline and follows an outer skin in early invertebrates but later follows the internal cell pattern and extracellular matrix in vertebrates. It is the appearance of a shaped, genetically inherited, outer organic 'skin' with internally organised cells, which gave rise either to a hard plastic case or a hard mineralised structure. Immediately we see that while the plant-like cells, algae, induced mineralisation (no shape), there are numerous cases of shaped, controlled mineralisation amongst different animals with different shapes somewhat later in time. (It is the animal cells of the coral combination which make the mineral of coral.) The shapes of biomineral shells and of the animals themselves then developed continuously from those in the Cambrian Period to those of today both in the sea and later on land. All the overall shape is known to be under some genetic control but as far as the crystals are concerned their shapes are defined by crystal geometry restrictions.[9–11] There was added, at about 0.4 Ga, a new dimension in the making of mineral support both internally and externally. The controlled minerals of land plants and related photosynthesising single-cell organisms occurred largely only after 350 (horsetails) and 60 million years ago (grasses). A second wave of new shaped inherited mineralisation mainly in the sea, now of single cells mostly plant-like, coccoliths, dinoflagellates and diatoms, is clear in the period 225 to 140 million years ago (see Table 3.3). It is noteworthy that they evolved after an extinction due to volcanic activity (Section 3.3) and there is the possibility that their appearance at this time is due in part to environmental conditions. The parallel controlled development of hard shaped 'plastic' (no mineral shells) is from around 0.6 Ga for animals and 0.4 Ga for plants and trees. The remarkable feature is that all these innovations occurred after the 'boring billion' years during which time there was little change in biological surface chemistry. Hence the gene products at these times indicate that the DNA sequences expanded to generate the new network of proteins outside the cytoplasm. One question we ask is why did biominerals form in these periods and not earlier, and why are plants so different from animals? The major timing does correspond to the time of the second rise of oxygen around 0.75 to 0.50 Ga and we shall therefore enquire, in Chapter 4, into changes of organic chemistry at this time. A secondary question is what caused silica to be used in some cases, but calcium carbonate and calcium phosphate in others in various animal and plant species? We shall not be able to answer this question fully but we can tackle the possible ways in which all these questions related to biomineral formation can be connected to environmental change and changes in DNA with time (Chapter 7).

We wish to impress upon our readers that in Sections 3.6 to 3.8 it is the strong influence of physical/chemical properties together with genetics that are controlling features of organisms. Note especially the numerical importance of Solubility Products, Section 1.3.

3.9 Molecular Fossils

A second kind of fossil information refers to organic molecule deposits in shales. The molecules of interest are cyclic, aromatic or chain compounds shown to be degradation products of molecules found in organisms today. Great care is necessary to avoid contamination as the material has to be obtained from drilled cores in previously undisturbed dated shale. Dating is done by isotopic analyses as described in Chapter 2. The materials from very early Archaean dates before 2.5 Ga have been checked against more modern deposits from the Permian Period, around 0.3 Ga. Although for many years, the dates associated with fossil chemicals were not regarded as thoroughly reliably associated with fossil organisms, the extent of detailed recent studies, illustrated here by one major reference[20] and references therein, has removed many doubts. The major findings are those of bitumens and kerogens, macromolecules, found in a variety of sideritic ($FeCO_3$), pyretic (FeS_2), various carbonates and BIF shales. The major smaller organic molecules have been classified under saturates, aromatics, steranes, hopanes and cheilanthanes. Of these, most could be from a great variety of organisms from prokaryotes to modern eukaryotes. Some even in the early Proterozoic Period are attributable to unicellular eukaryotes, namely the steranes, as opposed to the hopanes which are likely to be from prokaryotes. In Chapter 4, when we describe biological organic molecules arising in a deduced sequence of prokaryotes and eukaryotes of today, we refer to the molecules leading to these degraded molecular fossils.[20,21] In particular, the steranes are attributable to the degradation of eukaryotic steroids such as cholesterol and hence are indicative of some oxygen production by 2.5 Ga. The chemistry is confirmed by the finding of these steranes in molecular fossils of much later periods and their absence in some of the earlier periods. The next sections refer to massive degradation of living organic matter in a much more modern period. While generally biominerals are formed away from the reducing cytoplasm in oxidising conditions the fossils in Section 3.10.1 are due to the compacting of reduced organic material from all carbon-based parts of the organisms. The implication is that in the cycle of synthesis and degradation plants grew with lignin but animals failed to degrade the plants which died to become fossils.

3.10 Carbon and Carbon/Hydrogen Deposits

Apart from a fraction of the minerals, silicates, carbonates, sulfides (pyrite) and oxides (ferrite), three further minerals, sulfur, sulfates and carbon (coal with oil and methane to some extent) are produced indirectly by biological action, but of the three only some sulfur and some sulfate are in biological

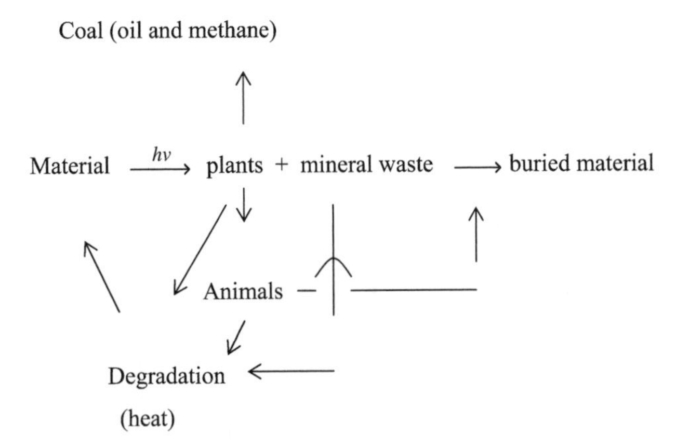

minerals, *e.g.* $SrSO_4$ and $BaSO_4$. We describe carbon first although its formation from carbon dioxide is complex and happened much later but it has become a much more significant mineral than sulfur or sulfate.[12]

Consider the way in which coal is produced in a simple scheme

Organisms, animals, were not able to metabolise certain novel protective products, plants, easily if at all so they have remained for hundreds of millions of years, in fact until man devised methods of using them for fuel. There is then a considerable hold-up of carbon plant material in this scheme. This leads to a pile-up of organic matter in debris which when buried and trapped under pressure gave rise to trapped carbon: coal, oil or gas (methane). In the Carboniferous Period (358–300 Ma) this build-up of carbon (fossil fuel) came about most strongly through the evolution of land plants, especially trees, which made compounds not digestible by animals at that time. Note that carbon, like sulfur, sulfate or Fe_2O_3, is an energised fuel and so are, for example, the oxides of nitrogen, possible fuels, but phosphate and carbonate are not fuels. All these fuels have an origin from the action of light yielding oxygen.

Much of today's burial of carbon is seen as peat, a degradation product of reduced chemicals from vegetation. It represents a huge reservoir because 2% of land today is covered by peat, and some frozen peat holds large quantities of methane, *e.g.* in Siberia. Most of today's peat is relatively young – it was formed less than 9000 years ago after the last Ice Age. Large amounts of this carbon return to circulation under conditions which expose it. Unfortunately for the creation of any steady state even today the failure of animals to digest all the available carbon sources from plants is still present and is included in the scheme. It may well be that organisms will evolve to utilise carbon of peat.

With peat, buried organic matter mixed with mineral sediment formed the basis of all soils. In soil there is a further mineral reservoir but now most of the elements required for life, mainly in silicates, are more available. Soil effectively retains moisture too. Thus when we come to cycling involving life we must remember that today there are a variety of element sources in the sea but also on land in soils of very different degrees of accessibility. Thus there

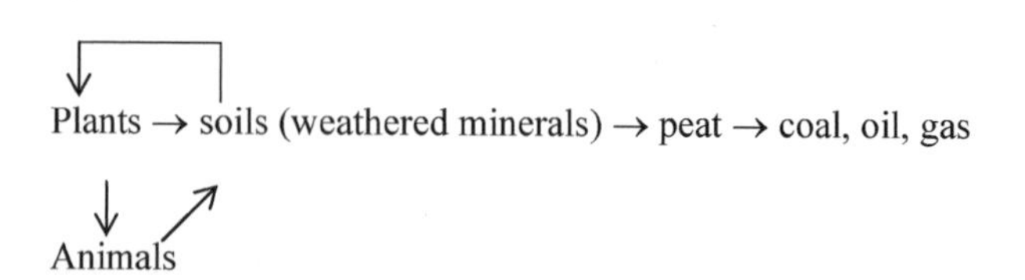

may be no simple rate constants covering cycling. These sources are not directly linked despite the obvious progression of degradation.

The further along the chain the more the carbon sources can become trapped out of circulation and away from catalytic action. It is temperature change and changes in foraging and digestive ability of land animals which can bring these stores into circulation. The obvious examples are global warming, which can set off fire in peat, and release of methane, and man's mining followed by use of them all as fuel. On the other hand movement of land due to erosion, ice coverage and tectonic activity lead to burial. In this discussion, as in the production of all organic chemicals, it is useful to keep in mind the nature of energised trapped chemicals (see Figure 1.8).

3.11 Sulfur Deposits

Sedimentary sulfur is due to oxidation of H_2S outside the cytoplasm. One mechanism of formation is from the action of light, bioenergetics, as described briefly in Chapter 2. Photosynthesising organisms may well have evolved to give sulfur very early in evolution, say by 2.2 Ga. Margulis[4] has considered that certain bacteria which use mineral sulfur as an energy source were of great consequence in early life. We describe later the formation of sulfates, especially of strontium and barium (Chapter 5). Although much is due directly to biological activity the earliest barite dates from an earlier time (see isotope studies in Section 2.4). Sulfur, being an energised mineral, is able to support organisms by its reverse reduction to sulfide.

3.12 Conclusions

This chapter has used the knowledge of fossils to show how we can date parts of evolution of organisms with increasing certainty. Information concerning the very early development of life supposedly from around 3.5 Ga to say 1.0 Ga is flimsy, based on soft organism imprints, but we can be reasonably certain that life was unicellular throughout the period and that there was no biomineralisation. Our knowledge from 1.0 Ga to 0.54 Ga, the beginning of the Cambrian Explosion, is better,[2] but much of it still concerns imprints of soft tissue until biomineralised multicellular organisms evolved on a large scale leaving much information concerning their shape and internal structure. At close to the same time, certain unicellular eukaryotes evolved with hard, shaped biomineral shells. All the biominerals have been characterised so that we know much of the ability of organisms to control precipitation and we know too that organisms could manipulate concentrations of some of the inorganic elements, especially Ca, Si and P apart from organic chemicals H, C, N, O and S in compounds. We also know that these structures are dependent on organic chemicals. It is the chemical ability to form relatively rigid crosslinked organic matrices which is essential for hard structures and they can form hard plastics by themselves with no mineralisation. This chemistry is largely oxidative, as is revealed by comparison with the same structures in organisms of today. The record around 0.75 to 0.55 Ga needs great extension and amplification if we are to test precisely ideas as to how this chemistry of organisms arose. After 0.54 Ga the preservation of imprints of soft-bodied as well as those of hard-bodied organisms improved until today. The purpose of the next three chapters is to show how we may explore this history by using extrapolation from the study of the different chemistry of the diversity of organisms alive today to their probable origins as the environment changed. This does involve some assumptions about which organisms came first.

Because of the diversity of shapes it is easily overlooked that these underlying general principles are applicable within groups of organisms. There is then a logic to some shapes, *e.g.* spirals broadening with growth.[7] In what follows therefore we wish to examine whether there was a logical progression to the organic as well as the inorganic chemistry of evolution as described in Chapter 2. It led to the groups of species which we have noted with special reference to the late advent, closely coincidental, of fossils of many kinds of multicellular organisms together with biomineralisation of organisms in all its forms and shapes, analysed in this chapter. In particular we need to explain why biomineralisation arose only after say 0.75 Ga, strongly after 0.54 Ga and again near to 0.30 Ga.[6] How is it that both silica and calcium carbonate minerals *both* formed biominerals in these periods, such that even the two minerals appear together in one organism, the sclerosponges, around 0.6 Ga, and again between 225 and 140 million years ago? How did they become made of crystals? Why is the first period of biominerals also of multicellular and eukaryote evolution in so many genera almost simultaneously? Why was evolution so late in finding these novelties, that is only after 85% of all time? That is to say we seek systematic chemical change which led to circumstances advantageous to cell development and then organic-matrix and mineral formation particularly that with shape. There is one clear possibility, the rise in oxygen (see Figure 3.3), and with it the changes in the chemical elements in the sea by weathering. It is then the environment which leads to the new possibilities and especially due to changes in oxidation state of inorganic elements and to changes of weathering. The fact that the overall chemical changes in the environment were certainly systematic (Chapter 2), and that the fossil record has an obvious sequence which is parallel in many different groups of organisms, gives us strong reason to believe that there is a continuous system to the organic and inorganic chemistry in organisms and that this has led to a systematic evolution of life. This we believe is supported by unavoidable genetic varieties.

The details are of comparative organic chemistry including the changes made in DNA, RNA and protein sequences with details of those of elementary metabolism will be given first in Chapter 4, followed by similar comparison of the inorganic chemistry in modern organisms of the bulk elements in them, such as Na, K, Cl, Mg and Ca, in Chapter 5 and of the trace elements in Chapter 6. At the same time, we shall note the changes of elements in the sea from the analysis of sediments (Chapter 2). We stress that the evolutionary paths are deduced and have limited degrees of certainty especially before 0.54 Ga. We bring all the chemistry together with genetics in Part A of Chapter 7. We shall not use genetics until Part B of Chapter 7, where we ask how could DNA control evolve to direct chemical development. Or did it?

3.13 Note

We have assumed that during the whole period from 4.5 Ga to today the major changes can be seen as a continuous evolution. This means that we have disregarded short-term fluctuations of the physical and chemical conditions which have certainly given rise to progress being quicker before falling back. Again we know that at certain times of extinction some species were lost.

An excellent example of a change in the ratio of phosphorus to iron and which may have affected oxygen levels and metazoan evolution around 0.7 Ga. The ratio is again high today.[22] Another example is that of a possibility that silicon limitation at a certain period caused loss of some sponges.[23]

Another interesting recent finding is of microfossils, probably of calcium phosphate, which appear to have existed for 0.75 to 0.60 Ga.[24] This is close to the time of the first appearance of biominerals but probably a little earlier. The fossils are cysts of certain metozoans. It may be that they were formed in an atmosphere of adequate oxygen but were later in an anoxic environment for some time. Cysts are a resting stage of many organisms and a shell is advantageous for protection in such a period.[25]

References

1. M. J. Benton, *The Fossil Record 2*, Chapman and Hall, London, 1993.
2. M. Brazier, *Darwin's Lost World*, Oxford University Press, Oxford, 2009.
3. M. A. Fedonkin, J. H. Gehling, K. Grey, G. M. Narbonne and P. Vickers-Rich, *The Rise of Animal*, John Hopkins University Press, Baltimore, 2007.
4. L. Margulis, *Symbiotic Planet*, Basic Books, New York, 1998.
5. K. Simkiss and K. Wilbur, *Biomineralisation: Cell Biology and Mineral Deposition*, Academic Press, San Diego, 1989.
6. S. Conway Morris, *Philos. Trans. R. Soc., B*, 2006, **361**, 1069.
7. R. J. P. Williams and J. J. R. Fraústo da Silva, *The Chemistry of Evolution*, Elsevier, Amsterdam, 2006.
8. J. R. Wilcock, C. C. Perry, R. J. P. Williams and R. F. C. Mantoura, *Proc. R. Soc. London, Ser. B*, 1988, **233**, 393.

9. A. Sigel, H. Sigel and R. K. O. Sigel (eds), *Biomineralization. From Nature to Application*, John Wiley and Sons, Chichester, 2005.
10. S. Weiner and P. M. Dove in *Reviews in Mineralogy and Geochemistry, Volume 54, Biomineralization*, ed. P. M. Dove, J. J. De Yoreo and S. Weiner, Mineralogical Society of America and the Geochemical Society, Washington, , DC, 2003, p. 1.
11. S. Mann, *Biomineralization: Principles and Concepts in Bioinorganic Materials Chemistry*, Oxford University Press, Oxford, 2001.
12. E. Sjöström, *Wood Chemistry: Fundamentals and Applications*, Academic Press, San Diego, 1993.
13. C. G. S. Phillips and R. J. P. Williams, *Inorganic Chemistry*, Oxford University Press, Oxford, 1965.
14. D. F. Shriver, P. W. Atkins and C. H. Langford, *Inorganic Chemistry*, Oxford University Press, Oxford, 1994.
15. J. H. Lawton and R. M. May, *Extinction Rates*, Oxford University Press, Oxford, 1995.
16. J. Erez in *Reviews in Mineralogy and Geochemistry*, Volume 54, *Biomineralization*, ed. P. M. Dove, J. J. De Yoreo and S. Weiner, Mineralogical Society of America and the Geochemical Society, Washington, , DC, 2003, p. 115.
17. R. M. Brown and I. M. Sazena, *Many Paths up the Mountain: Tracking the Evolution of Cellulose Biosynthesis*, Springer, Amsterdam, 2007.
18. D. T. Price, M. J. Schoeninger and G. J. Armelgos, *J. Hum. Evol.*, 1985, **14**, 419.
19. W. D'Arcy Thompson, *On Growth and Form*, 2nd edn. Cambridge University Press, Cambridge, 1968.
20. There is extensive information on molecular fossils in a series of papers in *Philos, . Trans. R. Soc., A*, 2010, **368**, 3059.
21. J. R. Waldbauer, L. S. Sherman, D. Y. Sumner and R. E. Summons, *Precambrian Res.*, 2009, **169**, 28.
22. D. Bhattacharya, H. S. Yoon, S. B. Hedges and J. D. Hackett, *The Timetree of Life*, ed. S. B. Hedges and S. Kumar, Oxford University Press, Oxford, 2009, p. 116.
23. N. J. Planavsky, O. J. Rousel, A. Bekker, S. V. Lalonde, K. O. Konhauser, C. T. Reinhard and T. W. Lyons, *Nature*, 2010, **467**, 1088.
24. M. Maldonado, M. C. Carmona, M. J. Uriz and A. Griszadcolo, *Nature*, 1999, **401**, 785.
25. P. A. Cohen, A. H. Knoll and R. B. Kodner, *Proc. Natl. Acad. Sci. USA*, 2010, **106**, 6519.

CHAPTER 4

Cells: Their Basic Organic Chemistry and their Environment

4.1 Introduction

In this chapter we provide a coherent connection between the evolution of the geological inorganic minerals (Chapter 2), together with knowledge of fossils (Chapter 3), and the biology of organisms through the ages, and the basic chemical content and nature of cells as they arose and evolved. Unlike the previous chapters this chapter will therefore concern mostly organic chemistry,[1,2] for its molecules include the large majority of a cell's contents. The description here has to be very limited, because the organic chemistry of cells is so very extensive.[3,4] It will be immediately apparent that many different inorganic chemical elements and compounds interact with this organic chemistry and we shall emphasise their presence.[5] Of course all cellular chemistry has its origin in inorganic environmental chemicals such as H_2O, CO_2, N_2, HPO_4^{2-} and SO_4^{2-} and we wish to stress this. In particular we shall record also the use of other, now metal elements, because we wish to see if changes in their presence in organisms match changes of these elements in the environment. Our analysis of the organic chemistry will not be orthodox then because it takes place largely in aqueous solution, much of it is coded, and we stress the concentration and changing roles of some 20 elements essential for it (see Figure 1.1). There are effectively two connected types of molecule, one of small organic molecules, the metabolome, the other the long, coded molecules, the proteins of the proteome, and the DNA (RNA) of the genome. We shall not need to use any diagrams of organic molecular structure, of mechanisms of reaction or of details of pathways of synthesis because these are very well detailed in biochemistry textbooks in much expanded chemistry and which we only outline in this chapter.[1,2] Our emphasis will be on the connection of

Evolution's Destiny: Co-evolving Chemistry of the Environment and Life
R. J. P. Williams and R. E. M. Rickaby

Published by the Royal Society of Chemistry, www.rsc.org

the organic chemistry to the environment. We have then to uncover the times of the evolution of cellular organic chemistry in life. Because organic chemicals have largely disappeared we shall use information from what is known from the organisms alive today which have the closest similarity to those and subsequent organisms from 3.5 Ga. Our objective is very different from, though clearly linked to, that of the analysis of systems biology of present-day organisms.[5] Only if we recognise that there are certain pre ordained features of the origin of cellular life described below, shall we be able to describe it and its evolution. Parts of this chapter may be difficult for a reader unfamiliar with biological organic chemistry and it could be helpful to read Sections 4.2, 4.3, 4.6, 4.7, 4.8, 4.15 and 4.25 first, before going back to more detailed aspects of it.

The organic chemistry in cells must have started from the reaction and energisation of very simple (inorganic) compounds such as H_2, CO and CO_2 in the environment before 3.5 Ga. It led to the formation of small reduced organic molecules and then to the major biopolymers, RNA/DNA, proteins, saccharides and lipids, using all the six major non-metal elements (see Figure 4.1; Table 4.1) by 3.5 Ga. To summarise: the chemistry had to be reductive, dominated by the need to reduce CO, CO_2 and perhaps N_2, so as to make the known essential small organic molecules and then the biopolymers mainly by condensation reactions needing energy, Figure 4.1. Condensation is by the removal of H_2O, in part from the ends of two molecules so as to join them together, *e.g.*

$$R{-}COOH + H_2N{-}R^1 \rightarrow R{-}CO.NH{-}R^1 + H_2O$$

as found in proteins and parallel synthetic steps leading to nucleotides and polysaccharides. The chemistry had to remain soluble in water but to be contained by the synthesis of primitive membranes and cell walls, made from reduced insoluble organic molecules, lipids. The limited number of metal elements that could have been involved in essential catalysis, structure and energy capture at first has been noted in earlier chapters and will be noted again here (see also Chapters 5 and 6). The next five sections of this chapter, 4.2 to 4.6, then cover the physical and chemical features of the first anaerobic cells, some of which is a simplified account of the material in texts on biochemistry including bioenergetic metabolism, large molecule structure, and the connection between different anaerobic cells.

All such reductive anaerobic in-cell chemistry had to be compensated by rejection of oxidised inorganic chemicals, perhaps sulfur originally but, unavoidably oxygen later. The later release of oxygen inevitably changed the oxidation of the environment in stages of various simple chemicals, both the non-metals and the metal ions, in the order of their chemical redox potentials (see Figure 1.11). The order is H_2, NH_3, H_2S and sulfides of metals. It is this oxidation that drives later evolution following the creation of anaerobic cells of micro-aerobic cells, by providing externally new energy sources, in turn oxidised Fe^{3+}, S_n and NO, and then low oxygen itself, for many later syntheses, and also slow changes in element availability. There followed, after anaerobic cells, the sequence of aerobic cells, and then of larger organisms of increasing complexity. We have seen the abiotic parallel of this chemical progression in the mineral

Table 4.1 The Seven Ages of Life's Organic Chemistry

Period (Ga)	*Section*	*Organic Chemistry Characteristics*
4.5-3.5	4.1 and 4.2	Beginning of energised reductive synthesis of molecules from H_2, CO, CO_2, (CH_4), NH_3, H_2S. Energy from unstable inorganic molecules or the sun acting in a restricted volume, bring first coded reproducible cells with simple shape. Anaerobic prokaryotes.
3.5-3.0	4.4 to 4.6	Cells increase in chemical diversity using same cytoplasm but range of coenzymes and porphyrins. Oxidation of H_2S to S and Fe^{2+} to Fe^{3+}: H_2, (CH_4) reduced. CO_2 incorporation by photosynthesis generating oxygen. Microaerobic cells. Metal ions as catalysts restricted to mainly Fe^{2+} and Mg^{2+}.
3.0-2.5	4.7	Increasing use of oxygen to make partially oxidised chemicals and as a source of energy, aerobic metabolism plus anaerobic reductive reactions. Aerobic reactions mainly in periplasm. H_2, CH_4 largely lost to outer space.
2.5-3.0 (1.5)	4.8 to 4.12	First single-cell eukaryote cells with compartments for reductive chemistry and the nucleus: peroxyzome and vesicles for digestion. First cells with mitochondria: chloroplasts added, for useful energy production. Calcium messengers. Membranes have cholesterol and are flexible. Endo- and Exocytosis. Some use of Cu, more of Zn.
(1.5) 2.0 to 0.75	4.14 to 4.15	Small improvements and increase in many cell types but still no biomineralisation. Some evidence for cell/cell interactions symptomatic of multicellular constructs from 1.2 Ga.
0.75 to 0.4	4.15 to 4.24	Multicellular eukaryotes. Connective external tissue with partly oxidised cross-linked large organic molecules. Tubular system for internal distribution of chemicals. Na^+/K^+ nerve messages. Organic messenger coupled calcium for message transmission between cells. Later bones and brain in animals. Vascular systems in animals and plants. Biominerals. Much increased Cu and Zn.
0.4 to today	Appendix 1	Fluctuations in organic populations but little new chemistry in cells. Many varieties of animal and plant life in considerable numbers. Very readily man exploits new elements of the Periodic Table. Much new organic synthesis. New sources of energy. New inorganic material, metals, silicon. Agriculture. Manufacturing.

N.B. Overall notice that during the whole time from around 3.5 to today energy is used to drive coded reductive chemistry in the cytoplasm and then increasing external oxidative chemistry in vesicles and extracellular spaces. There is an increasing organisation with a variety of communicating molecules and cooperative organism/organism cooperation, symbiosis. Cellular life increases further only interrupted for a time by extinctions.

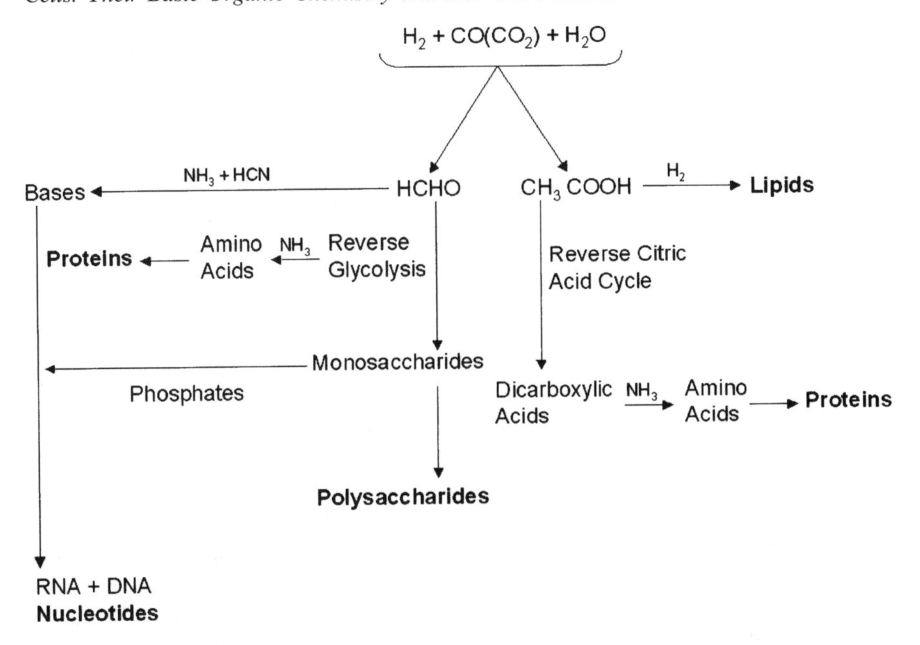

Figure 4.1 An outline of some basic features of the probable chemistry of the first cells.[1,2] Energy may have been required even in the first step and certainly in later steps.

chemistry of Chapter 2 (see also Figure 4.2). We believe that it was inevitable, as most inorganic reactions in the environment move rapidly toward equilibrium as oxygen is added. The feedback from the environment with energy and the above syntheses, suitably modified, created the possibility of more effective modes of new organic chemistry and use by organisms. The extra novel chemistry then

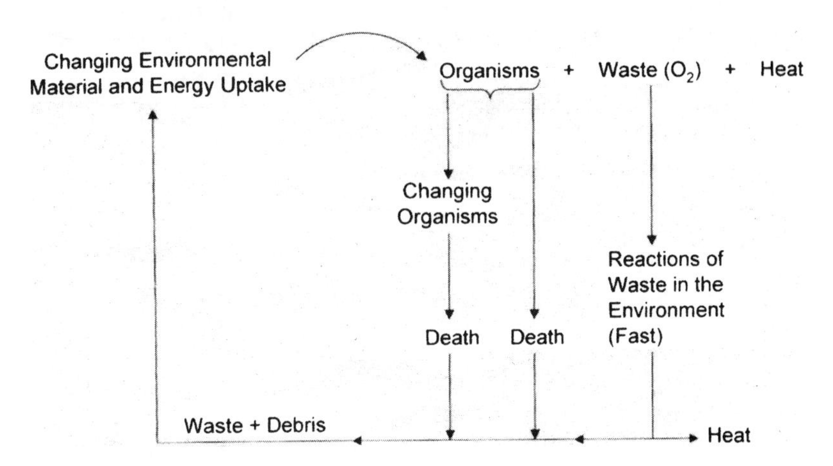

Figure 4.2 The unavoidable interaction of all metabolism and the environment which forces their mutual dependence during chemical evolution. Note that energy is required.

required greater complexity of chemical organisation. These inorganic chemical oxidative changes led, with a considerable degree of inevitability, to cellular organic chemistry changes. This is the theme of our analysis of organic chemical evolution. We shall examine complexity in the stages, anaerobic, micro-aerobic (Sections 4.2 to 4.6), aerobic prokaryotes (Section 4.7), aerobic single-cell eukaryotes (Sections 4.8 to 4.14), multicell eukaryotes and more complex organisms up to man (Sections 4.12 to 4.22). No matter the whole combination of organisms and the environment advanced together in what we shall indicate is a systematic chemical way initially by cellular reduction to give the earliest cells but later by oxidation to give the extensive forms of life today. The connection with genetic structure will be touched upon in several sections but, as no final description has yet emerged, much will be left until Chapter 7.

Before we start this description of the basic organic chemistry we elaborate a special feature of it, the nature of the catalysts which activate it. The catalysts were proteins that often contain inorganic ions, but their power is augmented vastly by holding substrates and ions in particular pockets. Whether they contain metal ions or not the catalysts are called enzymes. Enzymes are a special feature of the organic chemistry of organisms. Compared with the potential of simple ions, complex ions or organic acids and bases, these enzymes have vastly greater catalytic power, exceeding that of the simple catalytic entities, often by more than 1 million times. They, unlike the simple catalysts, by very selective binding and activation of substrates, direct cellular chemistry in very limited essential pathways. Although the enzymes that utilise metal ions are no more than one-third of the total their role is frequently in energetically difficult but highly important reactions, including energy transduction,[6] and they are directly connected to the evolving environment. Apart from the ease with which we can relate them to the environment we shall observe that these enzymes are deployed in different cell types and in particular parts of cells. Different elements have been potent in different types of reaction at different times in evolution, especially associated with increasing oxidation. We shall turn again to some of their properties in Section 4.4. The very presence of proteins in most metal functions forces us to search for special synthesis steps at the very beginning of cellular organic chemistry. Before we begin this analysis the reader should have in mind some basic questions (Table 4.2).

4.2 The Proposed Beginnings of Life: Anaerobic Prokaryotes

The prokaryotes known today are Eubacteria and Archaea. The non-metals required for them and which give rise to complex molecules are H, C, N, O, S and P. They are the essential elements of organic chemistry.[1,2] A way to consider the beginnings of life is then to look for chemical pathways and energy to organise molecules from the available compounds of these elements in the environment. The forms in which they were initially available were, in the above order of non-metals, H_2, H_2S, H_2O; CO, CH_4, CO_2; NH_3, N_2; SH^- and HPO_4^{2-}. The further

Table 4.2 Questions concerning the Evolution of Bio-Organic Chemistry

1.	When did certain reagents, substrates, become available, for example O_2?
2.	When did certain inorganic elements become available and useful as catalysts, controls, etc of organic chemistry, for example Cu and Ca?
3.	How was energy captured by the organisms?
4.	What is known of the historical development of biopolymers, e.g. proteins with different folds?
5.	What is the connection between biomineralisation and bioorganic chemistry as seen in fossils?
6.	When and why did bioorganic chemistry split into different compartments, e.g. in eukaryotes, and become a cooperative system of organisms, e.g. as in man? Man is a combination of many organisms.

requirements for the beginnings of cellular life, as mentioned, were a limited volume, that is a containing cell membrane structure, a supply of energy, control over ionic concentrations (see Chapter 5), and a set of catalysts for the synthesis of the selected organic chemicals, reduced relative to CO_2/CO, and eventually for that of the major polymers of life.[3,4] The energy requirement and the need for catalysts have been explained in Chapter 1. We must stress that apart from the six non-metal element the content of needed metal ions was very limited at first but we shall see that the coming of oxygen later led to a considerable increase of them in the environment. Their use and later cell expansion was then unavoidable. Also unavoidable were the rejections of sodium and calcium which later gave through their gradients essential assistance to the expansion of life from anaerobes. Energy is clearly of primary concern and it had to be both captured (absorbed) and transduced into usable forms to drive chemistry. The transduced energy had to be used with a source of catalysts and controls so that concentration and synthesis could be managed. These are the major considerations of the organic chemistry of anaerobes and all later organisms.

The first step in the energy capture system we shall describe is in fact the only one of which we are certain. It needed a membrane containing a light-capture molecule, chlorophyll, in a protein and a mode of energy transduction in this membrane. The membrane enclosed a trapped chemical solution. We know that there may have been earlier sources and capture of energy but they are based on speculation and do not connect to transduction. We shall refer to them only after the description of the known system.

4.2.1 Energy Transduction and Use

In order to make the source of energy from light usable in chemical synthesis reactions or in making essential gradients of ions it had to be captured (absorbed), and converted (transduced). The light capture was largely by chlorophyll bound in a membrane as it still is today, although earlier there may have been more primitive absorbing molecules. The synthesis of chlorophyll evolved from that of porphyrin, starting perhaps from 'available' cyanide. From its conception, with a central Mg^{2+} ion, in a specially bound condition this molecule was excited by sunlight into a positive radical state (+) and an electron *e* (−) within the membrane. The two charges were separated across a membrane at this special chlorophyll called a reaction centre, R (Figure 4.3), by suitable arrangement of electron carriers, mostly iron compounds:[6] because iron was available at this early time (Chapter 6). The electron travelled toward one surface of the membrane while H_2S or ferrous ions, at an active site on the opposite membrane surface, sent electrons into the membrane to neutralise the (+) charge on R. This created separated proton and hydroxide ions, and/or oxidised insoluble sulfur or ferric ions and a reduced set of organic molecules respectively on opposite sides of the membrane (Figures 4.3 and 4.4). This splitting of reduction from oxidation is the essential first step in the creation of organic/environment chemistry which with the production of the pH gradient is the first step of energy transduction. The organic molecules proceeded internally into synthesis reactions. A primary need therefore was to reduce CO_2 so as to give the starting organic chemistry from HCHO.

Within another membrane of a separate cell some of the newly created external oxidising agents, when they had increased in sufficient concentration, reacted across that membrane with internal reduced organic molecules to establish a similar charge or proton/hydroxide gradient. The gradient at both types of membrane built up energetically sufficiently to drive the reaction of bound pyrophosphate formation, ADP + P → ATP, internally (see Figure 4.3).[6,7] ADP and ATP are adenosine di- and triphosphate mostly bound to Mg^{2+} ions. Where the proton/hydroxide gradient comes from light absorption, photophosphorylation, the form of life today includes all plants. When it comes from oxidising agents plus reduced organic molecules (oxidative phosphorylation), it is found in all aerobic species, animals and plants. Oxidative phosphorylation is a secondary source of energy and is to be contrasted with the primary source, energy release by high-temperature systems, light from the Sun or any sufficiently unstable inorganic sources on Earth, or from debris from the precursors of cells. The ATP formed can drive all condensation syntheses,

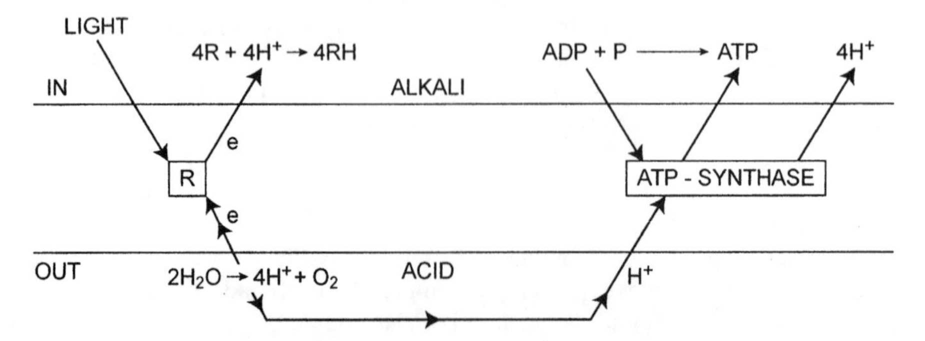

Figure 4.3 A scheme for energy, light, transduction via electron and proton transfer associated with membranes. It gives rise to internal reduction and external oxidation and to synthesis of the necessary energised molecule, ATP, for condensation reactions and waste oxygen.

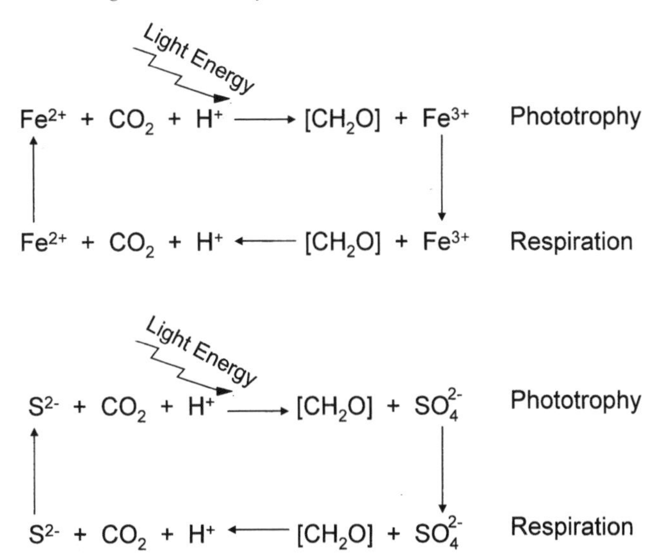

Figure 4.4 The coupling of photochemistry and respiration using sulfur and iron rather than oxygen. These reactions are likely to have been important from 3.0 to 1.0 Ga and are still ongoing.

leading to proteins, nucleic acids and saccharides, as it returns to ADP + P. A full account of this chemistry is given in Nichols and Ferguson[8] and the remarkable history of the discovery of the two phosphorylations is given by Weber and Prebble.[9] One essential feature was and is long-distance electron transfer in the membrane that was initially between Fe/S proteins.[5,6] Quite probably they were the only available catalyst for these steps. Condensation steps were usually reliant on magnesium bound to ATP. The central involvement of inorganic elements in living processes is stressed here, in the next section and in Chapters 5 and 6, because their roles, especially in catalysis of many reactions, cannot be brought about by the organic chemicals alone. Moreover the changes of inorganic elements can be followed in time and so used to note the times of the development of organic chemistry. The very limited number of original prokaryote inorganic catalysts is uncertain but Fe^{2+} and Mg^{2+} were certainly two of them. They were extremely common elements, for example in olivine, $FeMgSiO_4$, in the Earth's crust, and hence in solution in the sea (see Section 2.2). Additional components incorporated in proteins later were the small organic coenzymes and metal chelates of Fe, Co, Ni and Mg, note especially porphyrins, well described elsewhere.[3,4] There are more than 25 such coenzymes, many derived from amino acids and nucleotides. It is undoubtedly true that they arose early in evolution around 3.5 Ga but were they used in precursor chemistry?[2,4]

Further initial requirements were for osmotic and electrical stability and the avoidance of inappropriate bindings of organic molecules to elements such as calcium from the sea. To manage these conditions, even the first anaerobic cells had to control the internal concentrations of Na^+, K^+, Mg^{2+}, Ca^{2+} and Cl^- and did so using pumps energised by ATP or the proton gradients (Table 4.3).[5] These ion gradients are all then later subsidiary possible sources of energy for uptake. The pumps had also to give protection from certain heavy metal ions and some organic molecules (Table 4.3). We emphasise that the unavoidable selection or rejection of several ions was needed to sustain and to protect any possible organic chemistry which occurred in the sea. These choices of ions gave the possibility of certain essential steps in the evolution of organic chemistry later.

The reader will be aware that much of this form of the capture of energy and its transduction from light to give cell reactions and stability could not have been in existence at the time of the initial start of anaerobic organic chemistry in the environment. This requires the sophisticated collection of metal ions bound by proteins in an organised manner and of simple organic chemicals needed for the synthesis of proteins and other biopolymers, some placed in membranes, in cells. Unfortunately the real beginnings of bioenergetics and metabolism are unknown, probably unknowable, so that we shall only be able to discuss some possibilities.

We draw attention first to another pathway from the above for CO_2 reduction for initiating organic chemistry and it is found in both the earliest bacteria and Archaea. It is known under the names Wood-Ljungdahl. The striking feature of this pathway is that it catalyses the reaction

$$H_2 + CO(CO_2) \rightarrow HCHO(HCOOH)$$

without energy input. Two features of the pathway are of immediate interest. First the reaction is catalysed by what appear to be very primitive metal ion centres with Fe and Ni bound to small molecules, CO and CN^-. It is possible to make very similar small molecule Fe/Ni compounds which can act as catalysts of these reactions. Of course in cells such centres are in proteins. The second requirement in biological cells for these reactions is coenzymes related to folic acid. Both protein synthesis and that of folic acid require energy for their synthesis. It does not appear that organic chemistry in cells can be extended beyond the above first step without energy input from outside.

In the above account we draw attention to the need of external oxidising agents to drive oxidative phosphorylation. As far as we can tell the earliest organisms did not have an external supply of sulfur, sulfate or ferric iron available in the environment relative to any possible reduced states of elements

Table 4.3 Ion Pumps in Earliest Cells

Ion	*Pump*
H^+	Production of energised gradient
Na^+	Na^+ pumped out, K^+ replacement
Ca^{2+} (Mn^{2+})	Ca^{2+} removed to protect DNA Used to remove Mn^{2+} too
Zn^{2+}	Heavy-metal ATP-ases using RS^- centres
Cu	In and Out Pumps: copper ATP-ases
Fe^{2+}	Uptake systems
Mg^{2+}	Outward pump to sea

inside vesicles to give the energy supply, 'oxidative phosphorylation', needed to drive the initial organic reactions, for example further reactions of CO or CO_2 (see Figure 4.1). The earliest primary metabolism was possibly driven by energised inorganic materials, left by rapid cooling of the Earth (see Chapter 2). It may be that energy could have been obtained from NO, nitric oxide, due to lightning causing the reaction of $N_2 + H_2O$, but the main initial source was more probably the mineral (pyrite-forming) reaction[10]

$$H_2S + FeS \rightarrow FeS_2 \downarrow + H_2$$

Major problems are to manage and transduce the energy from this reaction so as to catalyse or drive especially the reduction of CO or CO_2 and the removal of water in condensation reactions in the next steps towards larger molecule synthesis. We have already explained the course of these reactions and that, as far as we can see, even the earliest energy capture had to make the creation of proton or charge gradients and then ATP. (Note a further possibility recently discussed is the action of a system based on zinc sulfide and UV light[11] which requires testing.)

Before we continue our description of the use of energy we point to the overwhelming and absolutely unavoidable features of the beginning and continued existence of organic chemistry leading to life: (i) the presence of certain elements and basic chemicals; (ii) the capture and transduction of energy; (iii) an enclosing membrane; (iv) the development of catalysts; (v) the rejection of sodium and calcium ions and of oxygen molecules. It is these features that limited the organic chemistry of life while making its evolution possible. They are not matters of chance but of necessity and led us to title this book 'Evolution's Destiny'.

The switch to the above use of light with the generation of energy and internal reduction with external oxidation is the only system known for certain. The initial external chemicals involved were H_2S and Fe^{2+} as already described, say, by 3.0 Ga. Their reactions supplied hydrogen until H_2S was lost or used up in the environment. The reaction which then arose in the evolution in cyanobacteria was

$$2H_2O \xrightarrow{h\upsilon} O_2 + 4H \text{ bound to carbon}$$

This gave the essential separation of redox reactions across membranes leading to reduced chemical products internally. It requires a complicated membrane protein in which the elements iron, manganese and calcium play a role. Models of this centre without proteins have been made (see Section 2.14). The origin of the complexity of the set of reactions in bacteria with that of light capture is unknown.

Now oxygen was a dangerous chemical for cells because it attacks all C/H compounds:

$$C/H + O_2 \rightarrow CO_2 + H_2O$$

and was therefore a pollutant, and a hazardous risk to all early anaerobic chemistry if it had become concentrated. Fortunately the Fe^{2+} and remaining H_2S in the environment reacted faster with oxygen, which protected the carbon compounds and as mentioned earlier also gave rise to novel secondary sources of energy:

$$H_2S + 2O_2 \rightarrow (S) + O_2 \rightarrow H_2SO_4 \rightarrow 2H^+ + SO_4^{2-} \text{ (with heat loss)}$$

where (S) is elemental sulfur, followed by (see Figure 4.4):

$$nSO_4^{2-} + 2(CHOH)_n \rightarrow S^{2-} + 2nCO_2 + nH_2O + \text{energy}$$
$$(S) + (CHOH)_n \rightarrow H_2S + CO + \text{energy}$$

Probably in an even earlier period extracellular Fe^{2+} would have been oxidised to Fe^{3+} by oxygen, as judged by the mineral BIF from 3.0 to 2.5 Ga. Fe^{3+} reduction could also supply energy to cells (see Figure 4.4):

$$nFe^{3+} + (CHOH)_n \rightarrow nFe^{2+} + nH^+ + (CH_2)_n + \text{energy}$$

All these cellular routes to energy capture are grouped under anaerobic or micro-aerobic conditions. After the first rise in oxygen this environmental oxidation of H_2S and Fe^{2+} buffered much of the environment against the free oxygen rise in the period somewhere between 2.5 and 0.75 Ga. At this time very few new organic chemicals could be synthesised and cellular evolution developed only slightly. During this time cells also evolved internal protection from the low levels of oxygen and from its products using other metalloenzymes in cells. Oxidised chemicals were likely to be greater in concentration in surface water than in the deep. It was only later that oxygen was used directly in cell chemistry, such as:

$$nO_2 + (CHOH)_n \rightarrow nCO_2 + nH_2O + \text{energy}$$
$$nO_2 + (CH_2)_n \rightarrow nCO_2 + nH_2O + \text{energy}$$

The importance of these oxygen reactions increased at the time of the second great oxidation event around 0.75 to 0.40 Ga (see Figure 2.5). On the other hand the supply of sulfate seems to have assisted cell energetics from 2.0 to 1.0 Ga. Notice how energy supply for the chemistry of organisms used environmental changes as an initial source, that is ferric ions, oxides of sulfur and then oxygen sequentially, and we shall observe that a sequence of organic chemical oxidations gave rise to the novel evolution of organisms.

4.3 Major Features of the Original Anaerobic Organic Chemistry

Before we consider the uses of oxygen in evolution we need to outline in a simplified manner the main probable features of the organic metabolite chemistry central to the original anaerobic prokaryote cells and to the production from those metabolites of large molecules from 3.5 to 3.0 Ga (Sections 4.3 to 4.7).[3,4] It will be assumed that energy was available, as described, to give a supply of ATP. This reductive chemistry is within the membrane in a space called the cytoplasm and is retained there even today in all cells (Section 4.7). The initial incorporation of carbon from CO_2/CO (Figure 4.5) must have led first to formate and acetate

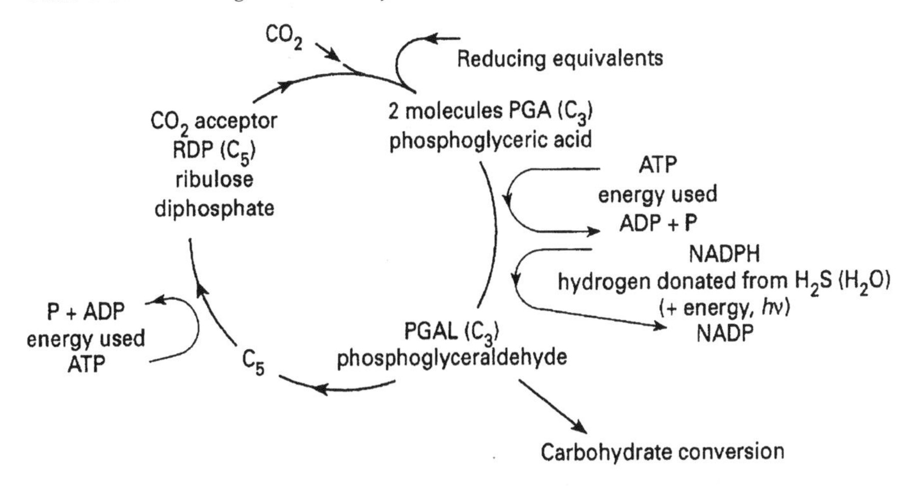

Figure 4.5 The Calvin cycle for CO_2 uptake[3,4]: note the energy, ATP, requirement.

and then through the (incomplete) reversed Krebs cycle (Figure 4.6), to many of the small molecules leading to macromolecules. Notice that eight selective enzymes and both Fe and Mg are involved in the Krebs cycle, see Figure 4.6, or the originally incomplete version of it. The iron enzymes were at first all Fe/S cluster proteins, for example aconitase with one dissociable Fe^{2+} ion. This enzyme is at the entry to the Krebs, citric acid, cycle which also produced the intermediates for several syntheses. A second Fe/S protein, fumarase, controls the Krebs cycle directly through its dissociable iron. The control by free iron is then of critical importance to all cell chemistry. A very similar protein to aconitase is a transcription factor and it also has a dissociable iron atom from an Fe/S cluster. There is another simpler Fe^{2+} protein, Fur, which is also a transcription factor. The importance of iron, in catalysts and controls, is a central feature of the origin of life.

Another very important pathway involves saccharides started from acetate. The synthesis of the saccharide pathway, known as glucogenesis, requires energy. Its reversal, degradation, by a different pathway is called glycolysis which gives energy, ATP. Glucose became a major carrier of energy later in life generating ATP in cells. This production of ATP, mostly bound to Mg^{2+} is known as substrate phosphorylation. Later, after photosynthesis arose, the CO_2 uptake started from more complex C/H/O/P compounds as used today in all plants. These reactions are known as the Calvin cycle (see Figure 4.5), and they too used phosphorylation with Mg^{2+} ions as a catalyst. The uptake of CO_2 is now via ribulose-1,5-bisphosphate in a large enzyme, RuBisCO. We discuss this enzyme in more detail in Chapter 7A. It is probably the most extensively used enzyme throughout all biology. It is important to observe that both the Krebs and the Calvin cycles are examples of the very great complexity of all organic chemistry, even in seemingly very simple cells. Many of the details of this chemistry are still found in all today's cells. They are well described in biochemistry textbooks. How did such sophistication arise so

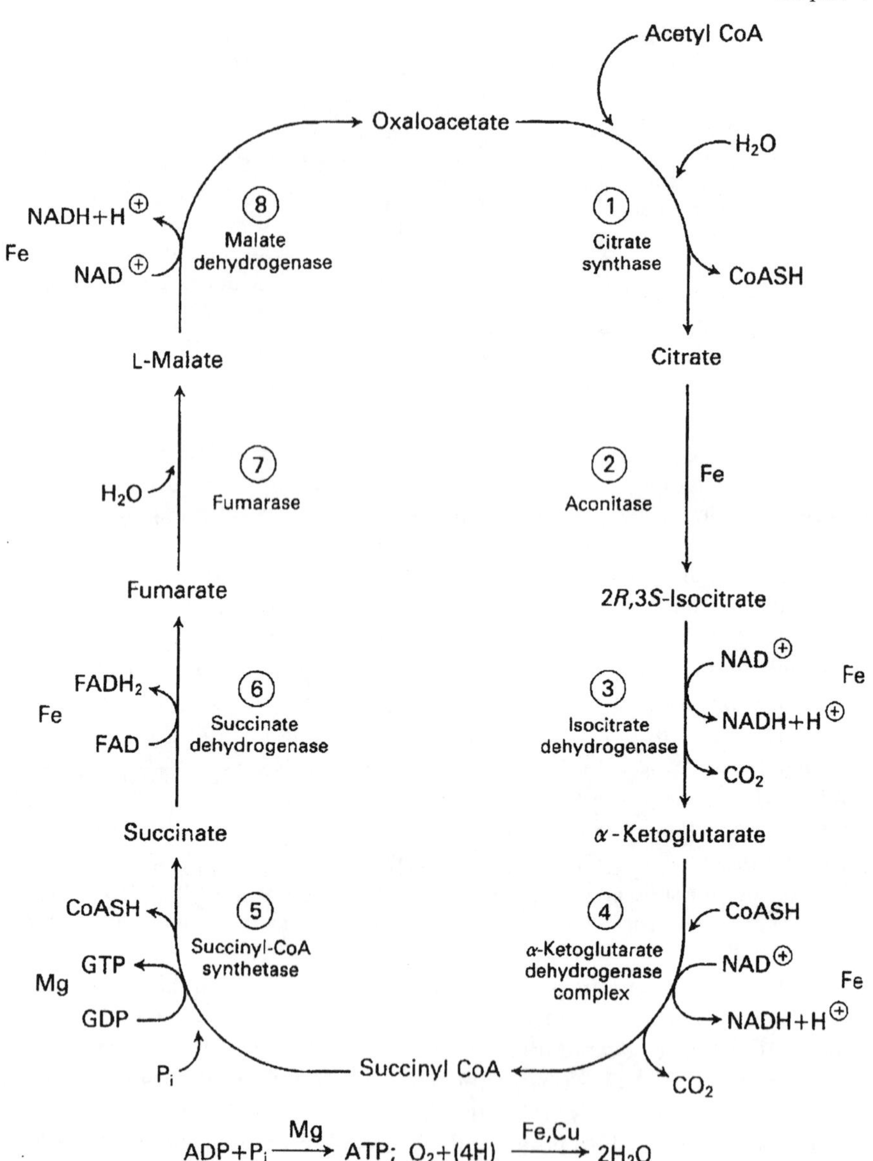

Figure 4.6 The Krebs cycle of oxidative phosphorylation but also providing starting chemicals for the synthesis of amino-acids and then other necessary larger molecules, see Figs. 4.1 and 4.8. It is believed that it was incomplete and ran backwards in early times to incorporate CO_2.

quickly and later become coded, controlled and coordinated in simple cells as illustrated in Figure 4.7? Some of the intermediate chemicals in these pathways have carbonyl groups that could have reacted with ammonia in side reactions to give the amino acids by condensation and further energised condensation of

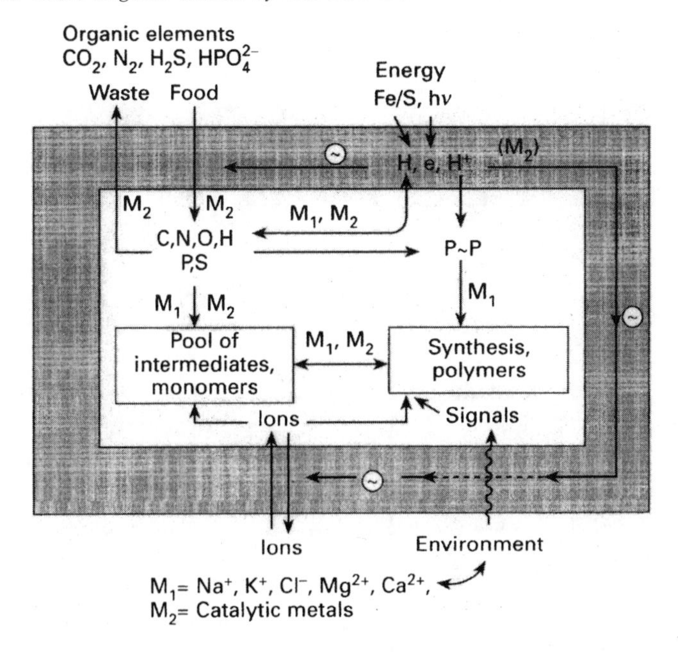

Figure 4.7 An outline sketch of a prokaryote cell, showing requirements and activities. The boxes are not isolated compartments. The shaded area is the outer membrane. Internal structure is not shown.

these amino acids could have given rise to proteins[12] (Figures 4.1 and 4.8). An alternative reaction of the first reduced carbon compound in cells, that is HCHO, with ammonia could have generated the nucleotide bases which together with a pentose monosaccharide and phosphate led to RNA/DNA.[13] Note that the conversion of RNA en route to DNA is by an iron enzyme which controls DNA synthesis. Many steps are further condensations driven by ATP with further reduction which produced lipids from saccharides.[12] The lipids are the molecules of membranes. These are all special organic chemical sequences. No metal ions other than Fe^{2+} (and possibly Ni^{2+}) are required for this reductive chemistry and for the oxidation/reduction reactions generally. As mentioned the only other metal of significance in this chemistry is Mg. Photochemical energy supply used Mg^{2+} in chlorophyll and Mg^{2+} acted in all ATP-driven condensations which were absolutely required. After the loss of the possibility of uptake of the available ammonia, lost by oxidation, cells had to take up N_2 and reduce it. Here a complex Fe/Mo set of enzymes evolved while Mo, molybdenum, was also used in transfer of oxygen atoms. (Mo increased in solution in the sea as the sulfide was oxidised.) Small quantities of other metal ions are found in certain special reactions, even in the earliest cells which we postulate from knowledge of today's organisms, such as Co, Mn, Zn and Se[5,14] (see Table 2.1). While all of these metals were certainly essential they were not concerned in more than one or two steps in any of the above syntheses. One or two of the metals, especially Zn, which are of great

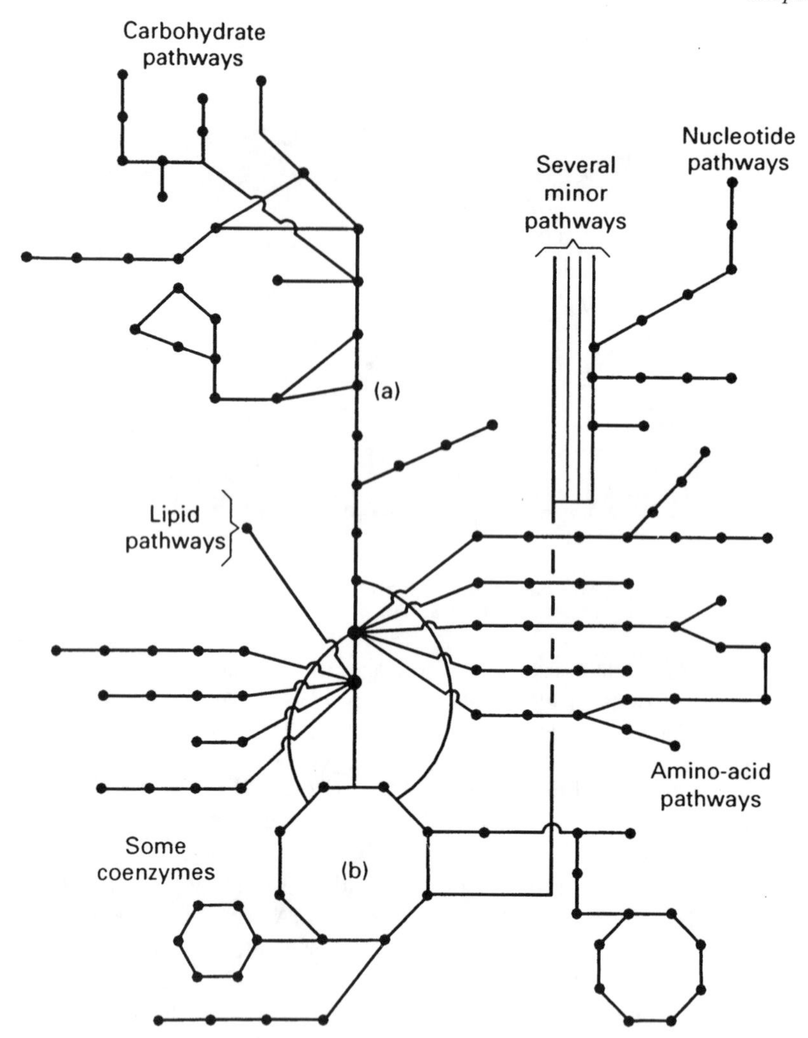

Figure 4.8 A diagram of the sophisticated connective pathways of basic metabolism. (a) is a part of glycolysis and (b) the Krebs cycle. The diagram is just symbolic. From Kauffman, ref. 12, with modifications.

importance in some hydrolysis reactions, were involved in the coded synthesis of DNA/RNA and proteins. We stress again that we do not know how any of these steps arose initially and became so selective and dependent on certain molecules and metal ions and especially on folded structures of catalytic proteins, enzymes. We consider that selection had to be based on the production of highly kinetically stable compounds which are reproduced in catalytic binding[13] (see Figure 1.8). (The value of metal ions, especially iron, was particularly important.) It is their evolution in novel environments which makes it possible for novel chemistry and hence novel organisms to evolve. No matter how the steps evolved there had to be dominant reductive pathways which used only the most kinetically stable molecules to produce an anaerobic cell. While we remember this, it is quite remarkable that much of internal cell metabolism from its very beginning is only a small part of the possible chemistry of small organic molecules. From the very first cell there had to be a code which expressed particular catalytic proteins which dominate the molecules selected in metabolism, and hence reproduction.

As well as drawing attention to the chemistry of the metabolism we need to describe the larger molecules necessary for reproduction of the cell before we ask how the integral cell might have arisen and evolved.

4.4 The Genome and the Proteome: Concentration Terms and Controls of Expression

(The introduction of the genome in a chapter on organic chemistry may appear odd. It must be remembered however that much of the organic chemistry in cells is coded by the DNA and therefore an understanding of cellular organic chemistry requires that of DNA, much though we wish to be engaged in it as little as possible until Chapter 7. This biological organic chemistry is then very limited and different from that described in current textbooks.)

The reductive metabolism of anaerobes has two major internal cytoplasmic polymer products, one built linearly from 22 amino acids, the proteins, collectively known as the proteome. The second is built linearly from four bases, nucleotides, and is known as the genome.[3,4] The nucleotides provided the reproductive code for the DNA, the genome. The DNA gave rise to RNA messengers, mRNA, which also has four nucleotides. We normally give DNA sequences by four letters, (C) cytosine, (G) guanasine, (A) adenosine and (T) thymine with (T) replaced by (U) uridine in RNA. Each group of three letters in mRNA is a code for an amino acid to make proteins. We need to see next how the three biopolymers interacted within a cellular system. The RNA (and the DNA?) as well as the proteins could have had a catalytic function at first as well as providing the code governing protein production, especially if the three were originally in separate vesicles. However they arose the three, DNA, RNA and proteins, were interrelated from the beginning of cells. The DNA is synthesised, copied in one machine, and mRNA is synthesised from it and expressed so as to make proteins in another machine, the ribosomes (Figure 4.9). There are carrier, transfer RNA (tRNA) molecules to take specific amino acids to the ribosomes. Each machine is composed of enzymatically active proteins and/or RNA and requires energy. Some RNA, iRNA, back-reacts with the DNA as a control on expression. All these nucleotides require bound Mg^{2+} and K^{+}. Certain proteins, transcription factors, are bound to DNA and cause particular mRNA expression. Although our aim is to understand primarily how the evolving environment chemistry affected organism chemistry it has to be the case that the DNA is also involved to maintain faithful reproduction. However we shall not need to discuss the problem of its connection until Chapter 7.

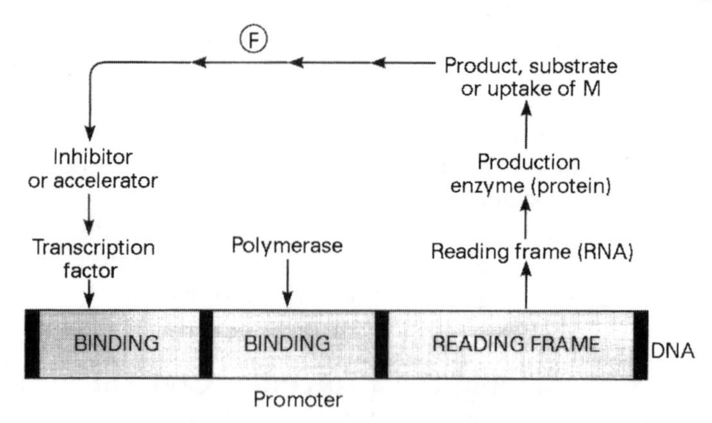

Figure 4.9 The essence of gene control through a series of genes and proteins providing uptake and enzymes in a unified system. (F) is feedback.

There are important system differences between the nature of the single molecule of DNA, the genome, in different prokaryotes, bacteria and Archaea cells, and the multiple number of molecules of each RNA and of each protein. The proteins are of greatest interest in this book as they are the dominant action part of genes. The nature of proteins is illustrated in Figures 4.10 to 4.12. They often bind metal ions in catalytic, machine-like or control functions linked to metabolism. The metal ions have, as we keep insisting, controlled concentration and provide a strong connection between cells (genes) and the environment. Metal proteomes will be examined in some detail in Chapter 5 (Mg and Ca) and Chapter 6 (transition metals).

The whole proteome is the complement of all protein concentrations produced in a cell under given external conditions. Unlike DNA but like those of RNA molecules the properties of the concentrations of proteins are not just those of single different molecules with different sequences as for DNA, but of considerable numbers of copies of each protein of a given sequence. Hence the synthesis of each protein (and RNA) and thereby its concentration must be

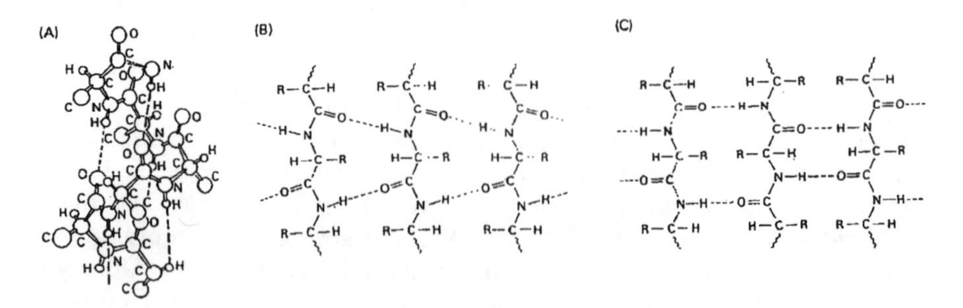

Figure 4.10 The structures of protein helices (A) and β-sheets (B) and (C). This and the next two Figures illustrate the unique nature of biological catalysts and structures in enzymes and energy transduction. See also Fig. 5.7.

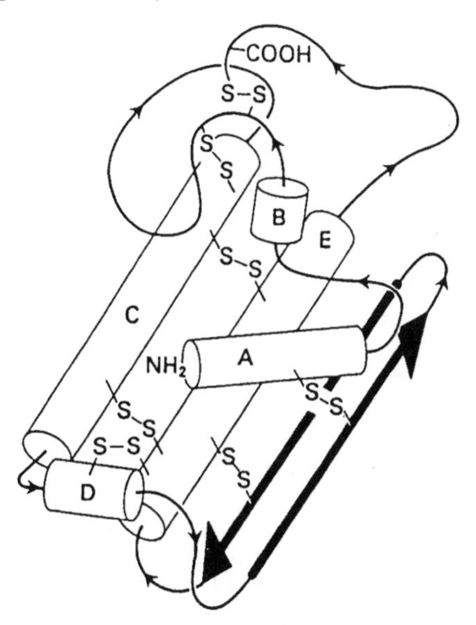

Figure 4.11 The structure of an extracellular enzyme showing helices as tubes and sheets as arrows. Note -S-S- bridges. The active site is at the top.

controlled. Moreover the proteins expressed are not necessarily all of those that are coded. The proteome therefore has, as well as a qualitative, a quantitative, selective description that is a complicated thermodynamic concept because some proteins are bound, not free, which increases the problem of the analysis. Concentration adds a variable to expression not obvious from a code. All such concentration dependencies have often been considered to be absent from any effect on DNA itself. However more recent genetic studies have pointed to modification in DNA and its structural proteins, histones, which affect expression in eukaryotes (see below). These modifications are brought about by enzymes present in controlled concentration, linking the genome to thermodynamic factors. Such effects will be described under epigenetics in Chapter 7B, linking the genome to thermodynamic considerations.

The proteins also find their way to particular positions in cells and most have particular sequence sections so as to be recognised by other selective structural regions or by compartmental membranes and proteins undergo trafficking. This is not the case for DNA or RNA. We can speak of the proteome of the cytoplasm, or of membranes, or of the exterior of a prokaryote cell. The expression of the proteins is under the control of promoters, transcription factors bound to DNA, which are also proteins leading, when activated, firstly to RNA synthesis and then to protein production. Now activation and prevention of the activity of protein transcription factors are usually by small molecular units, from the metabolome, or by ions, including phosphate, and metal ions from the metallome[5] (Figure 4.9). These are directly or indirectly obtained from the material of the environment and the quantitative expression of proteins is then

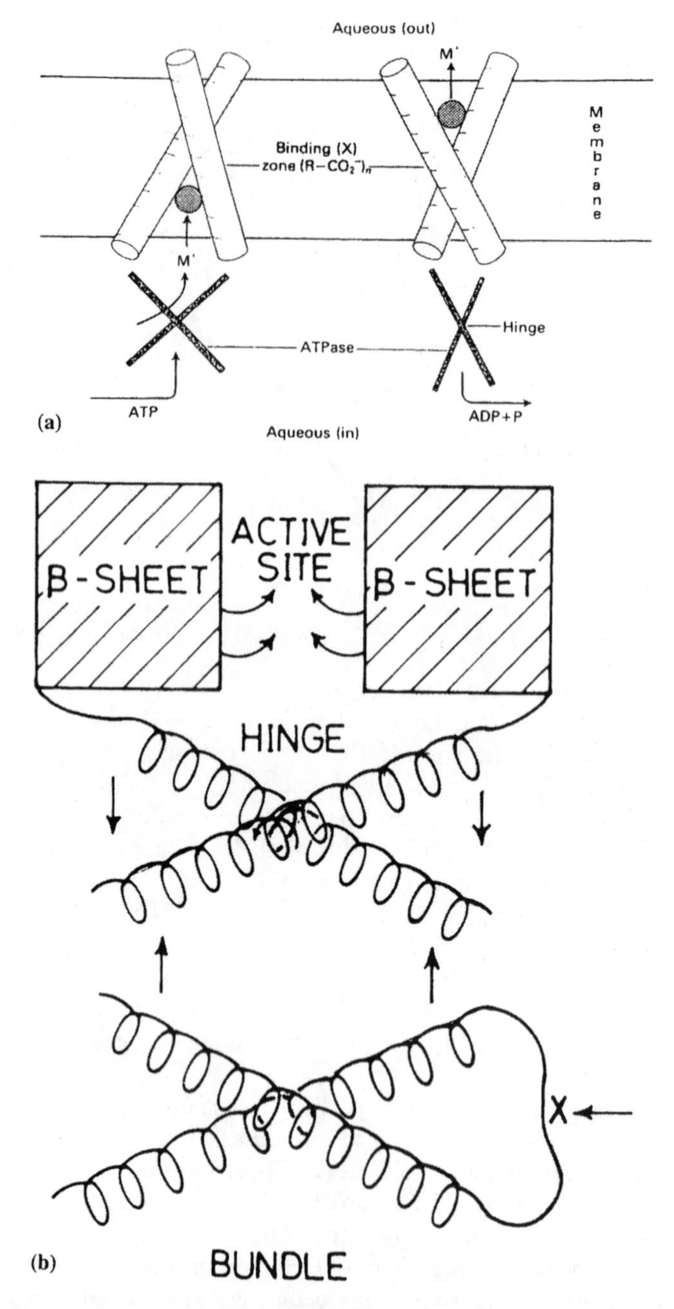

Figure 4.12 A schematic diagram of a biological machine in which the energy of the active site -ATP-ase based in part on β-sheets and a hinge is transmitted to movement of an ion, M^+, through a membrane to make a concentration gradient.

related to the environment. Recognition of this essential linkage between environment and expression in organisms, we shall maintain, is essential to an understanding of evolution.

Proteome expression is then dependent on a series of internal and external concentration-dependent factors. The concentration of each of these proteins together with the external environment decides the observable nature of features of cells, commonly called genes. Together with the DNA it is these concentration dependencies which emphasise the nature of cellular activity as a product of a complex feed-forward, feed-back system[15] involving the environment which cannot be revealed by DNA analysis alone. (We shall ask if the environment can feed back as far as directly to DNA changes in Chapter 7.) We shall make no attempt to offer a detailed explanation of how the system arose but proceed to a description of particular features of the proteome firstly of prokaryotes, anaerobics, because in this case the proteins are very largely contained in one compartment, the cytoplasm, and separately in its membrane until, mainly later, a limited number of protein kinds came to be located in the periplasm.

Other important features of the proteins are the different folds and their mobility but we cannot discuss these matters here.[16–18] We give an outline of their value in Figures 4.10 to 4.12 as they illustrate the complicated nature of proteins. The evolution of folds is not easily understood. It is the nature of these folds that gives enzymes substrate specificity as they create surface cavities which can exclude water. They have been likened to lock-and-key fittings. The cavity can also create energised substrates by straining them and 'energised' metal sites called entatic states.

The third polymers of interest are the polysaccharides. They are products of uncoded metabolism and can show a variety of structure including branching. One species can have individual characteristics generated by polysaccharides. Their study is called 'glycobiology'.

4.4.1 Differences between Anaerobic Cell Types

(We do not know how many different kinds of cell there were at 3.5 Ga. The fact that we describe only two classes, Archaea and bacteria, is based on the fact that these two have survived to today.) The essential original cytoplasmic anaerobic proteome can be characterised by a list of cellular activities related to a list of catalysts, machine-like enzymes, control and structural proteins and their variation with external conditions. It is easy to see how small changes have arrived through trial and error exploration of very small, mutational DNA changes, leading toward optimal activity of specific proteins in particular organisms. (Optimal activity is a systems requirement and individual isolated enzyme activity with a maximum efficiency is not the goal.) In later environments which provide different opportunities, such as a sulfur, sulfate, nitrate or ferric ion rich conditions, that is different sources of energy, it is found that different proteins for their uptake and use in metabolism are produced in different groups of prokaryote organisms. The groups are adapted and specialised for particular purposes. This being so all coexisting organisms did not need to have a protein apparatus for elementary sulfur or sulfate metabolism for example, much though they have a dependence on its products. This and the fact that organisms on death, or before, provided essential food (debris) for one another, a form of cooperativity (symbiosis?), means they cannot be said to be just in 'competition'. Many of the specialist functions can be transferred from cell to cell.

A striking different example is given by some differences between the two mentioned earliest prokaryotes, Eubacteria and Archaea,[3,4,14] which do not appear to be transferable via gene transfer. Many Archaea are 'extremophiles', capable of living at high temperature and in high salt situations. The most obvious chemistry, apart from the differences in membrane molecules, is the use of different coenzymes and cofactors. For example the Archaea are clearly very dependent on nickel, and its coenzyme F-430, and Archaea also often use tungsten rather than molybdenum as a catalyst centre. They differ too in many RNA/DNA features.[4] The Archaea have histones associated with their DNA. We consider that the two prokaryote classes belong to separate groups of organisms, which we call chemotypes, and it is this large difference which presents a puzzle for geneticists. To the authors' knowledge there need not have been a common origin to these two different groups of organisms. Could they have arisen in locally different environments? Despite these differences the very basic connected metabolic paths, shown in outline in Figure 4.8, are largely the same. So are many similar proteins and RNA as well as parts of DNA involved in reproduction. The simplest conclusion as given above is that this basic anaerobic organic chemical scheme is the only underlying one that could have generated reproducible anaerobic living organisms of high kinetic stability with a code. But how many times with variations did it arise? Maybe we have to consider that differences arose from different ways of coalescence of particular metabolic units in vesicles. An important question to be faced later is how did considerably different chemical capabilities arise in different cell types? We inspect possible ways of such evolutionary change in Chapter 7.

4.5 Internal Structure of Prokaryotes and Production of New Proteins

One important function of proteins, apart from catalysis, is to provide structure. Again such structure must have been present in the earliest cells. It provides a different way to control organic chemistry which is unknown in man's chemistry. There is now known to be a general internal structure within the cytoplasm of the rigid spherical, what we take to be original prokaryotes (see Figure 4.7).[19,20] Groups of proteins are arranged and held in, for example, multienzyme complexes and by various filamental structures. The implication is that small parts of metabolism are localised. Non-spherical prokaryotes, also rigid, have an actin-like protein[16] which maintains rigid shape, such as a spiral, underlying the inner membrane. Additional internal structure does appear in all prokaryotes during reproductive division when the DNA and the machinery for division are linked together. We shall not describe these structures in detail – or their origin – for they belong to the problem of the origin of reproductive life and they are well presented elsewhere.[21]

A general problem is the mechanism by which the DNA/RNA was able to expand to code for new activities as the environment changed, for example its element content, which brings us to an interest in different DNA structures in prokaryotes. One way to study the problem is to introduce poisons, organic molecules or metal ions, and check for DNA changes during cell multiplication. Remarkably the major DNA of the prokaryote cells is frequently not expanded at first but code expansion occurs in smaller separate rings of DNA called plasmids,[21,22] which are found in the earliest cells and are often present in more than one copy. The plasmids are curiously like certain viruses in that they can be transferred from one cell to another through the environment but interestingly there are now at least two coded molecules apparently necessary in one cell, the main DNA and the plasmid DNA. Is this to avoid complexity in one, the main DNA, and to allow adaptation at first in the plasmid? Moreover the DNA of the plasmid can be incorporated in the main DNA later by a process called conjugation. This transfer of bits of DNA indicates that DNA is a flexible and adaptable molecule open to different types of change.[23] We shall describe the relevance of this statement in Chapter 7. It also implies that bacteria can change chemotype and that there is a loose corporate life between groups of them.

The vast majority of internal compartments of prokaryotes are certain inclusions, membrane-surrounded spaces formed by invagination of the outer membranes, in which reserves, phosphates or 'waste' or calcium ions, or even Fe_3O_4 (magnetosomes), are stored, but where there is little or no chemical activity (but see Section 4.6.1). We shall see later that eukaryotes have greatly expanded the opportunity for change by using compartments for new activities.

4.5.1 Prokaryote Cell Walls and Membranes

(We are very interested in cell wall/membrane chemistry because, as it evolved, it gave rise to different types of fossil; see Chapter 3.)

The cell wall of Eubacteria is a rigid structure made largely from neutral N-acetylglucosamine and acidic N-acetyl muramic acid crosslinked by peptides and called peptidoglycosomes.[3,4] There are other small polymers, such as teichoic acid. Calcium especially helps to crosslink these very large, originally dense, polymeric anionic structures. (We return to the space-filling properties of such polymers later). The walls of their spores, their protected units for maintaining and distributing their genes and other necessities for reproduction, can be different and may include the calcium salt of dipicolinic acid. The prokaryotes may have some proteins and saccharides related to the extracellular structures of eukaryotes. Some bacteria induce and accumulate MnO_2 or Fe_2O_3 on their walls and use them as energy sources but none of them build any kind of controlled shaped shell including those from silica or calcium minerals. The bacteria of the stromatolites bind particles and can induce calcium carbonate precipitation but cannot build controlled shapes of minerals. (As mentioned earlier some others can make magnetosomes ((Fe_3O_4) of controlled shape internally.) It is the wall which we have to contrast later with the cell membrane and outer structures of eukaryotes, including those which are mineralised.

The membrane of Eubacteria is made from lipid esters of sugars and phosphates. It is not a particularly strong structure. It is unlike the membranes of Archaea which are made from long-chain ethers. This difference suggests again that they do not have a common cellular origin. The space between the membrane and the wall is called the periplasm and it contains proteins and activities especially for transport. The periplasm increases greatly in importance in aerobes (Section 4.7).

4.6 The Essence of the Chemistry of Anaerobic Life

Before we introduce any sense of an evolutionary drive we wish to draw attention to the very remarkable nature of the initial internal anaerobic organic chemistry which in essence has not changed in any cell cytoplasm for all of the 3.5 billion years to today.[12,14] It is this chemistry which is given in textbooks of biochemistry and which we cannot detail here but we have illustrated some cases by example. The evolution from the initial organic chemistry outlined in Section 4.4.2 is so remarkable as to throw grave doubt on this occurring by chance. However we believe it must have evolved without a code. This internal chemistry includes the metabolic pathways necessary to give rise to polymers. Later they became proteins with coded DNA/RNA for them. The question we pose is 'Is this complex set of energised reactions (see Figure 4.8) the only possible set of organic chemistry pathways which could have led to living organisms?' A second question is 'How did it arise?' Taking the second question first we repeat our description of the beginning of organic chemistry (Section 2.5). We know that biological organism chemistry overall adds to the degradation of energy at a low temperature to that generated by inorganic chemicals on Earth. This gives an increase in entropy. During the process of the initial absorption of light it is used inevitably to drive some intermediate thermodynamically unstable but highly kinetically stable organic chemistry syntheses only followed later by slow degradation. Though this activity might have been relatively small and piecemeal in different pathways in different containers, vesicles, it could only have lasted if it led to molecules which were catalysts when the contained chemistry in each vesicle could have become autocatalytic. Such systems are in self-sustaining cycles.[15] Clearly however these separate units cannot lead to the very sophisticated system we observe in the first organisms unless they can be combined and become reproducing in an organised way. The implication is that coalescence of compartments could aid evolution. We must accept that some such establishment of such special systems of organic reactions with attendant inorganic elements had to precede any production of large molecules for reproduction because reproduction needing these large molecules also requires the above system of energised reactions.

Simultaneously the cell had to acquire a defined content of many inorganic ions from the environment as catalysts, controls and stabilisers of cells. The requirement was met by pumps made from proteins (see Figure 4.12) and selective protein binding. The whole, with coded reproduction, is the initial prokaryote cell we know which existed at 3.5 Ga. (We find it difficult to accept an RNA-world with no proteins as the earliest form of life because we know of no RNA-based catalysts of redox reactions of such molecules as H_2, CO and CO_2.) This and previous more detailed descriptions of the anaerobic cell show it to be very complex but we cannot give any logical argument for its existence. The possibility that they arose through external action of some kind cannot be put in any rational context. Our answer to the second question is that we know how the organic chemistry of life started but we do not know how its later sophistication came about.

Our contention in answer to the first question is that it was possibly the most kinetically stable self-perpetuating energised kinetic system of any kind which could have arisen, given the initial environment and it is no accident that it has this particular complexity. It was, as we see it, the inevitable product of a very large number of collectively controlled reactions of considerable kinetic stability. The more we examine it the more we see that each pathway is linked of necessity to the whole cellular activity (see Figure 4.8). Cooperativity between the pathways increased sustainability before there could be coded reproduction. It is impossible to provide proof but it is reasonable to suppose that such a system could evolve from our knowledge of systems.[12,14,15] The probability of it arising by pure chance without guidance of specifically 'stable' kinetic factors in steps is extremely small. It must have had a high chemical probability which cannot be examined from studies of present-day organisms or of gene sequences because it could only arise from cycles which must be the first part of the creation of what is to be coded. It is clear that we do not know if it was the only possible system of reactions but we know of no other.

In addition to the nature of the pathways we must ask is the nature of the larger molecular products, which had already evolved by 3.5 Ga, the only possible set? We need to describe these larger molecules next. In Chapter 2 we discussed the mineral products of inorganic elements which are strongly chemically bonded structures overwhelmingly in three dimensions. There is much local symmetry in these structures. The organic chemicals of life are very different and are based on string structures, linear, in rings, often folded, even with branches in polysaccharides, all with no symmetry. Here we can only give the briefest introduction to them and refer to other books for more detail.[1–4]

The outstanding feature is that no other elements except C, N and O of the first row of the Periodic Table with H can build these structures with such variety and breadth of physical–chemical properties, including kinetic persistence and internal mobility. The organic molecular structures arrived through the limitations on the binding between C, N, O and H atoms (for details see organic chemistry textbooks[1,2]). Hence there is a variety of possible local small organic molecule structures from a small number of highly kinetically stable bonds between H, C, N and O. Now structures in small molecules went on to build proteins, nucleotides, polysaccharides and lipids. All these molecules have the potential to form weaker structural interactions internally, between themselves, or with other units. There are then very many ways of folding the long-chain molecules and combining them between themselves or with small molecules. Using weak interactions these structures also have both a degree of specific flexibility and a high degree of selectivity in binding to other molecules and metal ions. It is the functional value of the combination of strong internal chain links and weak external links from them and mobility that make for the unique chemical and mechanical character of the molecules of biological, organic chemical systems. In our view there is no other possible chemistry than this from these four elements knitted into these kinetically stable (energised) yet mobile compounds with their particular variety of properties which could have led to living systems.

Before closing this brief summary of early, anaerobic chemistry in prokaryote cells we must remove any implication that there is any sure knowledge of how their initial chemistry arose from abiotic inorganic origins. There is also no agreement about the origin of reproductive life, which must be later. Moreover there is no sure knowledge in how many ways such cells appeared. Our knowledge of how all the early chemistry evolved by 3.5 Ga is minimal. For example are the activities of plasmids indicative of how cells could evolve through coalescence? The simplest idea of a common origin of all prokaryotes and then a tree-like branching out of different species (see Darwin's picture, Figure 1.2), faces many difficulties, especially initially from the 'seed' and in its early branches. We know that gene transfer was relatively easy but limited between the Eubacteria so that the whole concept of 'species' (defined by their DNA) amongst bacteria is suspect. This transfer amongst cell types is sometimes called reticulate evolution (see Chapter 7). Their speciation may be more or less an artefact of laboratory culture practices, which has no place in the description of wild types. However there are certain groups of prokaryotes that differ markedly from others in their metabolism, even initially, while maintaining the general outlines we have described.

In this description of prokaryote and anaerobic chemistry we have constantly stressed the essential roles of inorganic ions. This allows us to point to the very strong link to the environment which is usually missing in conventional accounts of cell biochemistry. This is especially true because the availability of these inorganic ions can evolve rapidly both spatially and in time, thus affecting cell biochemistry. A second striking fact is that oceanic prokaryotes including many anaerobes (with aerobes) still make up 90% of the total weight of living things in the sea and there may be 10^9 of such cells in 1 g of soil on land. Most are, however, aerobes (see the next section). They make the largest contribution to interaction with the environment and cause much of its changes.

4.6.1 A Note on Prokaryote Diversity

From the beginning of this book we have referred to two kinds of prokaryotes, Archaea and bacteria. However we have qualified this by stating that these are the only early anaerobes we know of by comparison with the prokaryotes of today. It is not known how many, if any, other strict anaerobes existed. We observed that later there were different divisions of prokaryotes which depended upon the products of the oxidation of the environment. They included those using sulfate and nitrate as an energy source. They are traditionally included under anaerobic prokaryotes but are probably better referred to under the title of 'micro-aerobic cells'. As far as we know all of them have the same single membrane construction and metabolic system as the well-known earlier bacteria. There are a few exceptions which have separate internal vesicular compartments. Amongst them we mentioned those which have storage departments. We leave further description of them to the next section because they have a dependence on the oxidation of the environment by oxygen.

A most strikingly different class of micro-anaerobes is that of the Planctomycetes, *e.g.* anammox. They have internal compartments with complex enzyme activity. The anammox are particularly peculiar in that they have no outer membrane, no periplasm, a very different lipid in the inner membrane and three compartments all with different enzyme activities. Their metabolism is unusual in that they oxidise ammonia and use nitrate to supply oxidising power. It is their very different cellular characteristics that make us wonder if they belong like Archaea to a quite different group of bacteria. This would support the idea that new forms of prokaryote cell could arise, even after the atmosphere contained very low concentrations of oxygen. Because they are a minor and peculiar group of uncertain date and no connectivity which we know of we shall leave them to one side in this book. A good source of information about them is in a recent meeting report led by Gori et al.[54]

4.7 Resources and the Coming of Oxygen: Micro-Aerobic and Aerobic Prokaryotes

We have not been able to do more than describe speculatively the earlier organic chemistry which led to anaerobes. The first and greatest, yet least understood, part of evolution from abiotic beginnings is this reductive organic chemistry observed in anaerobes and in all later organisms. While we have stressed the basic reductive anaerobic prokaryote metabolism which had to be maintained to produce any cell we must now observe the advantages and problems raised by the appearance of oxygen and the environmental products of oxygen.[4] We shall make special reference to bioenergetics and basic environmental resources. Oxidative chemistry was at the very heart of the next step of evolution of chemotypes and we shall show that in general outline it was inevitable. It is understandable in essence but not in detail. (There is some discussion of the possibility that the oxides of nitrogen were used as oxidants before oxygen (see Chapter 6), and they certainly can oxidise organic molecules.) Fossil evidence (Chapter 3) suggests that cyanobacteria, associated with stromatolites, first gave rise to oxygen around 3.0 Ga. Confirmation of this time of the introduction of oxygen has been made by the observation of oxidised iron in sediments of iron oxide, BIF (Chapter 2). The major problem for cells was that slowly oxygen oxidised all the basic non-metal elements in the environment which were needed in organic chemistry so that they became more difficult to obtain. Importantly the land would have been oxidised too and weathering of it, run-off into the sea, would have introduced novel elements and new forms of elements unavailable earlier (Section 2.5). However due to poor mixing of the surface and deep water any such evolution could only apply to the surface layers at first with the introduction there of oxidised elements such as oxidised sulfur and iron from 3.0 to 0.75 Ga (see Section 2.5). Toward the end of this period, 1.0 Ga, all carbon became CO_2, all nitrogen N_2 or nitrate, all hydrogen H_2O, and all sulfur sulfate. The original prokaryote cells then needed new catalysts especially for reduction of sulfate and nitrate. However as both sulfate and nitrate and Fe^{3+} generated energy new micro-aerobic chemotypes appeared. Only after 0.75 Ga did oxygen itself eventually become the major source of energy for non-photosynthesising organisms which clearly can survive far from light. We have already illustrated the Krebs cycle in earlier organisms which now became an oxidising pathway and we need to note also the oxidation of lipids. All of the changes are departures from equilibrium by life as more energy is absorbed, transduced, used and later degraded. Many of the initial steps of metabolism of the basic novel oxidised substrates had to be outside the cytoplasm in the periplasm or outer membrane of aerobic prokaryotes to avoid the cytoplasmic reductive chemistry (Figure 4.13).

Before we proceed with the oxidative organic chemistry we must observe how dependent it was on inorganic elements. First manganese was the centre of the production of oxygen from water. The redox potentials of manganese are very well matched to those of oxygen production and manganese may well be the only possible centre for the catalysis of this reaction. Iron is the only element which can bind oxygen in the cytoplasm and it regulates oxygen use in all cells with links to transcription. Iron could not become useful in oxidases outside cells because it is precipitated there. Instead copper plays a major role. Another metal increasing its role was molybdenum. While we showed that inorganic elements were essential for anaerobic life their use in micro-aerobes

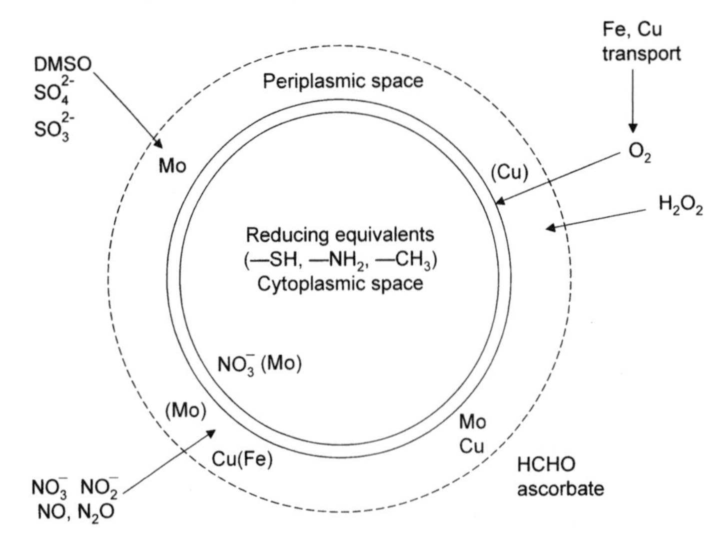

Figure 4.13 The modified prokaryote cell of Fig. 4.7 with a periplasm for many reactions of oxidised substrates, and particular metal ions as catalyst centres.

and aerobes is more extensive as oxygen and its environmental products can only be activated by inorganic chemicals.

After oxygen entered the atmosphere bacteria and Archaea in the surface also had to overcome the loss of external iron, oxidised to Fe^{3+} and mostly precipitated. They needed novel uptake methods for Fe (Section 6.4). (The progressive change of elements and of all environmental oxidative chemistry has been described in Chapter 2.) Despite this difficulty, they became equipped internally with additional non-haem and haem iron but very few new Fe/S enzymes because Fe/S centres are sensitive to oxygen. These new iron enzymes, included flavin and a few Fe/S dependent dioxygenases, many 2-oxoglutarate Fe^{2+} hydroxylases together with haem-dependent cytochrome P-450 hydroxylases. The P-450 enzymes probably acted at first in the cell to remove oxygen or used it to oxidise toxic organic molecules present in their cytoplasm. Later the new enzymes acted as catalysts in several new metabolic paths for the synthesis of hydroxylated compounds[24] and for preparation of poisons of other organisms which include: (i) a rich variety of organic compounds, (ii) the incorporation of halogen by oxidation of especially Cl^- and Br^-. (Some of these steps are common later to eukaryote cells, see below, but the use of several halogens is not common to higher organisms. Instead they oxidise and bind only iodide.) Note cells must not release O_2^- or H_2O_2 in the cytoplasm. The part played by iron, especially in early life, cannot be overemphasised.

The surface prokaryotes had to cope with an increase of certain metal ions released from sulfide minerals by weathering of the land such as Cu and Zn. Subsequently these metals became of great importance in useful enzymes. An outstanding novel use of copper is in high redox potential chemistry outside the cytoplasm but inside the periplasm (see Figure 4.13), and on the outer surface of the prokaryote cells' inner membrane (Figure 4.14) where iron is not very valuable externally due to the ease of its oxidation while Fe^{2+} is the state which interacts with O_2. Particularly striking is the employment of copper in denitrification (Figure 4.15), in energy transduction (see Section 6.7), and in the making of –S–S– bridges in proteins. Meanwhile molybdenum and then a haem plus an Fe/S centre were still used to reduce sulfate.

Many bacteria are facultative, that is able to switch to or from anaerobic to aerobic metabolism, maintaining the underlying fixed reductive metabolic network. The implication for cells is the need for control of expression by the presence of O_2. Interest then lies in novel O_2-sensing transcription factors,[24] themselves iron proteins which led to a new series of enzymes. The lack of further separate internal compartments within the cytoplasm for such steps and others in most bacteria made a decisive difference from eukaryotes in the exploitation of novel pathways as we shall describe later.

Overall the use made of Cu, Zn and Mo by the prokaryote cells made available by oxidation of their sulfides was not very great and we only see their full value when we describe eukaryote cells (Section 4.8).[5] There was continued use of manganese and magnesium but no novelty in the value of calcium. On the other hand the value of nickel, especially, and cobalt in H_2 and saturated

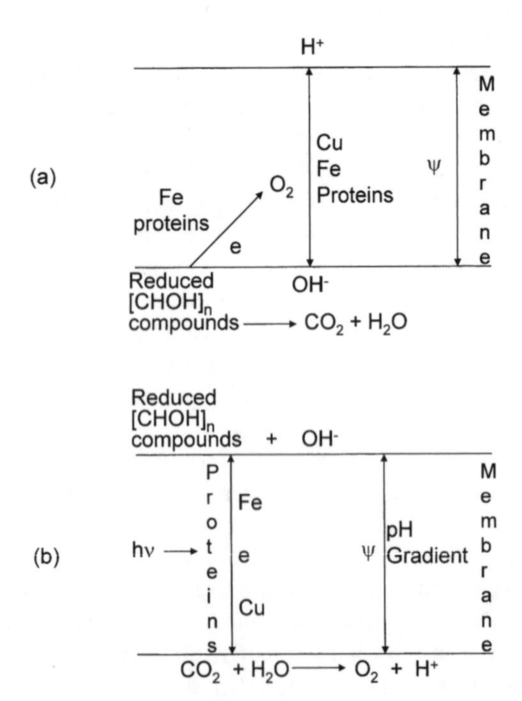

Figure 4.14 Two early functions of copper proteins in prokaryote membranes: (a) cytochrome oxidase and (b) in a cytochrome replacement in light conversion to electron gradients, Ψ.

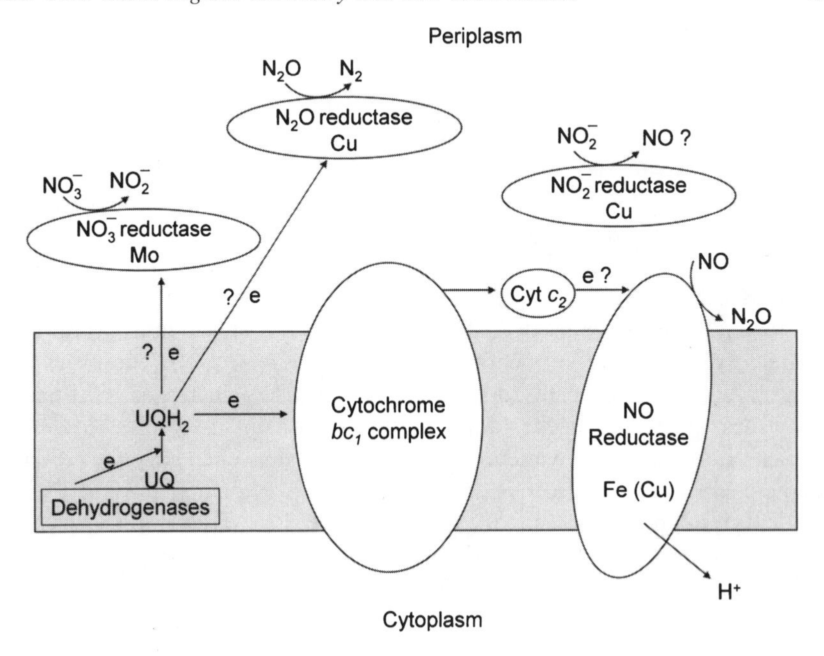

Figure 4.15 The enzymes in membranes of the reactions of oxides of nitrogen denitrification, mostly in the periplasm.

C/H chemistry in prokaryotes lessened as these substrate molecules became less available in the environment (see Chapter 6). These metal ions are not so useful as catalysts in oxidation (see Table 6.4) and their coding in DNA may have been rejected. Their 'rejection' was despite increased availability, although free cobalt in the environment remained low. Notice that changes to greater numbers of micro-aerobic and aerobic bacteria would have been very gradual with time and increased only slowly at depth of the sea from 2.5 Ga.

It is estimated that there were by 0.75 Ga various prokaryote life forms, photosynthesisers and non-photosynthesisers, anaerobes, micro-aerobes and aerobes, and divisions arose between these classes according to their abilities to use, for example, different pigments for light absorption and different oxidised chemical sources of energy, metabolites and metal ions generated in the environment.[14] They must have been unevenly distributed in the water column. While there was competition for many nutrients a striking feature is that this collection of organisms with different specialisms, different chemotypes, could clearly achieve greater use of the energy and material sources in the different environments than if any one organism had attempted to use all the resources as much as possible. We have seen this community life of different anaerobes earlier. We consider that the restrictions which generate specialisation are the inability to manage efficiently the chemistry of diverse processes given also the need for controls apart from access to different resources. The prokaryote gene complement is limited to about 4,000 expressed proteins which does not allow a very great variety of activities in one cell. Their combined debris gave communal access to all chemical needs. This cooperativity in a community, low-level 'symbiosis', is one side of evolution while competition for some resources needed by all is another. This period of the community of prokaryotes has lasted from 3.5 Ga to today, and remains the dominant form (90%) of all life in the sea. Full extension of life to the land must have evolved in soils but the fullness of life on land seen today did not begin until about 0.5 Ga.

In concluding the description of the mainly reductive and internal metabolic chemistry of micro-aerobic and fully aerobic prokaryotes there are three outstanding features.[3,4] The aerobic prokaryotes established the basic element capture, and use of oxidised C, N and S, from the environment; important internal organic oxidative metabolism, *e.g.* the Krebs cycle, energy capture using oxygen as in all cells to today which leads to increased synthesis of nucleotides, proteins, saccharides and lipids; and an essential, but limited complement of several new metal ions. Like anaerobic chemistry the oxidative organic chemistry is so selective as to be described as a singular system with a set of catalysts, often metalloenzymes, for selected required steps. Cells could only overcome the difficulties of the new external chemistry required by using extra divisions of space, in the case of aerobic prokaryotes, the periplasm. Aerobic prokaryotes established fourthly other basic requirements of this metabolism, especially control mechanisms of the concentrations of both bulk and catalytic trace ions in two different compartments, and lastly a maintained well-regulated internal average reducing medium at about – 0.1 V at a pH = 7.2. Many of these novel features are lasting conditions in the cytoplasm of all cells, prokaryotes and eukaryotes to today, but do not apply to activities and controlled extracytoplasmic spaces which arose and were greatly extended later. This means that these spaces, like the external environment, have different compositions and properties from those of the cell cytoplasm and offer the greatest chance for evolution and diversification.

4.8 The Single-Cell Eukaryotes

There was a considerable period of perhaps nearly half a billion years after prokaryotes produced oxygen before the probable first appearance of single-cell eukaryotes (Figure 4.16), around 2 to 2.5 Ga.[25,26] We consider their appearance as the second large step in the evolution of the chemistry of life. The great difficulty is in the dating of their evolution because the oceans were not homogeneous from 2.5 to 0.75 Ga (Section 2.2). Evolution would have been very slow in the deep reducing regions, due to the Fe/S buffer controlled by bacteria, which limited O_2 availability (see above). It could well have been more rapid in the more oxygenated surface exposed to light. The record is poor and large imprint 'fossils' of this period until 0.54 Ga are very rare and often classed under the name eukaryotic 'acritarchs'.[27] It is frequently difficult to distinguish between large prokaryotes, colonies of such cells, and eukaryotes, for example in the case of Grypania. The step to any eukaryotes must have required profound changes of metabolic activity and as stated above there are

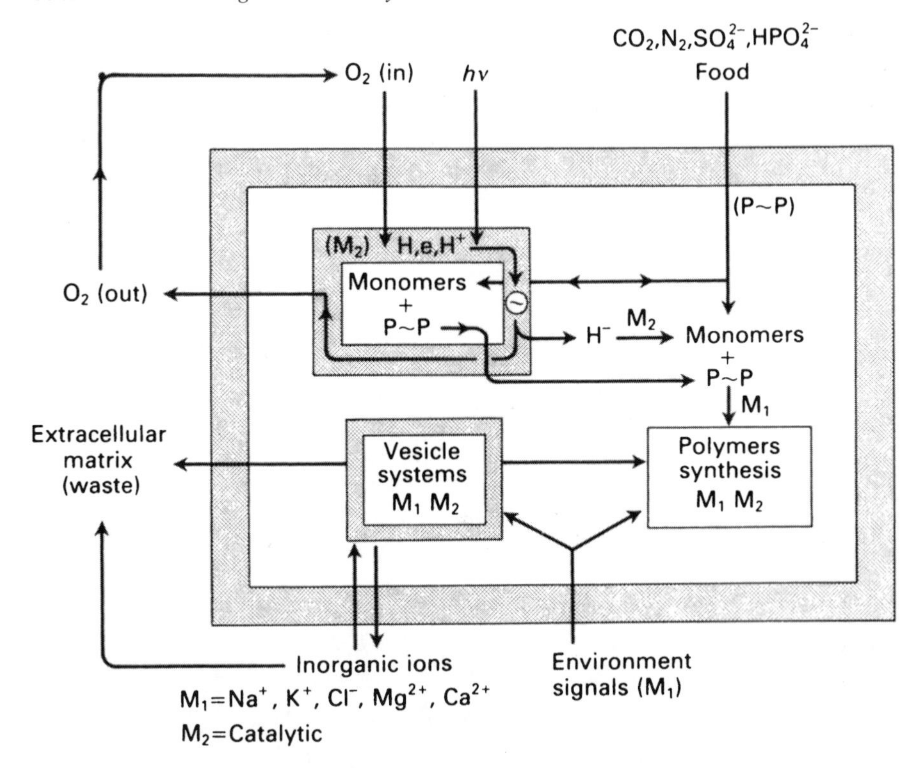

Figure 4.16 A basic outline of the chemistry of a single eukaryote cell with or without light capture. Compare the many compartments (vesicles) with the prokaryote cell's single compartment, Fig. 4.7. In what way was this cell derived from prokaryotes?

no really well-authenticated intermediate forms. We must now contrast the single internal compartment prokaryote cells with the complicated eukaryote cell of many internal compartments. It is often considered that, although we do not know with any degree of certainty how eukaryote cells arose, their origin was due at least in part to a combination of known prokaryotes sometimes called endosymbiosis, as shown in part by their DNA analysis.[23] Some authors consider that the more complex eukaryotes were in fact just combinations of the two simpler forms, Eubacteria and Archaea.[25] This combination by itself makes one addition to divided internal space. Biologists with detailed knowledge of today's organisms have described one or two apparently good possibilities of the combination of these two types of bacteria with transfer of the Eubacterial DNA to a basic Archaea nucleus.[25,26] But does all the DNA originate from one or the other of these prokaryotes?

The very basic novelties of all eukaryotes were for (i) use of a supply of oxygen, particularly for the synthesis of cholesterol from their novel flexible membranes which had no fixed shapes; (ii) the introduction of a number of internal compartments linked to the new endoplasmic reticulum and to the

Golgi apparatus; (iii) the creation of a nucleus compartment; (iv) the incorporation of captured and degraded prokaryotes, mitochondria (Figure 4.17), to bring about oxidative production of ATP. This is sometimes called symbiogenesis and includes the later inclusion of chloroplasts, again from bacteria, in some mitochondria-containing cells; (v) series of considerably

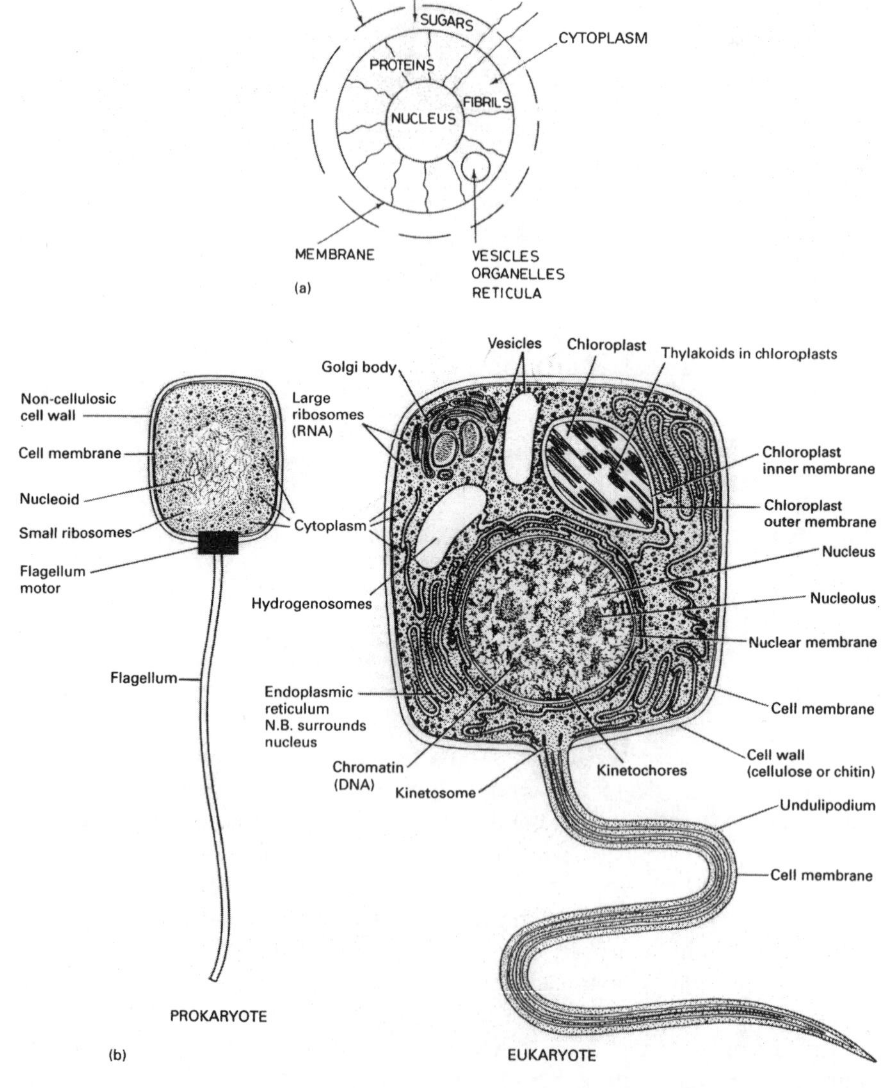

Figure 4.17 A schematic outline of the eukaryote cell (a) showing an outermost membrane (open wall if any) and connecting filaments to the nucleus compartment. (b) A prokaryote cell for contrast with a eukaryote cell of many compartments including both light capture thylakoids and O_2-using organelles (not shown) and a nucleus as in a plant cell.

advanced filaments positioning various compartments (Figure 4.17), and giving controlled mobility to the outer membrane and the whole cell; (vi) signalling systems for response to the environment using calcium ions. Here we see the realisation of the possibilities of use of the calcium gradient which was essential for the stability of life's organic chemistry; (vii) novel sexual relationships; (viii) large bodies, including prokaryotes which could be imported, endocytosis, and consumed, and vesicles could also export unwanted contents, exocytosis. While protection of the reductive cytoplasmic metabolism was maintained, often in novel ways, some quite new metabolism arose in these eukaryote organisms perhaps only in the surface at first. We give a list of some single-cell eukaryotes in Table 4.4. Given the buffering of O_2 in the sea together with the problems of the novelty of eukaryote chemistry, any further eukaryote evolution after, say, 2.0 Ga was slow for a long period. This pattern of evolution of the eukaryote revealed by the size of imprint fossils[27] and by molecular fossils (Section 3.9), can be partly confirmed and extended by genetic analysis but this is not straightforward in this period (see Section 4.15). We must note that several groups of single-cell eukaryote species arose at close to the same time in photosynthesising and non-photosynthesising cells. We have to note too that coalescence of organisms to give eukaryotes cannot fit into a simple branching Darwinian tree (Figure 4.18).

The single cell of the eukaryotes required a supply of fixed nitrogen because NH_3 was removed from the original atmosphere by oxygen. As they did not have the genes for NH_3 synthesis, they relied on bacteria, external symbiosis, for its supply. The only exceptional eukaryotes which were able to fix N_2 did so using plasmids derived from bacteria. The dependence on external bacteria adds an extra 'long-distance combination', simple community 'symbiosis' as opposed to coalescence (see Figure 4.18), again not captured by the idea of a tree of evolution. Other sources of nitrogen were from (a) oxidised nitrogen, other than by lightning activation of H_2O and N_2 and (b) cellular debris.

To understand in more detail the eukaryote cell and its organic chemistry we shall need to keep in mind the compartments because several are involved in internal and/or external uses of different elements and different metabolic processes from those in the cytoplasm and in prokaryotes (see Section

Table 4.4 Some Single-Cell Eukaryotes

Plant-like	*Animal-like*	*Fungi-like*
Euglena	Zooflagellates, e.g. trypanosomes	Some yeasts Sporozoa
Dinoflagellates	Amoeba	Slime moulds
(Coccoliths*)	Foraminiferae	Plasmodia
Acritarchs (algae and plankton)	Radiolaria, Ciliates, Acantharia*	

*Some of these species can carry out photosynthesis but they may do so by engulfing symbiotic algae. It has been suggested that Grypania found in iron oxide sediments could be the earliest eukaryote, while other very early examples are placed under Acritarchs.

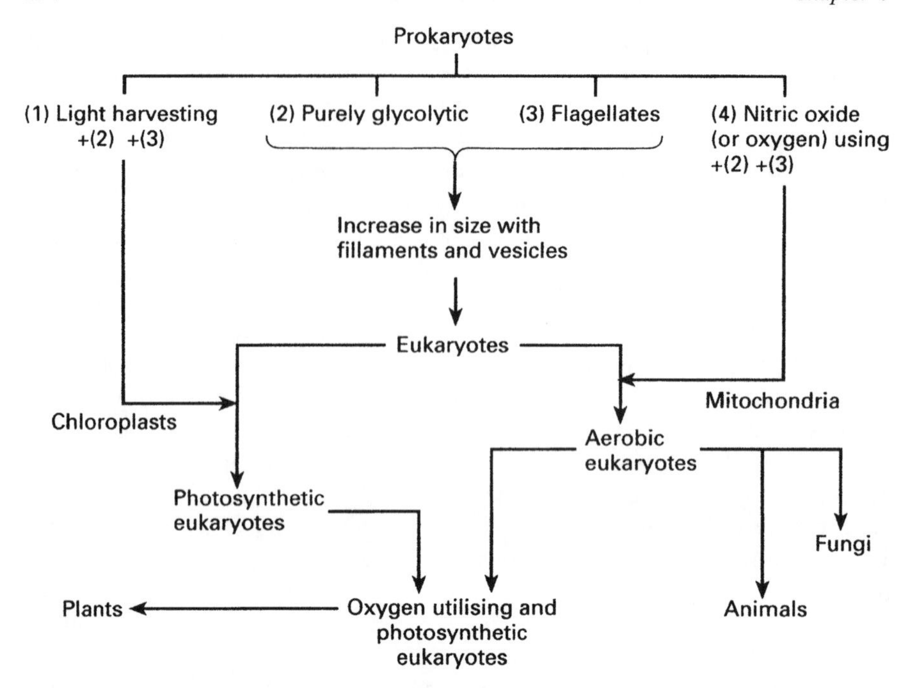

Figure 4.18 The progression from prokaryotes to eukaryotes but not on a simple branching tree. The initial eukaryote cell is not known.

4.11).We emphasise that placing enzymes and other activities which could conflict with cytoplasm reduction in separate containment is a clear logical evolutionary development. As was true for prokaryotes it is obvious that it is advantageous to separate oxidative from the basic reductive chemistry of the cytoplasm in any cell. We describe the compartments derived from bacteria, mitochondria and chloroplasts, here as they are very different from those from the endoplasmic reticulum (see Section 4.11). To a large degree keeping reductive cytoplasmic chemistry separate from interference from environmental oxidative chemistry correlates with the separation of reductive and oxidative chemistries within prokaryotes, eukaryotes and man's laboratory and industrial practices.

The oxidative degradation of some small molecules in the mitochondria is used for energy transduction in all eukaryote cells and provided a gain in efficiency of organisation and in removal of other oxidation risks from the eukaryote cytoplasm and the nucleus. Probably only a little later certain cells incorporated additional different photosynthesising bacteria and converted them into light-capturing organelles, chloroplasts, a further separate space. Thus all energy transduction and oxygen production remained in the organelles. The two classes of organelles lost most of their genes to the parent DNA but retained some independent genes. Note how easy uptake of symbionts and gene transfer seems to have been. Has it continued up to today in higher organisms? Why not? Clearly much protein transfer is also involved in both directions (and to vesicles where this trafficking is largely novel).

The photosynthesising eukaryotes with two types of organelle occupied surface regions of the ocean bathed in light while the non-photosynthesising, largely dependent single-cell eukaryote organisms could also occupy space in deeper water, often the seabed. In summary the overall turnover of chemicals in both regions was increased by the introduction of these different eukaryotes with organelles. It was driven by increased light absorption by the sum of prokaryote and single-cell eukaryotic syntheses and by greater degradation of synthesised material by all organisms generating heat. At the same time the cytoplasm maintained the metal cation salt balances as seen in prokaryotes, much though the total complement of ions and catalysts and of metal elements, the metallome, is not the same for all the different kinds of prokaryotes and eukaryotes, as we shall discuss in Chapter 6. We see also that in the single-cell eukaryotes the processes of effectively anaerobic synthesis of the major units, proteins, nucleic acids, lipids and saccharides from basic units such as CO_2 and N_2 and then amino acids and small C/H/O compounds (see Figure 4.7), takes place overwhelmingly in the same way and in the cytoplasmic space as in prokaryotes but in a protected way. The implication is that reductive catalysts, often employing iron, Fe_n/S_n, with Mg^{2+} dependent enzymes for acid/base catalysis as in prokaryotes, remained almost unchanged.

The eukaryotes had the great advantage and protection that they could engulf and consume threatening bacteria. However the complexity of the eukaryotes meant a slow reproduction rate and here the prokaryote still had great advantages. The two came to coexist and frequently actually assisted one another in the ecosystem at first in the surface of the sea.

The next sections will describe the important properties of single eukaryote cells in more detail in the following way: Section 4.9 outlines the nature of the cell nucleus, which is common to single-cell and multicell eukaryotes; Section 4.10 will describe their filaments; Section 4.11 looks at their vesicles and will describe their proteomes; Section 4.12 outlines their protective systems; Section 4.13 some features of their genetic structure; and finally Section 4.14 gives a quick summary of all features of unicellular eukaryote organisms. We must observe the closely similar timing of new cellular activities with the environmental change of O_2 and its oxidisable inorganic compounds, probably firstly in the shallowest levels of the sea only (Chapter 2). Much of this description depends upon the presumed parallel between today's and earlier organisms.

4.9 The Eukaryote Cell Nucleus

As we have stressed earlier the organic chemistry of cells is limited and unique in that it is limited by translation of a code. It is then essential that we always bear the code in our minds when we discuss cellular organic chemistry. Furthermore we must give a description of the novelties of the DNA of the eukaryote cell because eventually we shall have to consider how it, as an organic chemical, interacts with the environment in Chapter 7. The original prokaryote genetic instructions were provided by a single, double-stranded, often cyclic DNA plus plasmid DNA. The size of the main DNA in bacteria is usually small, up to around 4,000 genes, although very variable even in these cells but it has grown considerably in single-cell eukaryotes, to 6,000 up to 8,000 genes, which have their DNA in a nuclear compartment. The nucleus now contains extra genes transferred from the organelles. The eukaryote nucleus has non-coding regions, introns, not expressed as mRNA and therefore as proteins, but some are coded for micro-RNA, mi-RNA, which acts as a messenger or a control of growth and back-interacts with DNA.[3,4] We shall return to the value of introns in Chapter 7B. The reading of the DNA is made difficult in eukaryotes because the DNA is packaged with proteins, histones, in small units, nucleosomes, which are then organised together (Figure 4.19). (Note that histones and some introns are also found in Archaea.[26]) The special folding and packaging of the nucleus in eukaryotes requires extra binding of K^+ and Mg^{2+}. The activation of the nucleus is dependent in part in the eukaryotes on a few novel transcription factors, some of which are dependent on metal ions. One group of them, the zinc-finger transcription factors, which are rare in prokaryotes,[28] are of great interest as they have control features in unicellular eukaryotes. The zinc must be obtained from the environment in greater quantity. Zinc in the transcription factors, though not in large numbers, behaves differently from that in earlier zinc enzymes in that it is more easily exchanged (see Chapter 6). We observed in Chapter 2 that zinc in the environment increased as oxygen increased.

The histones, basic proteins of the nucleus, can be modified by reactions such as methylation or acetylation which can control expression. The dependence of such reactions, and hence of gene expression on external factors, introduces complexity in genetics. It is often called epigenetics,[29] for example in the immune system of man.[30] The inheritance of epigenetic changes is now generally agreed. We return to this subject in Section 7.8. As in prokaryotes

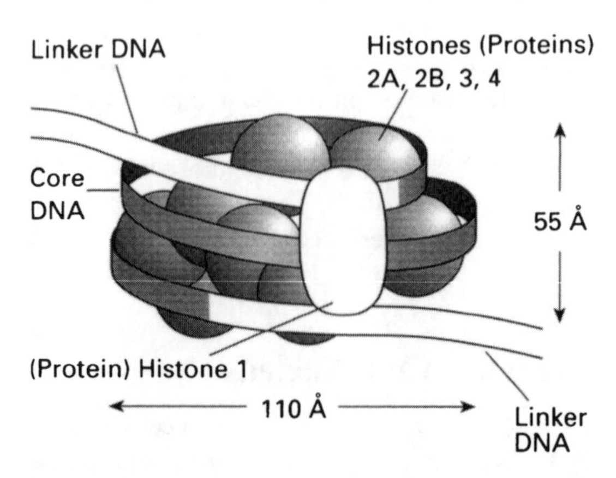

Figure 4.19 The structure of the packaged eukaryote DNA with histone protein-celled nucleosomes.

insertion of genetic material is observed, now by viruses, and movement of DNA sections within the whole DNA, transposition, is not uncommon. In passing observe that the complications of the nuclear compartment added to the discontinuity from any single prokaryote. It may well be said to be a missing or certainly an unsure link in the branching of the 'tree' of evolution. The complexity as well as the limitations of the chemistry of DNA itself make the understanding of it difficult.

4.10 Filaments in Single-Cell Eukaryotes

The major filaments inside the single-cell eukaryote cytoplasm are the tubulins, intermediate filaments and the actomyosins.[3,4] They control the cell shape and the distribution of compartments, management of space, just as in prokaryotes, but now they are adjustable, extendable or controllable in the cell on receiving a signal (see below). In eukaryotes these filaments have additional properties. The actomyosin units are linked to the new protein cadhesin, a combination of a calcium binding site and an adhesive structure for binding to surfaces, which extends outside the cells, as well as to several other extracellular proteins. They are bound temporarily to an inner calcium-activated protein, a calmodulin[31] called troponin which switches on/off contractile action. The trigger for such action is the influx of calcium activated by environment sensors in the cells' outer surface.[2,6] As the calcium binds to a large number of different calmodulin centres of different metabolic pathways and is in fast exchange it integrates cooperative action to external novelty. The activity is a pulse followed by rapid recovery because the calcium is rapidly removed from the cytoplasm by the action of outwardly directed ATPase pumps. It does not make long-term changes in the cell unless it is connected to certain protein phosphorylations. Clearly the apparatus is the father of modern muscles. We can then relate the calcium ion properties to the beginnings of controlled motion in animals as well as a messenger activating many cell internal responses to environmental change. In Section 4.2.1 we observed that it was essential for life that calcium was rejected from the cytoplasm and that this was then one of the unavoidable conditions but it gave rise to certain possibilities for the evolution of cellular organic chemistry. In this section we see the realisation that the possibilities given by calcium rejection have given rise to the evolution of eukaryotes and, as we shall see, to that of all animals. Compare the specific roles of calcium with that of internal exchangeable Mg, Fe and Zn ions. The tubulin in eukaryotes is a guide for vesicle movement and all filaments and vesicle positionings. There is no evidence of extracellular structured filamentous proteins for controlled biomineralisation in early unicellular eukaryotes.

4.11 Vesicles in Single-Cell Eukaryotes

As listed in Section 4.7, a new feature of the single eukaryote cells is their internal enclosed spaces, vesicles, other than the nucleus and organelles mentioned above, including the weaving reticula of the endoplasmic reticulum connected to the Golgi body or vesicle (see Figure 4.17). Some large vesicles are more or less permanent though they may bud off small vesicles for transport of material. Other vesicles are formed from the external membrane, bringing in food particles, and then entering or forming internal lysosomes for digestion. Some of these vesicles are at a different acid pH from the cytoplasm. These and other vesicles have uptake pumps for protons and/or ions such as Ca^{2+}, Mn^{2+}, Na^+ and Zn^{2+} but not Mg^{2+} and Fe^{2+}. The last two, the very ions of greatest importance in the cytoplasm of all cells, appear to be excluded from very many vesicle spaces but this is not so for the organelles. Uptake of iron is of Fe^{3+} ions into special vesicles where it is reduced to give Fe^{2+} to the cytoplasm. Vesicles containing calcium are of immense importance as they can be stimulated to provide amplification of the calcium-signalling system between the environment and the eukaryote cell activity. We describe this stimulation in Chapter 5. There are yet other vesicles which hold enzymes for oxidation, the peroxyzomes, which must have a higher average redox potential than the cytoplasm. They contain haem and copper oxidases but no Fe/S proteins. There are also vesicles which store organic molecules, mainly small but also some large proteins, which can be ejected into the extracellular fluids or the environment as poisons of other organisms. We return to the value of them in Section 4.22. Finally there are vesicles in which proteins are glycosylated, often by Mn^{2+} enzymes, aided by calcium proteins. They control some protein folding and are called chaperones. The position and movements of these vesicles as stated are controlled by novel filamentous structures inside the cell cytoplasm. Observe that the vesicles differ from those of the prokaryotes generally in that they contain activities associated with proteins, enzymes, but see Section 4.6.1. We must be aware too that the single-cell eukaryotes can live like some bacteria in multicellular organisms and can be agents of disease, parasites, or as required symbionts. Could they be like organelles and give DNA to the nucleus? What is an independent organism?

4.12 Protection in Single-Cell Eukaryotes

The introduction of oxygen and then many heavy metal ions from the oxidation of sulfide into the environment was a risk to prokaryotes and eukaryotes. The worst feature to this day was the partial reduction of oxygen, catalysed by metal ions, to give superoxide, O_2^-, and peroxide, H_2O_2, which attack most organic molecules. Prokaryotes devised protection from these reactive species, using an Fe/Mn superoxide and a nickel dismutase in the cytoplasm, still seen in eukaryote organelles and a catalase. Eukaryotes have a catalase, now in peroxyzomes, which is very like the prokaryote protein but their superoxide dismutase is a quite different Cu/Zn enzyme and is in the cytoplasm. They do not have a nickel superoxide dismutase. As mentioned earlier some bacteria also have such a Cu/Zn enzyme but its origin is much debated and is probably due to gene transfer from eukaryotes.[32] Note how the newly increased Cu and

Zn have been used. A new protection against organic peroxides is provided by a selenium peroxidase only found in eukaryotes. Selenium as selenate also became more available with the oxidation of environmental selenides. Protection in prokaryotes against new environmental Cu and Zn themselves is by polypeptides containing cysteine, which are conserved in plants, but in eukaryotes it is due to a new protein, metallothionein. The maintained levels of both free cations in all cells are also kept low by outward selective ion pumps and carrier/buffer proteins to avoid poisonous excesses.

In summary, for the purposes of this book, as we have stressed, linking the environment to the organisms during evolution in the origin of prokaryote life, is a dominant theme and it is oxygen, oxidised small substrates and the released metal ions, which bring this out most clearly. To emphasise this point the evolution of single-cell eukaryotes is notable for: (i) The rise in calcium-signalling to the cell of features of the environment which link not just to physical cellular activity, *e.g.* contraction, but also to metabolic organic pathways of phosphorylation/dephosphorylation in turn connected to gene expression.[3–5] These phosphorylating enzymes require Mg^{2+} bound to ATP and ADP. Some 30% of the eukaryote proteins in the cytoplasm may be phosphorylated dependent on conditions. (ii) The somewhat increased iron and the value of additional copper in oxidative enzymes. (iii) The use of a few novel zinc proteins in transcription factors. These zinc proteins are associated with signals, especially concerning growth, zinc fingers, and with zinc peptidases/proteases. The importance of the last two will become greatly strengthened in multicellular eukaryotes where there must be connection and communication between cells as well as with the environment and between compartments. (iv) The novel value of selenium. We must not forget that single-cell eukaryotes, mainly algae, with prokaryotes are the major organisms of the sea surface except at its fringes even today.

4.13 Genetic Analysis of Unicellular Eukaryotes: Algae and Metazoans

We shall not refer to the value of genetics greatly until we come to Chapter 7. However genetics is a study of the organic chemistry of cells.[33] We use genetics here to show how this branch of organic chemistry helps us in the comparative analysis of the sequences and the number of proteins in evolution. Earlier we have used fossil evidence but this does not give any picture of a certain connected series, especially before 0.54 Ga. This is also the case with evidence based on the comparative chemistry of organisms using today's examples. In principle genetic analysis could give a continuous sequence from the earliest cell of a particular kind to today. Unfortunately we have to base the genetic analyses largely on present-day organisms for an assessment of evolutionary relationships. This is very successful from about 0.50 Ga to today. However where experimental DNA data are inadequate mathematical methods of projecting the mutational data back in time are used.[34,35] Here we give a basic account and some observations on a part of organic chemical evolution reserving detailed discussion to Chapter 7. (Before introducing the topic our thesis is that long periods of evolution, say from 3.5 to 0.5 Ga, can be followed most easily using trace elements in sediments to guide us to evolution generally but this has not yet yielded a well-defined continuous line of evolution.)

By making certain assumptions as to rates of mutational base changes any small novel step in evolution of organisms can be assigned to a place in time.[35] The procedure is at its best using the sequences of DNA or RNA related to proteins involved in well-conserved activities, such as slow modification of an internal enzyme as opposed to the introduction of an apparently novel one. Before we begin this analysis we have to see that it is one of substitution in or extension of organic chemicals, DNA, without any reference to the inorganic nature of the environment and the cellular content of inorganic chemicals. We describe here the algae and single-cell metazoa which are related to the present-day plants and animals as examples as they illustrate the methods and the problems of this particular branch of organic chemistry.

Berney and Pawlowski[36] show that the RNA of a group of photosynthesising single-cell eukaryotes, diatoms, dinoflagellates and coccolithophores of today (plant related), can be traced to convergence in the chromalveolates not too far back from 0.8 $\pm$ 0.2 Ga. They are all red/brown algae, two or more types, and both the uncertain impressionistic fossil record, imprints and these genetic data suggest then that the parent algae evolved some time before but not long before 1.0, say 1.5 Ga. A striking fact is that the brown algae are distinct genetically from the photosynthesising green algae, probably of a greater age as the green algae are simple symbionts while the red/brown algae are doubly symbiotic with a simple symbiont inside an earlier large cell (but see reference 25). In particular the brown algae have the different organic pigment chlorophylls *a* and *c* (the red only *a*) from the green which have chlorophyll *a* and *b*. This is indicative of different sets of genes inherited from different captured prokaryotes, at least four. Hence coalescence with photosynthesising bacteria was not uncommon but is not a single genetic event among algae, unlike that of mitochondria, giving eukaryotes and leading to animals in which there is only one kind of mitochondria. The genetic differences between the algae are seen also in the different mineralisation, in trace element content (Section 6.8),[37,38] and in efficiency of protection of CO_2 uptake from O_2 (Figure 4.20). There are also considerable general metabolic differences. There are clearly difficulties in giving one clear, timed evolutionary sequence of all the organic chemistry. The later multicellular or colonial seaweeds (algae) are of all three genetic kinds, green, brown and red/brown algae and the last two are the most dominant (in the sea), and their record is better defined. The green algae later evolved to give the multicellular algae (plants) on land from around 0.4 Ga including vascular plants. This explains the fact that the sea is brown but the land is green.

Given that genetic studies give the date of the evolution of all photosynthesising algae cannot be far from 1.5 Ga and they all have mitochondria, the evolution of single-cell non-photosynthesising eukaryotes with mitochondria

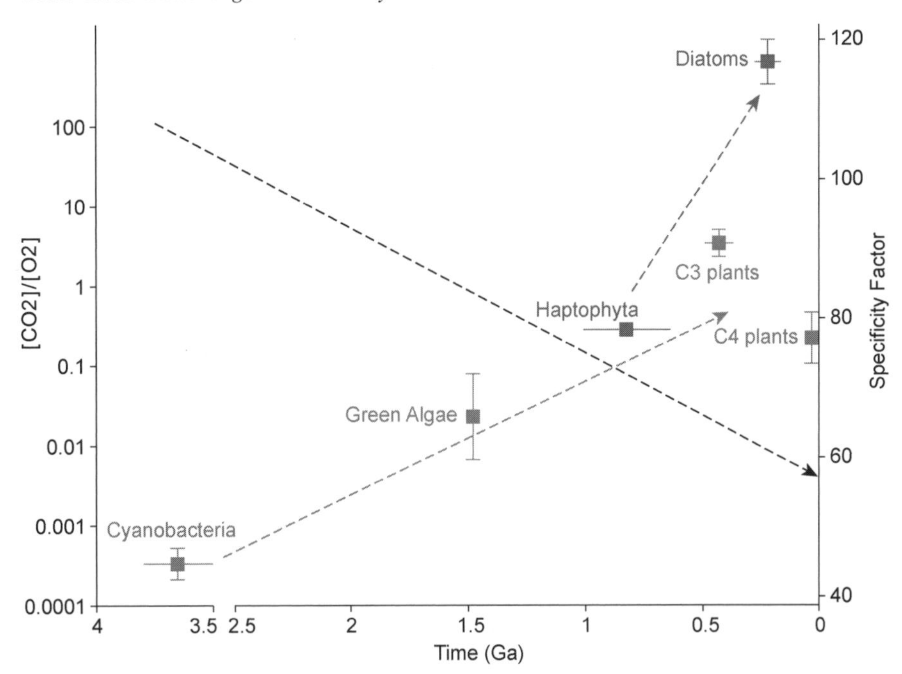

Figure 4.20 A plot of RubisCo specificity factor (τ), that is CO_2/O_2, against geological age for some single cell organisms.

alone must be earlier.[33] Yet the extrapolation of some genetic data to their origin of the earliest of these cells, amoebae, appears to give an origin of 600 to 750 Ma. Various lines of argument from fossils have led to the proposal that their origin was in fact around 1.5 Ga,[39] which is indicative of a 1 billion year gap before fully multicellular eukaryotes arose (see Section 4.15). Clearly before 0.75 Ga much is uncertain in fossil findings and in the comparative biology of organisms and in the genomic analysis of unicellular eukaryotes and greater confidence should rest in chemical analysis of the chemical elements in them and in sediments.[40] The difficulty here is that sediments may reflect the nature of the deep not the surface water which we believe were poorly mixed from 2.5 to 0.75 Ga. A much greater effort is required to increase knowledge of this chemistry.

In conclusion genetics is of great value in the understanding of advances and errors in parts of purely organic chemistry. The further back in time we go the harder it is to use the method with confidence and especially so when there are no well-classified fossils which give definite knowledge of evolution.

4.14 Summary of the Evolution of Unicellular Eukaryotes

To help the reader we give a brief survey of the above changes in structure in unicellular eukaryotes which are common to many if not all their branches.

These changes came about in large part from the way in which oxidation provided opportunity based on novel organic chemistry.

1. The cells are larger than those of prokaryotes.
2. Their outer membranes contain cholesterol and are flexible; their shape is adjustable, hence they require oxygen.
3. They have a nuclear structure in a compartment.
4. They have a variety of compartments separating activities and including enzymes. Some new compartments are derived from the new endoplasmic reticulum.
5. All members have internal degraded Eubacteria, mitochondria, and some have chloroplasts. These two organelles supply much of the energy of eukaryotes.
6. Their activities are controlled by external/internal messengers such as calcium ions and phosphate esters.
7. They import (endocytosis) and export (exocytosis) large particles, vesicles and even prokaryotes.
8. They undergo sexual reproduction.
9. They have novel complements of metalloproteins, including a few of zinc in transcription factors, a small number of copper enzymes for oxidation and with zinc for protection. They also have a new selenium enzyme also for protection.
10. There is diminishing importance of nickel and cobalt catalysis of organic chemicals.

All these features of the character of cells apparently rose after the first marked changes in the environment, the first great oxidation event of about 2.5 Ga with the changes in environmental element composition. We do not know which of these features were unavoidable. We do know that many are related to the much earlier unavoidable production of oxygen with its inevitable oxidation of the environment and the calcium gradient. In Chapter 7 we shall ask whether the changes in organisms were due to the random searching by mutation, finding the organisms of greatest fitness, or do the novel environmental chemicals impose local genetic changes? The same question will arise with the evolution of multicellular organisms, which we describe next. We stress at each stage that the increases in variety of the chemistry in each of the three steps is only possible by increased physical size and compartments of organisms. Remember that initial enclosure of chemistry was not a genetically controlled event but an unavoidable consequence of the need for retention of energised molecules of certain kinds to be contained.

4.15 The Multicellular Eukaryotes

The next development of concern in the description of organic chemistry is that of the multicellular eukaryotes which have all the features listed in Section 4.14 with the following additional characteristics (Figure 4.21).[3,4] We wish to note

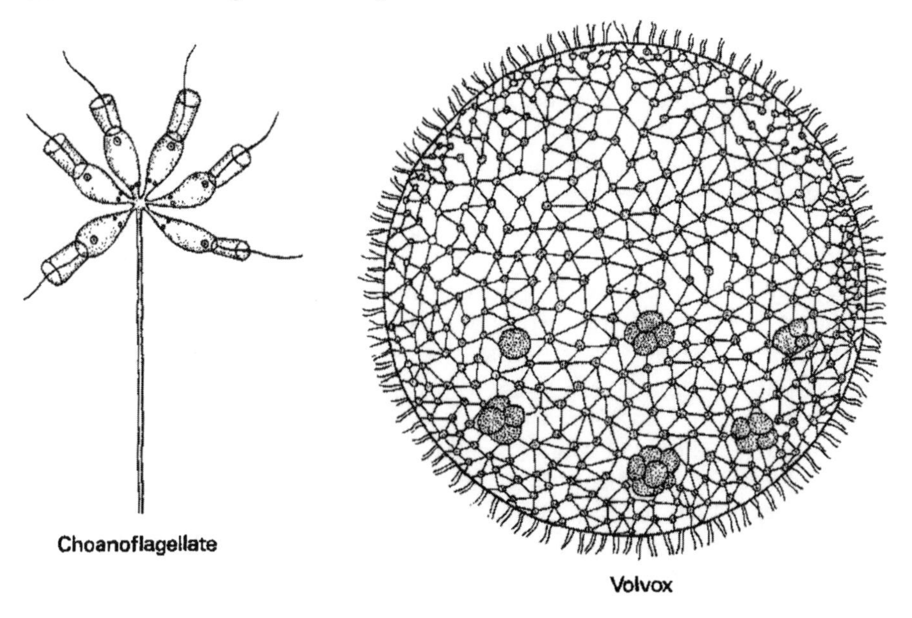

Figure 4.21 The multicellular organism, Volvox, which could typify early forms of such organisms. It has internally differentiated cells held by a matrix within a "skin". For comparison a later flagellate is also shown.

most strongly the chemical changes due to coincidence with further oxidation in the environment.

11. Cells are linked by an extracellular, partially oxidised matrix (see Table 4.6), which, while the organisms grow, must be open to degradation to alter their space yet, at any one time and in an adult, the proteins of the matrix must be crosslinked to maintain shape and relative rigidity of the multicellular construction. The role of metal elements Cu and Zn is very considerable here. The crosslinking is by oxidation of organic chemical large biopolymers, *e.g.* proteins, often using copper enzymes. The opening up of the crosslinks to allow growth is due to zinc enzymes.
12. Cells differentiate as organisms grow and in early organisms can dedifferentiate but in later organisms cells do not dedifferentiate easily. Differentiated cells multiply independently in separate organs. Clones of whole organisms can be produced today from many differentiated cell types. During growth the genetic complement generates many different appearances, forms, either nearly continuously or by metamorphic jumps. Genetic expression is under changing, timed control with age from conception of the organism. Special chemicals, hormones, control stages of growth.
13. The chemistry and metal ion content of different cells in the same organism are now different.
14. The differentiated cells must communicate with one another and therefore require a new set of messengers internal to the organism in their

extracellular space. The metal ions play a large role in these organic messenger syntheses, by catalysing oxidation of small molecules.

15. The most sophisticated organisms control the immediate extracellular fluids of organs within an outer skin which has novel crosslinked proteins, keratins in animals and lignin in plants, again partially oxidised using metalloenzymes.
16. The cells require that extracellular fluids circulate within the outer skin carrying oxygen, salts, nutrients and waste and this fluid flow is often contained in tubular structures, connected in later animals to a pump, the heart. It is very different from the mode of circulation in vascular plants (Figure 4.22) from the arteries and veins of animals. In many most recent cases the ionic content of the whole extracellular fluid is controlled by hormones, new chemical messengers. Many of these hormones are partially oxidised organic compounds made earlier in an aerobic condition. As we stressed in Chapter 2 shape is often related to tubes.
17. The cells often have an organised extracellular biomineral structure, a shell mostly of calcium carbonate or silica which is shared between cells in multicellular organisms and which is contained in the outermost structural region. The matrix supporting the mineralisation is of cross-linked organic polymers often due to oxidation. Later skeletons appeared inside this outer region, now of deposited calcium phosphate (bone) only, not calcium carbonate, in vertebrate animals and silica in plants.
18. Dependence of the more advanced organisms on lower organisms of all kinds has increased greatly so that as well as the internal symbiosis there is considerable external cooperation. The dependence of symbionts is in the supply of chemicals mainly.

We note that all the different types of organisms, unicellular bacteria, unicellular eukaryotes and multicellular eukaryotes coexist. The tree of

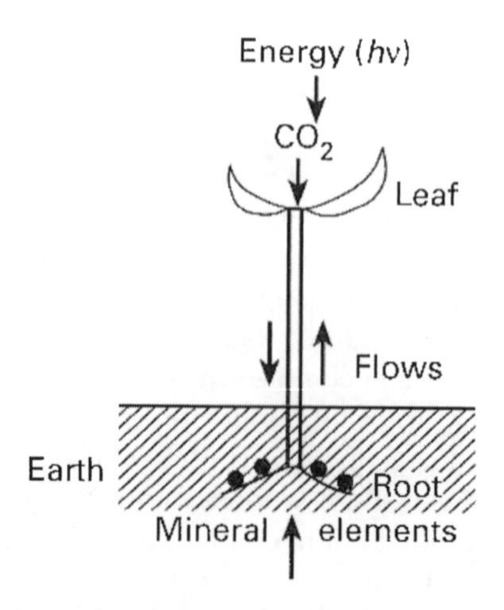

Figure 4.22 A very diagrammatic sketch of a plant to illustrate the simple flow compared to the complex networks of arteries and veins in animals.

evolution (see Figure 1.2) apparently connects independent series of animals, fungi and plants and prokaryotes with many variations and a vast variety of earlier prokaryotes and single-cell eukaryotes along lines of branches (see Chapter 7). All these organisms are in fact chemically collaborative to a considerable degree between branches. Especially the later multicellular organisms depend heavily upon earlier (now associated) organisms for some essential chemicals, *e.g.* vitamins. The advantages of this organisational separation of function were enhanced by the increase in the large-scale exploration of space by specialist cells: (i) those reaching to catch more light, plants, especially on land, while being dependent on fungi and prokaryotes in the roots; (ii) others, fungi, having extensive spreading 'filaments' in earth to take in minerals and to supply roots of plants, especially minerals, in exchange for saccharides; and (iii) those able to move in a deliberate, controlled way and were able to engulf others for 'food', scavenging animals. Both fungi and animals are dependent on plants for energy and many chemicals derived ultimately from light and the environment. All these developments of multicellularity required that extensive controlled, organic cellular chemical activity was extended to space outside the cell cytoplasmic membranes as well as within it, while also increasing the number of internal vesicles for further novel reactions. Both activities again required new or an increasing number of inorganic catalysts, especially for handling oxidative and hydrolytic steps of growth. The number of these organisms increased enormously from 0.54 Ga so that the surface of Earth, especially land, came to be a true biosphere, but notice how intimately it is involved with the mineral chemical environment.

Before we describe further details of multicellular organisms we must note the remarkable speed of the introduction of cells with many novel chemical features. Somehow before 3.5 Ga the basic reductive anaerobic cytoplasmic chemistry arose before there were genes. Anaerobic cells were present after this time, followed by micro-aerobic and aerobic prokaryotes around 3.0 Ga. From 2.5 to 0.75 Ga there were only unicellular organisms, all of which had some additional oxidative chemistry but their beginnings are difficult to date. The unicellular organisms included the first eukaryotes. Remarkably quickly after 0.75 and before 0.54 Ga a large number of multicellular organisms arose as seen in the Cambrian Explosion. It is certain that the complete basic chemistry of evolving organisms including much novel oxidative chemistry developed between 3.5 and 0.4 Ga and especially in two periods around 2.5 to 2.0 Ga and 0.75 to 0.40 Ga. We can follow some of it by traditional means: deduction from that of extant organisms through steps of introduction of new organic chemicals and by the chemistry of fossils and of molecular fossils. However we consider that the best of all methods is by following the proven inorganic chemistry changes in the environment and in sediments and noting their inevitable interaction with organisms. The final phase of organisms' chemical development by the novel oxidative chemistry was virtually complete around 0.54 to 0.40 Ga in the environment and in multicellular organisms. Evolution has undergone a marked change in emphasis from that time, say from 0.4 Ga to today. We agree with Darwin that in this recent period the changes were largely random with some dependence on very small environmental differences (see Chapter 7B). We now wish to look at particular features of multicellular development.

4.16 The Evolution of the Divisions in Space in Multicellular Organisms

We describe in this section how the observed spatial features of multicellular organisms evolved[27] before looking at the physical–chemical development which gave rise to them. The general premise is that multicellularity arose through the containment of cells in a roughly formed surrounding skin. Depending on position in the container the cells then develop their own particular differentiated character. Differentiation is a way of separating chemical activities in groups of cells within large numbers of cells in one organism, all with the same DNA, but different environments and chemistry. This reduces local complexity in any one cell while being able to increase it within whole organisms. The construct is to be likened to modern-day Volvox (see Figure 4.21). The break-up of these early multicellular organisms leads to new organisms as each cell remained able to dedifferentiate and multiply. Notice that the formation of an outer skin to enclose cells parallels that of the original cell membrane before 3.5 Ga to enclose the initial chemical reactions. Later groups of internal cells of a given kind in multicellular organisms came together to form fixed gross structures, organs which are new enclosures, in organisms of particular shape within an outer skin (see Chapter 3). It is quite hard to picture how organs for such different purposes as digestion, mobility (muscle) and reproduction arose with vastly different novel protein contents but the same total code, DNA. Clearly expression is differentially controlled. Regardless of the complications the DNA became very large (30,000 to 40,000 bases). Structures of fixed differentiated cells followed on from those in Volvox above, to one-layered organisms such as sea sponges; two layers in animals such as jellyfish and three-layer organisms with true organs with an increase in numbers of different types of cell. This development is viewed as an increase in divisions of spherical space when further advance came from dividing the space by x, y and z planes giving bilateral and then trilateral organisms. The reader can map this simple description onto an evolutionary tree of multicellular animals. This is called a cladogram. The ability to form a controlled pattern in space allowed differentiated cells in organs to be arranged in an advantageous way. This involves extracellular connective constructions both between cells, organs and of the 'skin'. An example is the extracellular filamentous structures that were also used to position constructed tubular structures for flowing chemicals. These structures also gave means of distributing food and waste and passing information. A new series of channels in cell membranes was needed and devised. Through them and the contained external fluids (Figure 4.22, for example), the basic requirements, including the mineral elements, could be transferred so that all internal cells, organs, of animals and plants were in long-distance contact with sources of food and even with environmental factors. Different organs were made aware of these different external changes by chemical messenger systems in the fixed, flowing, interior fluid spaces. These messengers are in addition to the earlier Ca^{2+} in transfer. We shall describe the chemical changes in cells and their fluid contents with their communication systems in Sections 4.23 and 5.5. It is very important to observe that all of this evolution is that of organic chemicals developed in a united manner.

As mentioned above a further novel point is that, just as single eukaryote cells evolved using greatly modified internal symbiosis of bacteria, organelles, so now multicellular organisms evolved with external prokaryotes and eukaryotes. These symbionts used the external environment to transfer essential chemicals with their host, so multicellular eukaryotes came to use both organelles in cells and internal to their own contained environment, novel symbionts. This growth of collective 'organisms' within an overall unity is a further use of space to avoid complexity and possible confusion. Increasing cooperation as well as competition is clearly a feature of evolution[14] and is readily followed using mineral elements as markers of life forms, so-called chemotypes. It is the rise in oxygen with the changes in the environmental chemicals, especially of oxidising catalysts, which made this use of space possible.

4.17 Control of Growth and Shapes

One of the greatest problems in the understanding of multicellular organisms is the appearance of shape which is controlled during growth. The previous section has described the final shapes of eukaryotes but has not considered the controls as shape develops. The outstanding theoretical study which remains the basis of all present analyses is called the Reaction/Diffusion Theory of Turing.[41] It depends on the presence of gradients of electrical fields or chemicals and has linked to them a feedback mechanism of the gradient components to cells defining their positions. These components are called morphogens. In this book we are particularly interested in metal ion effects on organism chemistry including its organisation, because they are easy to study. A very powerful morphogen is the calcium ion.[42] Its gradients are easily recognised by experimentalists using fluorescent dyes. Other morphogens are the hormones which bind to zinc fingers known to be factors in growth.

An interesting observation on multicellular organisms is that one DNA must control many shapes during growth. This leads to an inquiry into the causes of shape changes in organisms with time. The obvious example for study is metamorphosis. Here for long periods one shape is observed and then after a short period the organism can make huge shape changes. While metamorphosis is common to all kinds of multicellular organisms, even to some single-cell algae, it is most striking in insects and reptiles. A well-examined case is that of the salamander, axolotl, which does not metamorphose as it should to the full adult stage. One clear feature of these changes is that messages inside organisms are required to stimulate them at certain moments of the animal's life. It has been found that the final stage of development can be reached if the axolotl is injected with the hormone thyroxine or even with the element iodine. Thyroxine acts on zinc finger transcription factors. In humans it is known that the metamorphosis known as going through puberty, as well as features parallel with it, can be induced by zinc in food. In fact there is good evidence that the growth patterns in all multicellular organisms are due to changing hormones, produced by oxidative metabolism and interacting with zinc fingers. A further example is the action of light needed for vitamin D synthesis which interacts with zinc transcription factors and then calcium of bone. For the purposes of this book, however, the major evidence of interest is that introduction of elements such as copper or zinc which arise in the environment and the oxidation of iodide also from it. Both relate to DNA expression. It is evidence such as this which will make us ask in Chapter 7 whether there is a mechanism by which changes of elements in the environment can induce evolution in a hereditary fashion as well as immediate expression by interacting with DNA.

4.18 Building Larger Structures: Internal and Extracellular Tissue Proteins

We turn now to the chemistry which structured the extracellular compartments and shape of organisms. We have shown that the prokaryotes have some structural proteins that organise internally the protein complexes, the nucleus and contacts with membranes (Section 4.5). Their magnetozomes, containing small magnets of pure Fe_3O_4, are to be compared with inclusions mentioned above and they are known to be organised by in-cell filaments, but not internal to the vesicles. The larger, single-cell eukaryotes have different membrane proteins and many connecting sets of internal structural proteins, including tubulins, linking their various compartments to the membrane, as described in Section 4.8. Tubulins are related to prokaryote proteins.[16,17] Cell vesicles and cell compartments such as mitochondria are also held in juxtaposition with the endoplasmic reticulum membrane, no doubt positioned there by protein extensions from membranes in the cell. Here we concentrate especially on proteins outside or on the exterior of cell membranes, as opposed to these internal structures, so as to demonstrate the connection with increased organisation and biominerals in multicellular eukaryotes. All the structures require quantitative control of protein and often of polysaccharide production.

An example of development of external structure is provided by the analysis of membrane proteins.[43] This showed that the eukaryotic membrane proteins from the unicellular and multicellular eukaryote organisms have much larger external regions than those of the prokaryotes. Moreover these extracellular loops have a higher content of oxygen-containing amino acids, *e.g.* glutamate and aspartate, and the proteins are usually glycosylated. These are properties associated with protein flexibility[44] related to function. Mobility is desirable for detecting and responding to the environment.[45] These external surfaces are again quantitatively expressed, described under glycobiology. Their increased acidic (oxygen-containing) proteins give them the general ability to bind saccharides to other proteins, *e.g.* to lectins and cadherin using calcium. They then play a part in the formation of crosslinked multicellular tissue, with links to the environment. Is it possible to trace the rise in these exterior proteins and the cellular content of acidic proteins to a cause based on the rise of O_2? We consider this possibility under an analysis of calcium uses in cellular systems in Section 5.4.

An additional way in which to build larger structures is to grow large elongated cells. The simplest are large vascular plant cells, mostly full of vacuoli, placed along the stem of plants which act in transport. Animals evolved quite distinctly different large muscle cells which have little but acto-myocin filaments which act to control tension. They are energised by calcium input from outside, stimulated by nerve cells (see below), and inside by calcium from stores in their sarcoplasmic reticulum. The muscles control the movements of the skeleton through connections with ligaments. They are supplied with sugars from arteries made of connected cells which are also able to contract and together with the heart muscles drive the blood carrying 'food'. The importance of calcium both in its binding capabilities and its function as a messenger are emphasised here to draw attention to the integral role of inorganic ions in almost every activity in multicellular organisms. It was the very necessary rejection of calcium from cells and its quantitative presence outside cells together with the rejection of oxygen to back-react and give oxidised surfaces to cells which made the uses of calcium possible. We turn next to extracellular proteins which connect the cells of multicellular organisms.

The first multicellular species do not appear to be very large,[27] are soft-bodied and flexible, but already they have novel large extracellular proteins joining cells together (Table 4.5), for example collagen and cadherin proteins linked through calcium ions.[46] An additional way of creating rigidity of structures is by H-bonds or chemical crosslinking organic polymers. It appears that the first structures of considerable rigidity in plants were made from cellulose (or a similar linear polysaccharide) plus chitin. Both polysaccharides were synthesised earlier in evolution in association with the outer membrane of

Table 4.5 Extracellular Matrices

Organism	*Basic Polymer*	*Cross Links*	*Enzymes*
Higher animals	Collagen	Oxidised amino acids	Cu oxidases
Vertebrates	Elactin Keratin	- S – S - bridges	Oxidases
Invertebrates Arthropods	Chitin	Proteins + phenols	Cu oxidases Heme oxidases
Plants	Cellulose Chitin	Hemicellulose Xylo-glycans	Mn enzymes?
Late plants (Trees)	Lignin	Oxidised phenols	Cu oxidases

Table 4.6 Examples of Oxidative Synthesis of Small Signalling Molecule Messengers

Molecule	*Enzyme*
Sterols (hormones) slow-acting	Fe-oxidases, P-450 cytochromes
Adrenaline (messenger, fast-acting)	Fe- and Cu-oxidases
Hydroxytryptophan (fast-acting)	Fe-oxidases
Amidated Peptides (fast-acting)	Cu-hydroxylases
Nitric oxide (hydrogen peroxide)	Heme-oxidases, peroxidases

bacteria. Through them molecules of all kinds are passed out directly to the periplasm but there they are not chemically crosslinked. In eukaryotes the synthesis of polysaccharides is in the Golgi vesicles but only fully assembled and crosslinked in the extracellular matrices. These polymers are not of equal flexibility and chitin, N-acetylglucosamine, has much stronger crosslinking H-bonding than cellulose. With added agents, *e.g.* keratin, both can form very strong plastic outer membranes or skins with well-sculptured rigid shape. Keratin forms an oxidised, strongly crosslinked –S–S– bridged network. Note that it is oxidising conditions outside cells that maintain the stability of the –S–S– crosslinks. Amongst the earliest examples is the keratinised 'skin' of fungi related to yeasts but this also forms the outer skin of vertebrates. Somewhat later is the inclusion of chitin in the frameworks of certain animals, arthropods, insects and crab-like animals. Some of these frameworks are of tightly packed proteins, leaving no room for minerals but others, such as those in the trilobites (see Figure 3.4), do contain some crosslinking calcium.

Relative rigidity is also achieved by oxidative crosslinking other organic polymers, as in modern hard plastics. In plants the strongest crosslinked polymer is a polyphenol, lignin, and its use in bark enabled the production of large upright plants such as trees, which evolved strongly just before and in the Carboniferous Period, 0.4 to 0.3 Ga. The lignin is dependent on copper phenol oxidase enzymes and oxygen for its synthesis. The arrival of these large plants, largely on land and fed by fresh water, required novel ion pumps. The largest are the novel vasculated plants. Their abundance and the size of their leaves led to a vast increase in light absorption and incidentally food for animals. On the other hand it is very probably the inability to break down lignin structures by animals in more rigid structures, such as that of tree bark, which led to their burial. In turn this produced the formation of oil and coal in the Carboniferous Period and caused the further modest drop of atmospheric CO_2 with a modest rise in O_2 around 0.4 to 0.3 Ga (Section 2.6). We return to the modern use of coal by man in Section 7.14. In many animal polymers, such as certain novel proteins, *e.g.* collagen and glycoproteins, are also chemically crosslinked by oxidised side chains. They became common in 0.54 Ga, the Cambrian Period (see Chapter 3). Quite possibly they or at least some of their proteins, evolved first around 0.7 Ga, *e.g.* in sponges. Of great consequence therefore are the catalysts which could bring about this oxidative crosslinking

(Section 6.7), glycosylated side chains, and novel polysaccharides. We shall see in Chapter 6 that the catalysts, enzymes, for oxidation and glycosylation are mostly dependent on life being able to gain access to novel trace elements, Cu and Mn.

Finally there are the sulfated and branched polysaccharides which are so plentiful in vertebrates, as well as invertebrates and later plants, forming open structures, allowing easy flow of small components in extracellular fluids. Sulfate is of course an oxidised form of sulfur. Generally the extracellular structures gave rise to the vast increase of eukaryotes, especially on land. Notice how many different developments arose between 0.54 to 0.35 Ga, after little change from 1.5 to 0.75 Ga. We shall turn to the reasons for this in Chapter 6.

Once again we emphasise that the necessary rejection of oxygen by prokaryotes generated the increased availability of elements such as copper as a catalyst, which provided the possibilities of new extracellular organic chemistry described in this section.

4.19 The Evolution of Biominerals and their Associated Structures

Given that they are major sources of the fossil record, and hence a novelty in the chemistry of evolution, we are particularly interested in the origins of biominerals which we have associated with particular proteins or polysaccharides. As we described in Chapter 3, starting from the beginning of the 'first great oxidation event' around 2.5 Ga, the initial remarkable advance in size and shape of organisms, shown only by fossil imprints of soft bodies, occurred from perhaps as early as 2.0 Ga with the appearance of single-cell eukaryotes with external structures. There is no suggestion that controlled biomineralisation was present in any of these single-cell organisms. The second accelerated increase in oxidation lies between 1.0 and 0.5 Ga when the multicellular organisms arrived, many of them being mineralised in a controlled manner (Chapter 3). This required a controlled manipulation of inorganic elements, such as calcium, in fluids, to form the mineral fragments attached to extracellular proteins. The peculiarity is that both multicellularity and biomineralisation appeared in several lines of organisms at almost the same time. Knoll[45] describes the evolutionary scene as follows: 'A conservative estimate is that carbonate skeletons evolved twenty-eight times within Eukarya', and 'We estimate that silica skeletons evolved at least ten times in eukaryote organisms'. In the same period external calcium phosphate shells also appeared in a further group of organisms. Now a necessity for any controlled mineralisation is a well-defined organic matrix. There is little doubt that the organic matrix requirements, which control the mineralisation, must be closely similar, but they are not identical, in all the organisms for production of external (shell) calcium carbonate, calcium phosphate and silica deposition, respectively. The implication is that very similar groups of external crosslinked proteins (glycoproteins and collagens) and polysaccharides together with the complex handling of ions arose at close to the same time in some 30 to 40 different phyla. All these evolutionary changes will be shown to have also common novel simultaneously acquired trace metal requirements, especially in addition to zinc's new organic chemistry (see Section 6.7), as well as bulk inorganic elements (Chapter 5). It is surely apparent that the study of biomineralisation, and of the cellular apparatus required to bring it about, provides an indication of a link between DNA, the proteome and the environment. Backed by the direct evidence of fossils, this must be one of the best ways to date a major part of the organic chemistry required and hence of evolution as a whole.[46]

The most amazing development is that of the bone,[47] segmented internal mineral skeleton, around 0.4 Ga, which combines advantages of the mobile, skin-limited soft invertebrates with the unadaptable rigid structure of external shell-limited organisms. We described the adaptability of bone shape related to its piezoelectric properties in Chapter 3. Bone is a composite of calcium hydroxyphosphate with various phosphorylated proteins.

4.20 Extracellular Fluids

The evolution of extracellular fluids appears to be by modification of the sea within a 'skin' in multicellular organisms and today in higher animals there are very carefully managed flows of such fluids in special tubular vessels, and in equally well-maintained volumes in special organs.[47] While the whole organism is held in an outer skin of keratin-like proteins there are also barriers between some gross internal compartments such as the blood/brain barrier. The flows of fluid are ultimately from fresh water. The mineral composition of the fluids is increasingly accurately controlled in the evolution of complex organisms and is very different in groups of organisms.

The composition of the extracellular fluids of many seawater animals can be closely related to that of the sea in mineral content but the later more sophisticated animals have more controlled flow systems in fluids in tubular structures. They are somewhat modified with respect to Mg^{2+} and Ca^{2+} and Na^+, K^+ and Cl^- concentrations. The composition is much the same for vertebrates in fresh water and in those which move on land, and we return to their quantitative composition in Chapter 5. This implies tight control of the composition by regulatory devices.

Just as in the case of growth the strict control of both sodium and calcium in higher organisms is through the action of hormones, here mineral corticosterols, themselves the products of oxidation (see Section 4.21).

To maintain circulation of essential oxygen and trace elements these animals also have carrier proteins, *e.g.* haemoglobin and albumins. However as in the case of fungi, only more so, they have a limited organic chemistry independent of other organisms. Clearly animals must intake and digest proteins from plants and circulate this food to all their cells. They also are required to obtain and to circulate minerals, some 20 elements, in their diet as they have no direct source. As we have said before, the demand of animals for vital chemicals that they do not synthesise is clear evidence of cooperativity of organic chemical resources in ecosystems. In multicellular organisms the food and elements are in chemical flows to a local or distant internal environment from digestive organs. These flows also carry messages between cells (Section 4.21).

Most higher plants on land have a fluid of low Na^+, K^+, Cl^-, Mg^{2+}, Ca^{2+} and SO_4^{2-} content circulating from the roots via their vascular system containing the xylem to the upper reaches and returning via the phloem (see Figure 4.22), with a very different composition of waste and some 'food' for the roots. A living modern plant is not an isolated species but very dependent on cooperative species for nitrogen, Eubacteria, and on fungi for minerals. The uptakes of minerals and bound nitrogen (NO_3^-) are transported in the xylem while the symbionts obtain sugars by the opposite flow from the leaf in the phloem. Carbon compounds are synthesised by the leaf and sent to the root. All such tubular structures require some crosslinked polymers and quite possibly that could not have arisen in the absence of considerable concentrations of oxygen.

4.21 Signalling with Organic Molecules and Electrolytic Gradients in Multicellular Eukaryotes

The major signalling between the cell's environment and the interior of single-cell eukaryotes remained in the cells of multicellular eukaryotes via the pulsed input of calcium to each individual cell triggered by external events.[48,49] These new differentiated cells also required additional, often selective, complex signalling for cell–cell (organ–organ) communication across extracellular space in the tubular structures. The complexity of such signalling lies in that it must generate not just controlled growth while maintaining differentiation and organ integrity, all relatively slow processes, but must also convey rapid awareness of the external environment of the whole organism to the interior cells in organs.[50] There had to be many different extracellular cell-to-cell messengers activating rapid calcium input channels in selected organs, and others connected to slower response to give novel protein expression and growth selectively of different organs. The two sets of organic messengers, for awareness and growth/differentiation (Table 4.6) have very different timescales of action, as the first has to be fast (millisecond) while the second, on the timescale of growth, can be of various slow time constants, even of years.

We treat the slow irreversible growth messengers first. They are called hormones and have their own receptors as transcription factors of the DNA and must have a sustained effect through protein production. The slow messenger receptors for the hormones, transcription factors, are often zinc metalloproteins, which are up to 5% of the proteome in many of the most recent organisms, and initiate certain protein syntheses. They are found in all eukaryotes but develop largely and differently, in kind and in numbers, in classes of multicellular organisms, *e.g.* plants and animals (see Figure 6.13).

Many of these hormones are partly oxidised sterols, made from squaline oxidised by iron haem proteins. One of these proteins is cytochrome P450, an enzyme already known in prokaryotes. We have maintained (see Chapter 6) that binding of many hormones to zinc proteins and zinc exchange between these zinc proteins gives a cooperative control of growth, mineral (Na/K and Ca) uptake, and of bone biomineralisation.

Fast messengers are released from donor cell stores by the action of triggered calcium entry. They act very differently on membrane receptors on the outside of cells so as to cause calcium entry followed by cooperative protein (enzyme) action and/or structural chemical modification such as phosphorylation or dephosphorylation. These are fast chemical changes and can be followed by mechanical movement, not involving any slow carbon, nitrogen, and hydrogen synthesis chemistry. They are rapidly stopped as calcium is removed by pumps described in Chapter 5. We note that many of the enzymes involved in the syntheses of these messengers are oxidative, use oxygen and have copper at the active site (Section 6.9). This was also shown to be the case for the synthesis of connective tissue (Section 4.20). Both arose together with extracellular fluids, within each and every multicellular plant, animal and fungus, with or without mineralisation, in the period just before the Cambrian Period, or in it, or just after it. It is the historical suddenness of several, similarly timed developments which involves many genetic and chemical structure changes in many organisms at close to the same time after, say, 0.75 Ga but before 0.4 Ga, which we have now seen several times. This is the changed chemistry of the time around the Cambrian Explosion and we must address its linkage to DNA in Chapter 7. How did it happen? We shall see that in all those developments the two elements, copper and zinc, released from sulfides by oxygen in small part with the early rise of oxygen around 2.0 Ga – but more so after 0.75 Ga – played a major role.[14] Copper outside the cytoplasm became the major oxidising site in several enzymes, acting in the synthesis of organic messengers and to crosslink matrices in part for biomineralisation, while zinc in other enzymes catalysed the hydrolysis of many of the molecules to destroy messengers and structural proteins to make space for growth. The outstandingly obvious feature to consider is that the quick changes in the environment, due to higher oxygen content and its consequences, are in some way responsible for the novel organic chemistry of a wide variety of multicellular organisms. Following the beginning of anaerobic cells and the evolution of single-cell eukaryotes the appearance of multicellular organisms is the third most striking event in evolution.

The last development of cellular organic chemistry was the nerve cell and, following from it, the brain. The nerve cell has three regions: a normal central body, long thin tubular protuberances and closing the ends of the tubes, synapses. The synapses have vesicles and they contain the organic transmitters (messengers). Their release is by influx of calcium ions stimulated by the arrival of an electrolytic message along the tubes. The electrolytic message is a depolarisation of the membrane of the tube which carries charge due to the outside/inside distribution of Na^+ and K^+ respectively. Here we see the realisation of the potential value of the rejection of Na^+ mentioned in Section 4.2.1. We indicated that this rejection was a necessity for the maintenance of organic chemistry within an enclosed space from the very beginning of life.

This takes us to the end of the major chemical changes of the environment and so to the period from, say, between 0.50 and 0.40 Ga to today. The cell chemistry has not stopped but in the period it is totally dependent on mutational change to create fitter organisms. It is visible to us in the variety of plants, fungi and animals in our present-day natural environment. Outstanding developments are in the varieties of vascular plants, including flowering of many. Amongst animals the outstanding advance has been in the brain, leading to man. A further step of environmental chemical changes, now not natural but brought about using the brain, is man's chemical industry and the applications in huge constructions, foodstuff and medicines. We return to this topic in Chapter 7C (Section 7.13), as the activity has not been finalised.

4.22 Genetic Analysis of Multicellular Animals

In Section 4.15 we have described considerations relating to the origin of unicellular eukaryotes and stressed especially the poor agreement before 0.54 Ga between fossil and genetic analysis, especially in the case of photosynthesising single cells.[26] We considered that they preceded the multicellular plants in the sea by about a billion years. We had poorer information about non-photosynthesising single cells but we estimated that they evolved probably around 2 billion years ago. We turn now to a comparison of genetics with the fossil evidence given in Chapter 3, which led us to conclude that both the multicellular non-photosynthesising and the photosynthesising organisms arose to some degree just before the Cambrian Explosion but mainly with and after it, 0.54 Ga.

Utilising cooperative RNA/DNA sequences, the evolution of non-photosynthetic multicellular metazoans, animals, initially in the sea appears to be from the sponges of the species porifera which can be traced back through their genes to a possible source of them all in the so-called urmetazoa close to 0.65 Ga; see also Müller *et al.*[51] However it may be that this date is too early as it could refer to colonies of single eukaryote cells. The date is reasonably consistent however with that from the less certain fossil record before 0.54 Ga. It is generally agreed that fully multicellular algae and animals probably arose at about the same time of around 0.75 to 0.55 Ga. The fascinating further evolution which follows has been of variety not of chemotypes. We observe many millions of species on Earth and the evidence is that they belong to Darwinian branches of the development of these varieties by very small incremental changes and not by any change in chemistry if we take into account cooperation not just competition.

Much of this and other possible recent new protein chemistry has been followed by detailed genetic analysis of the proteins of organisms relative to their DNA which we describe in Chapter 7B, but we must be aware again that expressed RNA and proteins are concentration-dependent in their properties unlike the DNA of genes (but see Section 4.4) and linked to environmental factors. The need is for controls of translation by small molecules and ions from the environment to stop as well as start activity. Our particular interest will lie in the genes for metalloproteins as through these proteins we can discover any linkage to environmental geochemistry, adsorption in sediments, and to its timetable (Section 2.11). Again we shall observe that many very different classes of organisms underwent parallel if not identical evolution of novel transcription factors and messengers at close to the same time.

Before closing this section we must note the remarkable change in reproduction from simple doubling of DNA and division in a prokaryote to the variety of features of reproduction in more complex organisms. Apart from the different rates of increase of organelles (note also plasmids) there is the introduction of sexual reproduction. Here an egg and a sperm DNA come together in one cell to create an organism. The process is one of complete coalescence. This step is reminiscent of the introduction of bacteria as organelles into certain cells.

4.23 Loss of Genes and Organism Collaboration: Internal and External Symbiosis

As well as gains in genes such as is apparent in adaptation to aerobic conditions and the general gain in numbers of genes in novel organisms over long periods, there is systematic gene loss of certain metabolite pathways with increasing complexity in different organisms.[26] This is a loss of organic chemistry. The loss emphasises importantly the way evolution overcame complexity by cooperation. Even the aerobic Eubacteria have lost the ability to synthesise some amino acids and rely on cooperative activity in the environment. The original aerobic eukaryote cell (or was it a kind of prokaryote?) was apparently not able to or had lost the ability to photosynthesise or to do oxidative phosphorylation and relied on incorporation of bacteria as organelles for these functions. This is the beginning of full cooperativity, symbiosis. This original cell must have been aerobic because it made cholesterol. What other functions were completely lost from the cells? An extreme and very significant case already mentioned is the absence of N_2-fixing genes from eukaryote DNA leaving eukaryotes dependent for reduced nitrogen upon symbiotic prokaryotes (or on organelles in a very few examples) or the uptake of nitrogen from compounds ranging from nitrate to more complex molecules. Another extreme example of loss is that some eukaryote cells seem to have even lost later mitochondrial integrity and function, perhaps generating hydrogenosomes.[50] Again some unicellular eukaryote organisms such as coccolithophores lost the ability to synthesise essential chemicals using coenzyme B_{12} containing cobalt. This loss is also apparent in man for whom B_{12} became a vitamin. It is a general trait in evolution that increasing complexity in organisms is accompanied by increasing loss of essential chemicals and their genes and hence dependency upon simpler organisms for the most essential biochemicals such as certain coenzymes (vitamins), amino acids, sugars and fats. One such example is the inability of all animals to synthesise the three essential aromatic amino acids phenylamine, tyrosine and tryptophan, through loss of the genes of the shikimic acid pathway. At the same time all animals or related non-photosynthesising bacteria are dependent on photosynthesising organisms, plants, for capture of carbon combined with hydrogen. The plants themselves are often dependent on fungi to supply minerals and animals return nitrogen to the soil. Animals get much of their minerals from food, not directly from soil. There can be no clearer example of the cooperation rather than competition between organisms, in order to gain effectiveness of the ecosystem, than the sharing of the products of the total gene pool. What then is the selfish gene (organism) in life?[51] Another peculiar observation is that bacteria and parasites which live in large complex organisms have lost many genes and have a much reduced genome. This allows them to survive very economically. We consider that while there is competition between similar 'selfish' members of one group of related organisms to a large degree, very unlike organisms belonging to different groups frequently cooperate in an ecosystem to achieve the optimal absorption and degradation of light energy. The first of these tendencies, local competition, has a random character and leads toward survival of the fittest amongst similar organisms but this is within the larger-scale direction of the whole of life toward the greatest uptake and degradation rate of energy by many organisms, which requires cooperation for its efficiency. The more systematic and inevitable chemical character probably finished by 0.4 Ga.

Genetic analysis looks for a way of following sequences but there are now seen to be six difficulties which make a sequence study going back a long way in time difficult: (i) faster, punctuated rates of change of genes in particular periods; (ii) sequences rooted at a particular date with no apparent antecedents; (iii) sequences which are lost from an organism and yet their products are essential and the genes are found in a different group of organisms, *e.g.* genes for vitamins, often coenzymes; (iv) as studied genetic analysis does not relate to rates of physical changes or of chemical changes in the environment; (v) gene transfer which is now known to be very considerable, even from remote branches of a 'tree' to a given branch (this is called reticulate evolution and we will refer to it in more detail in Chapter 7);[34] (vi) recently attention has been drawn to the appearance of genes from non-coding (junk) DNA, and two ways of changing DNA sequences, not just mutation alone, to which we shall refer together with epigenetics in Chapter 7.

4.24 Summary of the Distinctive Features of Biological Organic Chemistry

At the beginning of this chapter we observed that the bioorganic chemistry of cells is quite different from the organic chemistry as taught and practised in academic and industrial laboratories. Due to the extent of this chemistry we did not describe the structure of its molecules, their synthesis or the mechanisms of their formation. Additionally we could not give details of most pathways of synthesis or degradation. We indicated that all this chemistry was available in textbooks of organic chemistry and biochemistry. We turned to the unavoidable character and limitations of biological organic chemistry imposed largely by the environment.

1. For the most part it takes place in water and is trapped by a membrane.
2. It requires energy capture from outside for both synthesis and controls.
3. The medium had to be an aqueous solution of fixed basic salts inside the trapped membrane, including Na^+, K^+, Mg^{2+}, Ca^{2-} and Cl^- contents. The medium is managed by energy-driven pumps.
4. Some oxidised chemicals had to be rejected to gain access to reducing equivalents. After a period this rejection had to be oxygen from water. Also rejected from the very beginning were sodium and calcium ions.
5. Organic compounds in the cell are synthesised and degraded along highly selective pathways. Many pathways are organised in space by structures.
6. The pathway chemistry is directed by enzymes, some fixed in space.
7. About one-third of the active sites of enzymes contain catalytic metal ions and many controls are also by trace metal ions.
8. The controls gave at first internal homeostasis in resting states in prokaryotes. Later controls in single-cell eukaryotes extended to response of cells to the outside environment. The expansion of inorganic controls was from Mg^{2+} and Fe^{2+} internally and to Ca^{2+} responses in pulses from the outside. Later still cell–cell interaction in higher organisms became controlled. Then more generally internal responses depended on Zn^{2+} transcription factors. Much controlled chemistry is in directed flow.
9. After an initial period evolution became unavoidably more and more dependent upon oxidation by oxygen in the presence of newly available catalytic elements, especially dependant on copper.
10. There are also different classes of structural proteins, required to retain ordered activity in both space and time.
11. Much response of organic reactions is linked to messengers from the external, especially calcium, or later to the internal environment of multicellular organisms by oxidised organic molecules and finally to Na^+/K electrolytes.
12. The combination of the elements described under (1) to (6) indicate the unusual nature of the cooperative, biological, organic chemistry.
13. All of the above changed with the changing environment because the evolution of organic chemistry is very dependent on it from 3.5 to about 0.5 Ga.
14. From the very earliest cell a coded molecule DNA (RNA) limited the expression of both RNA and proteins including the enzymes that limit metabolite synthesis. Were DNA/RNA and proteins the only biological organic polymers which could have evolved to give present-day life?

15. The DNA code has evolved but the mechanism of change and links to the environment are not known (see Chapter 7B).
16. The development of nerves and the brain was the final development of biological organic chemistry. They use Na^+/K^+ electrolytic messages. The brain evolved to give rise to man who has created a non-biological chemistry.

This organic chemistry in cells can only have started from the injection of energy into very simple inorganic C, H, N, O chemicals in a trapped aqueous volume, giving rise to their reduction to C/H, N/H by H_2, H_2S and then by H_2O sequentially. Their subsequent incorporation in the metabolites required to make the molecules necessary for the eventual synthesis of proteins (plus sulfur), nucleotides (plus phosphate), saccharides and lipids (see Figure 4.1). To this scheme we added the introduction of waste in Figure 2.4. We included later the steps of reproduction so as to give the overall picture, illustrated in Figure 4.23. We have suggested more partially a plausible, and no more than that, scheme of basic reactions. The scheme depends, as it must, on energised intermediates of high kinetic stability to degradation but it also depends essentially on the presence of molecules of catalytic capability. The energy required was obtained initially perhaps from unstable compounds (minerals) found when Earth cooled but later from the use of sunlight, using chlorophyll to absorb light to make charge or proton gradients across the cell membrane. We have indicated that these gradients were used in the production of redox reagents on uptake of reducing equivalents. The proton gradient made pyrophosphate, ATP, to drive condensation reactions. It is only possible to get sufficient catalytic power by taking into the trapped volumes certain available inorganic ions, particularly at first iron and magnesium then manganese ions bound in proteins. It was also essential that other ions of the environment were rejected, sodium, chloride and calcium, while potassium was taken in and

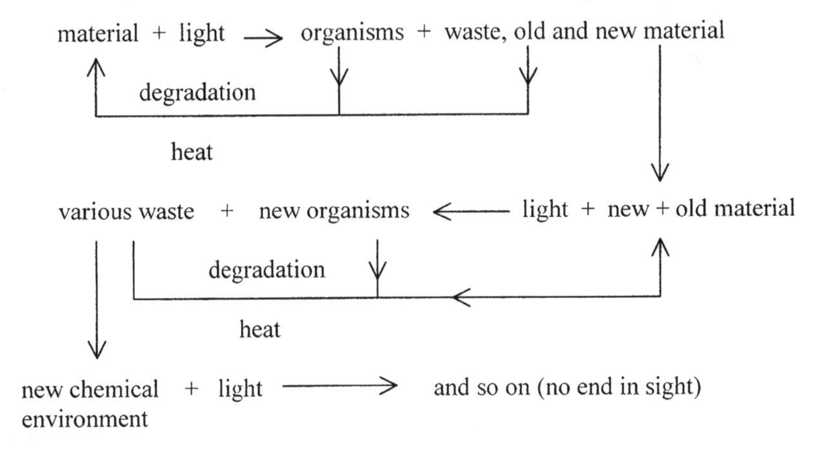

Figure 4.23 The sequence of the development of new organisms using new materials and energy to create novel organisation.

magnesium was largely retained. These are the conditions in the cell cytoplasm in which organic chemistry began and has continued in the cell cytoplasm. We pointed out in Section 4.2.1 that the necessary rejection of sodium and calcium and the release of oxygen to the environment very early in evolution gave rise to many possibilities. Much of the chemistry which we have observed in this chapter in fact developed from these possibilities. Given these limitations of the environment there was an inevitability to this beginning of the organic chemistry of cellular life and it is extremely difficult to devise any alternative chemical system for evolution. It had to evolve before or with a coding molecule, genes, as control centres for biological activity. We stress that our knowledge of how this proposed chemistry gave rise to the complexity of the first anaerobic cells hardly exists.

As this cytoplasmic (in cell) organic chemistry was reductive of necessity, because the ingredients were made from oxides of carbon, it was inevitable that oxidising compounds, which became oxygen, would be released. There followed in the environment an unavoidable and predictable sequence of oxidation of minerals and non-metal elements in solution, limited by diffusion but generally following sequentially equilibrium constants,[52] redox potentials, since the reactions of them are relatively fast (Chapter 2). A big advance was that of single-cell eukaryotes with essential calcium messengers. The limitations on diffusion were seen particularly between 2.5 and 0.75 Ga while external S^{2-}/Fe^{2+}, redox buffering of the sea was dominant.[52,53] This buffering restricted the change of cell organic chemistry because there could not be much increase in oxidation in this period. Before the first time and after the second time buffering was not effective and change was more rapid.[53] The inevitable effect on the required organic chemistry of the oxidation of the original non-metals forced a change in the methods of incorporating and reducing nitrogen, sulfur and selenium compounds which were now in oxidised forms in the environment. In turn it made possible a sequence of new oxidative organic reactions catalysed by the newly available inorganic elements, especially Cu, Zn and Mo. It is by using these elements that the possibilities of oxygen release became exploited, driving evolution. These reactions could not be carried out extensively in the reducing medium of the cytoplasm and therefore had to be mainly in extracytoplasmic and the extracellular space. To take advantage of the two incompatible parts of organic chemistry the number of spatial cellular organic structures increased with increasing oxygen released. Those in the initial anaerobic and then aerobic prokaryotes in the periplasm were followed by those of unicellular eukaryotes 2.5–2.0 with inner vesicles and organelles, and later still in the spaces between the multicellular organisms, enabling oxidative chemistry free from the reductive cytoplasm. In all the eukaryotes compartmental separation of reactions involved the uptake of bacteria, mitochondria and in some cases an uptake of photosynthesising bacteria which became chloroplasts. Once again there is no complete understanding of the steps often looked upon as coalescences of prokaryotes to give unicellular eukaryotes. It is at this stage that the unavoidable calcium gradient in the earliest cells gave rise to an essential messenger system and control of mechanical movements which are so central to the life of an animal.

Multicellular organic structures followed especially around 0.54 Ga in which much more of oxidative chemistry took place in extracellular but internal fluids in these organisms. It provided extracellular connective and message systems between now differentiated cells within an enclosed volume or skin. Major catalysts in these later changes were enzyme-based as before but especially on newly increased released copper and zinc in the environment. We consider this progression of organic chemistry was unavoidably dependent at all times on changes of elements in the environment. It is the only way in which organisms could develop as they are dependent on novel resources, that is on inevitable sequential redox inorganic chemical changes of the environment. These changes provided the possibilities and the limitations of biological organic chemistry. Much of the organic chemistry (80%) remained that of prokaryotes in the sea but eukaryotes flourished later as multicellular plants and animals, especially on land. Developments such as that of vascular plants required the new environment after 0.54 Ga. It would appear that all major innovations of the cellular organic chemistry occurred before 0.4 Ga.

The very complexity of change requires controls. At first there were internal mobile chemicals, small molecules and coenzymes, and ions such as Fe^{2+} and Mg^{2+}. When eukaryotes evolved there were extra necessary controls allowing them to change behaviour as their external environment changed. The most remarkable control was the use of calcium ion gradients, unavoidable from the earliest cells, in pulses. Other messenger molecules, later in multicellular eukaryotes, included sterols and oxidised molecules such as adrenalin and the terminal amino acids of peptides. Those related to sterols controlled growth and connected directly to transcription through DNA. They were slow-acting. The fast messenger molecules such as amidated peptides acted differently at membrane receptors followed by input of calcium. These small messenger molecules were synthesised with the aid of novel copper oxidases. The corresponding development of inorganic chemicals was notable for the evolution of Zn fingers for the slow-acting controls. The value of zinc was enabled by the action of oxygen in increasing its availability by release from its sulfide. Zinc also became of major importance in external hydrolases which aided growth.

A quite different oxidation was that of extracellular protein side chains to form crosslinks and connective tissue. The crosslinked proteins helped to give rise to biomineralisation. Plants evolved special polymers, phenols, chitin, for their synthesis of crosslinked 'skin' (bark). All these activities depended on an increase of environmental oxidation which released copper from its sulfides.

Now as this chemistry increased in complexity in advanced organisms they lost the ability to handle some of the oxidised environment. Advanced animals cannot reduce nitrate or sulfate and various essential organic compounds became vitamins while many elements became essentially required foods. Hence symbiosis between individual organisms increased continuously in evolution.

The evolution of this chemistry can be compared in part with the way organic chemistry has developed in industry and in research laboratories where the needs for inorganic catalysts, controlled conditions and the use of separate vessels for different steps of reaction are clear. However, much of this chemistry uses organic solvents, not water, and is carried out in isolated steps. If this account of the way biological organic chemistry evolved, coupled and in part forced internally and externally by inorganic elements from the environment and their changes is accepted, it has a marked degree of inevitability to it. This follows of necessity from the very essence of the original organic chemistry in the sea and in an enclosed volume. The removal of calcium and release of oxygen were intrinsic to the cell synthesis and stability. One final twist to the remarkable evolution of the action of oxygen in the chemistry of combined cells was the evolution of nerve cells and the brain with the possibilities arising via the unavoidable primitive rejection of sodium ion (see Section 4.2.1).

It is this that leads to man and his activities. There remains the puzzle of how it arose and became reproductive. Reproduction is essential to maintain each chemical advance and brings with it restrictions on the organic chemistry. We shall return to this major question of the understanding of evolution which has to be related to genetic studies of RNA and DNA, in Chapter 7. We wish first to present all of evolution of the environment/organisms system in chemical terms. Before we attempt to analyse genetic studies we shall give therefore the evolution of the interaction of the organic chemistry in organisms with the bulk elements, Mg, Ca, Na, K and Si, largely of unchanging availability in Chapter 5 before that of the developing organic chemistry associated with the trace elements, Mn, Fe, Co, Ni, Cu, Mg and Zn, and their changing availability, in organisms in Chapter 6. They are both essential for modern organisms, and a few were dominant in the initial cell while others have increased in importance with time and environment oxidation. In each chapter we note that the overall driving force of the evolution of the organic/inorganic chemistry of organisms following that in the environment is just that of the inescapable energisation by the Sun and the subsequent degradation of energy through intermediate physical and chemical steps. The flow of energy is seen to increase with the evolution of organisms and with weathering.

In concluding this chapter the emphasis has been on the increase in the energised, strictly limited, organic chemistry in cells in water but with the connection to the nature of and the changes in inorganic elements in the environment. We refer especially to the necessary rejection of sodium and calcium ions and the release of oxygen from water so that new organic compounds could be made by oxidation. The flow of energy and material, inorganic and organic, is unavoidable but in our view it is not understandable how the progression of organisms is based on complex reactions with their timing and containment in divided space. Through this emphasis we demonstrate the intimate relationship between the slowly changing biological chemistry and that of the environment, the source of fast-changing elements which became available. They drove evolution of the basic organic matter in cells which is central to our approach to evolution through analysis of chemistry in the whole ecosystem. This explains the title of this book, 'Evolution's Destiny'. To what degree was it in fact inevitable?

Around 0.5 to 0.4 Ga the rise in oxygen ceased and the environment for organisms became fixed. The evolution of a joint inorganic environment and organic/inorganic organisms ceased. This did not stop the process of random mutation which had always created variety of chemotypes and a slow progress in adaptation. The mutations allowed further slow development of random processes as in Darwinian progression. Only very recently has this development been upset by a return to a situation in which new chemical change arose. This is due to man and is not related to random processes.

References

1. J. Claydon, N. Greaves, S. Warren and P. Wothers, *Organic Chemistry*, Oxford University Press, Oxford, 2002.
2. M. B. Smith and J. March, *March's Advanced Organic Chemistry: Reactions, Mechanisms and Structure*, 6th edn, John Wiley and Sons, New York, 2007.
3. D. Voet and J. G. Voet, *Biochemistry*, 3rd edn, John Wiley and Sons, New York, 2004.
4. L. Stryer, J. Berg and J. Tymoczko, *Biochemistry*, 5th edn, W. H. Freeman, New York, 2002.
5. R. J. P. Williams and J. J. R. Fraústo da Silva, *The Chemistry of Evolution*, Elsevier, Amsterdam, 2006.
6. R. J. P. Williams, *J. Theor. Biol.*, 1961, **I**, 1.
7. P. Mitchell, *Nature*, 1961, **191**, 144.
8. D. G. Nicholls and S. Ferguson, *Bioenergetics*, 3rd edn, Academic Press, London, 2000.
9. B. H. Weber and J. N. Prebble, *J. Hist. Biol.*, 2006, **39**, 125.
10. G. Wächterhäuser, *Proc. Natl. Acad. Sci.*, 1990, **82**, 205.
11. A. Mulkidjanian and M. Y. Galperin, in *Structural Bioinformatics of Membrane Proteins*, ed. D. Frishman, Springer Verlag, Berlin, 2010.
12. S. Kaufman, *The Origins of Order*, Oxford University Press, New York, 1993.
13. M. W. Poner, B. Gerland and J.D. Sutherland, *Nature*, 2009, **459**, 239.
14. R. J. P. Williams and J. J. R. Fraústo da Silva, *The Chemistry of Evolution*, Elsevier, Amsterdam, 2006.
15. M. Eigen, *Steps towards Life*, Oxford University Press, Oxford, 1996.
16. A. M. Lesk, *Protein Architecture*, Oxford University Press, Oxford, 1991.
17. R. Diamond, T. F. Koetzee, K. Prout and J. S. Richardson (eds), *Molecular Structure in Biology*, Oxford, Oxford University Press, 1993.
18. R. J. P. Williams, *Eur. J. Biochem.*, 2005, **183**, 479.
19. P. L. Graumann, *Annu. Rev. Microbiol.*, 2007, **16**, 589.

20. J. B. Foth, M. C. Goedecke and D. Soldate, *Proc. Natl. Acad. Sci. USA*, 2006, **103**, 3681.
21. D. Nath, R. Dhand and K. Eggleston, *Nature*, 2010, **463**, 445.
22. S. Silver and L. T. Phung, *Appl. Environ. Microbiol.*, 2005, **71**, 590.
23. E. Kim and L. E. Graham, *PloS ONE*, 2008, **3**, e2621.
24. W. Gong, B. Hao, S. S. Mansy, G. Gonzalez and M. A. Gilles-Gonzalez, *Proc. Natl. Acad. Sci. USA*, 1998, **95**, 15177.
25. L. Margulis, *Symbiotic Planet*, Basic Books, New York, 1999.
26. E. R. Barry and S. D. Bell, *Microbiol. Mol. Biol. Rev.*, 2006, **70**, 876.
27. M. Brasier, *Darwin's Lost World*, Oxford University Press, Oxford, 2009.
28. L. Decaria, I. Bertini and R. J. P. Williams, *Metallomics*, 2010, **2**, 706.
29. L. H. Caporale, *Darwin in the Genome*, McGraw-Hill, New York, 2003.
30. M. S. Neuberger, R. S. Harris, J. H. Di Moia and S. K. Peterson-Maku, *Trends Biochem. Sci.*, 2003, **28**, 305.
31. R. H. Kretsinger and C. E. Nockolds, *J. Biol. Chem.*, 1973, **248**, 3313.
32. J. M. McCord and I. Fridovich, *Free Radical Biol. Med.*, 1988, **5**, 363.
33. T. Cavalier-Smith, M. Brasier and T. M. Embley, *Philos. Trans. R. Soc., B*, 2006, **361**, 848.
34. J. Deschamps, *Curr. Opin. Genet. Dev.*, 2007, **17**, 422.
35. M. P. Scott and N. J. Weiner, *Proc. Natl. Acad. Sci. USA*, 1984, **81**, 4115.
36. C. Berney and J. Pawlowski, *Proc. R. Soc. B*, 2006, **273**, 1867.
37. Y. Zhang and V. N. Gladychev, *Chem. Rev.*, 2009, **109**, 4828.
38. P. G. Falkowski, M. E. Katz, A. H. Knoll, A. Quigg, J. A. Raven, O. Schofield and F. J. R. Taylor, *Science*, 2004, **205**, 354.
39. P. D. Tortell, *Limnol. Oceanogr.*, 2000, **45**, 744.
40. J. N. Young, R. E. M. Rickaby, M. V. Kapralov, D. A. Filatov, *Philos. Trans. R. Soc., B.*, 2012, **367**, 483.
41. A. M. Turing, *Philos. Trans. R. Soc. B*, 1952, **237**, 37.
42. L. G. Harrison, *KineticTheory of Living Patterns*, Cambridge University Press, Cambridge, 1993.
43. C. Acquisti, J. Kieffe and S. Collines, *Nature*, 2007, **445**, 47.
44. R. J. P. Williams, *Carlsberg Res. Commun.*, 1987, **52**, 1.
45. A. H. Knoll in *Reviews in Mineralogy and Geochemistry*, Volume 54, *Biomineralization*, ed. P. M. Dove, J. J. De Yoreo and S. Weiner, Mineralogical Society of America and the Geochemical Society, Washington, DC, 2003, p. 329.
46. P. M. Dove, J. J. De Yoreo and S. Weiner (eds) *Reviews in Mineralogy and Geochemistry, Volume 54, Biomineralization*, Mineralogical Society of America and the Geochemical Society, Washington, DC, 2003.
47. D. Krogh, *Brief Guide to Biochemistry and Physiology*, Prentice Hall, New York, 2007.
48. S. A. Nichols, W. Dirks, J. S. Pearse and N. King, *Proc. Natl. Acad. Sci. USA*, 2006, **103**, 12451.
49. L. Qiao and X. Wang, *Biogeosciences*, 2007, **4**, 219.

50. B. Gomperts, I. Kramer and P. Tatham, *Signal Transduction*, Academic Press, New York, 2009.
51. W. E. G. Muller, J. Li, H. C. Schröder, L. Qiao and X. Wang, *Biogeosciences*, 2007, **4**, 219.
52. M. van der Giesen, J. Tovar and C. L. Clark, *Int. Rev. Cytol.*, 2005, **244**, 175.
53. D. T. Johnston, F. Wolfe-Simon, A. Pearson and A. H. Knoll, *Proc. Natl. Acad. Sci. USA*, 2009, **106**, 16925.
54. F. Q. Gori, S. Tringer, B. Kartal, E. Madison and M. S. K. Jetten, *Biochem. Soc. Trans.*, 2011, **39**, 1799.

CHAPTER 5

Other Major Elements in Organism Evolution

5.1 Introduction

The genome, the proteome, the glycosome and the metabolic products of a cell, the metabolome, are mainly based on H, C, O, N compounds with distant connection to the same 'organic' elements in the environment. Their biochemistry has been outlined in Chapter 4 and is very well analysed in standard biochemistry texts.[1,2] The underlying reductive cytoplasmic organic chemistry has not altered greatly since life began insofar as it has provided the foundation cytoplasmic chemistry of all organisms. There have been considerable additions to this reductive chemistry since oxygen entered and altered the environment, as described in the previous chapters. Especially the outer membrane and vesicle regions of the cell evolved in this new environment a new oxidative organic chemistry (and new genes internally) together with extracellular chemistry and with a strong link to the environment.

The analysis of the earliest cellular organic chemistry, with an essential accompanying inorganic chemistry, led us to propose that the system of precursor chemistry which led to the first living cells was a consequence of the availability of elements, of a very selected set of especially kinetically stable, energised and reduced organic compounds, and of particular gradients of ions, all energised. The unavoidable gradients of Na and Ca ions created certain possibilities for evolution which were realised only later as we shall see in detail. The further consequence of what we regarded as a singular beginning to life was accompanied by an inevitable development of rejection of oxidising chemicals from cells and of an increasingly oxidised environment which led by feedback to an added cellular oxidised chemistry of ever more complexity to

Evolution's Destiny: Co-evolving Chemistry of the Environment and Life
R. J. P. Williams and R. E. M. Rickaby

Published by the Royal Society of Chemistry, www.rsc.org

today. The environmental inorganic chemistry was prescribed on the oxygen release by equilibria (Sections 1.2 to 1.4).

In addition to the largely C/H/N/O organic chemistry in organisms there has always been the involvement of the major elements, P, S, Cl, (Si), Na, K, Mg and Ca, as ions or in compounds (Table 5.1; see Figure 5.5), which are equally essential in life's chemistry. They are often called 'inorganic' as they enter and exit in similar form from and to the environment. They have all been relatively plentiful in the environment while being of matching importance with the organic chemical elements in organisms. With the exception of S their environmental chemistry was not affected by oxygen evolution. We shall dwell only briefly on the role of P, as phosphate, as its functions in cells are also very well portrayed in standard texts.[1,2] It has little or no redox or C–P bond chemistry in organisms and it generally has not changed function greatly from that in the earliest cells. Exceptions are the roles in biominerals, in some external matrices, and in cell signalling by phosphates. We shall also only outline the early roles of sulfur as they have been described in some detail in Chapter 4 and its in-cell functions have not changed with time, but externally its chemistry has altered greatly, forced by the coming of oxygen. The roles of both P and S are kinetically controlled not at equilibrium. To what extent were the uses of these two non-metals inevitable (Sections 5.2 and 5.3)?

Na, K and Cl ions have always and inevitably been involved in cellular osmotic and electrostatic balances which involved rejection of Na and Cl ions. This use of them was inevitable as only they have always been sufficiently available in high free concentration in the sea and hardly interact with organic chemicals in cells. In each compartment they are in equilibrium between free and bound states if any (see Sections 1.2 to 1.4).Their use evolved but slightly until the gradients of the ions became messengers in the nerves of multicellular animals in oxidised structures (Section 5.10), but without change of oxidation states. Potassium has additional roles in a few enzymes and is part of the

Table 5.1 Composition of Cytoplasmic Fluid of All Cells

Element	*Form*	*Concentration (M)*
H	H^+	10^{-7}
Na	Na^+	$<10^{-3}$
K	K^+	10^{-1}
Cl	Cl^-	$<10^{-3}$
Mg	Mg^{2+}	10^{-3}
Ca	Ca^{2+}	$<10^{-6}$
Mn	Mn^{2+}	$\sim 10^{-6}$
Fe	Fe^{2+}	$\sim 10^{-7}$
Co	Co^{2+}	$<10^{-9}$
Ni	Ni^{2+}	$<10^{-10}$
Cu	Cu^+	$<10^{-15}$
Zn	Zn^{2+}	$<10^{-11}$
Mo	MoO_4^{2-}	$\sim 10^{-7}$

structure of DNA telomers, while chloride has acted in a minor way as a cofactor in the chloroplasts. Sodium gradients also have a general role in pumps for the exchange of small molecules and ions. Just as for organic chemistry we shall show there was an inevitable beginning in cells of the properties of these elements with extra energised development of value as the cell systems changed their structures (Section 5.8). Did all their evolved actions occur unavoidably due to the particular value of their availability and physical–chemical properties as ions?

In this chapter we shall look most closely at the changing functions of readily available Mg and Ca ions without oxidation state changes (Sections 5.4 to 5.7). Calcium was rejected from cells to permit a beginning to any life while outside cells and, using its energised gradient on the rise of oxygen, it was changed in use. The activity of Mg was an equally inevitable feature of its chemical properties and apart from incorporation in chlorophyll developed little from its extremely important initial value as a weak acid and a stabiliser of structure. The uses of these two elements were again due to their availability and binding at equilibrium in compartments. The changing roles of the trace elements Mn, Fe, Co, Ni, Cu, Zn, Se, heavy halogens and Mo and W (Table 5.1)[3] are examined in the next chapter. We shall show that their functions were directly unavoidable in evolution due to the coming of oxygen. An important bulk element, but only in some organisms, is silicon and we turn to it briefly in Section 5.8, as its most extensive role has been in biomineralisation, which has been described in essence already in Chapter 3 but here we give some more details. Throughout this chapter and Chapter 6 we must always have in mind that evolution with all its major environmental changes took place in the sea. Not until after 0.54 Ga did life invade the land by which time the basic chemical nature of living organisms had evolved.

We describe first the chemistries of phosphorus and sulfur as they resemble that of C/H/N/O in that in their compounds, they are less strongly covalently bound and their exchange, though faster, requires catalysts. Different states are not in equilibrium. We contrast the behaviour with those of the four cations at the end of Section 5.3.

5.2 Phosphorus in Cells

We have already met phosphorus in the diester linkages of DNA and RNA and in many roles in metabolism and membrane structures.[1,2] Here we draw this chemistry together briefly. Since the beginning of life it has been found as phosphate in cells as a free ion at relatively high total concentration (near mM). There have been very few observations on forms of reduced phosphate as the redox potentials of phosphorus are below that of hydrogen (H_2/H^+) (Figure 5.1). The forms of bound phosphate are still to be found in all organisms in essential substrates of many enzymes, in energy transduction and in bioenergetics as intermediates, pyrophosphates. They are also present in coenzymes and regulation factors and later in signalling, in combination

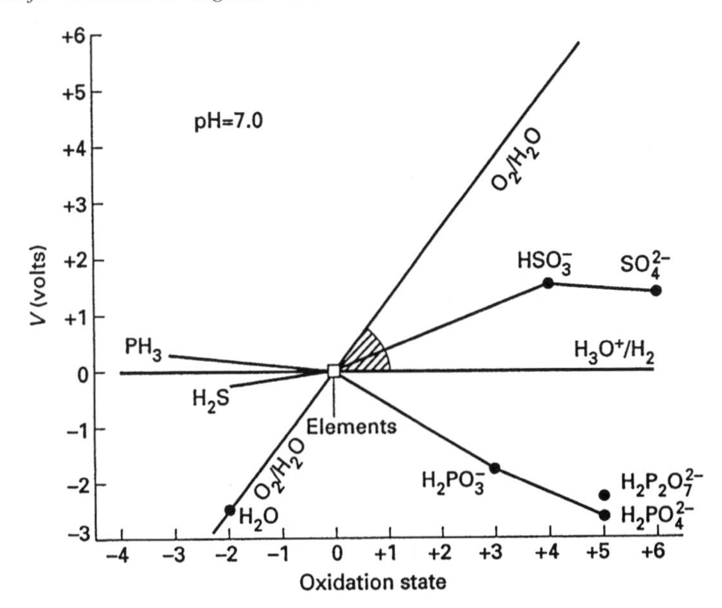

Figure 5.1 The oxidation state diagram for sulfur and phosphorus relative to the hydrogen and oxygen. Phosphorus cannot be changed from oxidation state five. Sulfur has a multitude of accessible states according to the availability of oxygen and hydrogen.

bound to lipids and to saccharides as well as to proteins and its role in diesters is an essential part of nucleotides. No matter how life started phosphate has always been essential and required in considerable quantity. Undoubtedly this followed from the kinetic value of phosphate in relatively fast covalent bond exchange and in its stereochemical flexibility compared with most C, N, O and H bonds in compounds. Interestingly the diesters of phosphate seen in RNA and DNA are much more kinetically stable than mono-esters. The difficult problem is to see how these valuable chemicals were acquired by the primitive initial system, because phosphate has never been plentiful in the sea and is still only present at below 10^{-5} M. It could be directly in exchange for any gradient, say, of protons.

There is now sure knowledge of how the steps of energy capture and transfer originated. It is often considered that energised proton gradients in and across membranes were formed first (see Figure 4.3), and after formation, were followed by synthesis of energised pyrophosphates (see Figure 4.2). The general initial uptake of material into cells used this energy carrier, linked to pumps and enzymes and in the synthesis of the cells' basic organic ingredients. In energy transduction using pyrophosphate or its derivatives, such as adenosine triphosphate, ATP, or other nucleotide triphosphates, the pyrophosphate linkage is bound and activated[3] by Mg^{2+}. Regulation of enzyme function still frequently employs direct, kinetically reversible phosphorylation and there are many enzymes and coenzymes for both forward and back reactions, phosphorylation and phosphate ester hydrolysis. Phosphate transfer is a major cellular control connected to transcription and linked to Mg^{2+} and to Ca^{2+} activation. It has been estimated that a quarter of all proteins in today's eukaryote cells are phosphorylated.

A more detailed relatively recent feature of phosphate biochemistry continues to be the use of a number of monosaccharide phosphates in signalling, generating or assisting in amplification of the calcium signal (see Section 5.5.1). In particular inositol triphosphate, IP_3, released from membranes by an external, catalysis coupled, event, has acted as an activator of calcium release from vesicles causing secondary calcium signalling in eukaryotes (see Figure 5.6).[4] This dates from around 2.0 Ga which is also the origin of eukaryote cells and is absent in prokaryotes.

In summary the somewhat low abundance and availability of phosphorus made it difficult to employ in cells. However it has unique features. Note that even as a monophosphate it binds protons and especially Mg or Ca ions. In this second capacity it has supported internal biomineralisation, such as that of bone from about 0.4 Ga, assisted by heavily phosphorylated matrix proteins. It remains a primary energy transporter in pyrophosphate and a controlling activator of reactions and in saccharide esters it has acted as part of signalling systems. A glance at the Periodic Table (see Figure 1.1) shows that it is one of the second row of non-metals with sulfur which had to be employed with catalysts, proteins, to augment the stability of certain crosslinks and to aid internal kinetics. It is generally essential in many coenzyme fast-transfer functions in the synthesis of compounds of slower reacting non-metal first-row elements. In this way it strongly assisted the essential overall dynamics of cellular biochemistry. Many of the special properties and uses of it in the environment and in cells are an inevitable consequence of the evolution of the elements in the universe and their availability on Earth. The essential chemistry of P, like much of that of S, cannot be replaced by any other element and they have both maintained their original functions in all cells to this day, which is part of very early evolution, that is of the inevitable requirements of the basis of life even from its precursors. Many cells, even prokaryotes, can store phosphate often as polyphosphate in vesicles, called calisomes, with a variety of cations including calcium.

An interesting peculiarity is the finding of calcium phosphate, apatite, in rocks of age close to 4.0 Ga, some of it associated with graphite which in turn is contaminated with sulfur and nitrogen.[5] The findings have led to speculation as to whether these usually small deposits belong to very primitive organisms. In fact phosphate may have been more available when Earth cooled but was then precipitated by metal ions due to weathering. Apatite is certainly a major mineral in some early multicellular eukaryotes, about 0.54 Ga, such as brachiopods but then appears to have only a relatively small part to play in exoskeletons. However it becomes of immense value internally in mineral skeletons of vertebrates, bone, but it is not found in plant biominerals. The late known use of it as a biomineral, not before 0.75 Ga, makes it improbable that the deposits of apatite of 4.0 Ga are anything but inorganic in origin.[5] The later special value of apatite in animals lies in its proton conductivity allowing skeleton bones to be adjustable in shape during growth. It is called a piezoelectric solid.

5.3 Sulfur in Cells

The very important diverse roles of sulfur in cells both in redox and hydrolytic enzyme reactions are described in detail in the book by de Duve.[6] It has maintained several major functions mentioned earlier which we regard as inevitable features of life's chemistry from its original availability and properties which we now collect together. For example inside all cells it is found as both –SH and –S–S– in small molecules in relatively high concentration. In particular the relatively fast exchange between –SH and –S–S– bridges of glutathione is concerned in maintaining the redox steady state of the cytoplasm of cells. (Note also the Fe^{2+}/Fe^{3+} couple of similar potential in cells and in the environment.) The slow kinetic rates of redox change of C–H, C–N, C–O and C– S make R–S–S–R exchange with RSH also extremely useful in the rapid handling of reducing equivalents in compounds. As we have stressed a valuable distinction between the elements of the first row of the Periodic Table and those of the second row, Si, P and S lies in their kinetics (see Table 1.5). P and S covalent chemistry coupled to Mg^{2+} and Fe^{2+} controls and catalysts, are of the very essence of the homeostasis of the moderately fast metabolic steps of the chemistry of cell cytoplasm to this day.

The R–SH group is a very effective base centre too, when ionised, and R–CO–SH is important in transport and as an energy carrier.[6] From the earliest times, while reduced sulfur as –SH had an important role in enzymes, coenzymes and cofactors, RS^- and S^{2-} also formed strong metal complexes, especially with transition metal ions and are found unavoidably in M/S clusters and complexes in the environment and in cells. It is especially associated with essential Fe, Ni, Zn and Mo in environmental minerals and in enzymes often as clusters. They are present in many very primitive organisms.[3] We shall return to the role of these complexes in Chapter 6. They provide a clear structural relationship between minerals, model complexes and cell sulfides. RS^- and RSH thus have roles in enzymes in hydrolysis, in energy and group transfer, and in redox buffering through all of evolution. Observe the very different metal ion association of Mg^{2+} and Ca^{2+} with phosphates and carboxylates but never with RS^-.

The first steps in the evolution of a different sulfur chemistry were due to the increasing redox potential of the environment. Initially sulfur was available as H_2S (HS^- in the sea) and was used as such in cells. From very early times H_2 was an energised source of hydrogen for all organic chemistry and formed elemental sulfur. Around 2.5 Ga more stable –S–S– bridges were formed outside cells or in vesicles due to the evolution of oxygen. Outside cells –S–S– bridges became a common stabilising non-exchanging feature of protein structure to today. Certain bacteria used elemental sulfur in vesicles as an energy source. (The reaction $FeS + H_2 S \rightarrow FeS_2$ (pyrite) was probably an initial energy-generating reaction). Though some sulfur was perhaps still available as H_2S after 2.5 Ga, it became less and less so, especially at the Earth's surface, until about 0.75 Ga when the sulfur of the sea, away from anaerobic regions, had become inevitably, overwhelmingly oxidised to sulfate (see Figure 5.1). During this period prokaryote organisms evolved uptake methods for it from the sea and then metabolic paths for sulfate, SO_4^{2-}, reduction. Some steps involved iron ions, *e.g.* reduction of sulfite. This pathway is not found in higher animals and they are dependent on symbiotic organisms for sulfur supply. Note also the inability of higher animals to reduce nitrate. (Intermediate by-products such as SO_2 and dimethylsulfoxide also arose in the atmosphere. The latter is metabolised by a molybdenum enzyme.) Sulfate, with metal ions, was and is extremely important in many oxidation/reduction steps in cells and on their outer surface (Section 2.7). Once oxygen was more available, that is soon after 2.0 Ga, sulfate could also have been used in certain eukaryote organisms for making internal shells, *e.g.* in Acantharia as $SrSO_4$,[7] and elsewhere as $BaSO_4$[8] as a gravity sensor. However these materials probably did not appear until much later, even after 0.3 Ga. Their evolution is difficult to follow as such sulfates are not stable in the sea. Outside cells many sulfated polysaccharides became synthesised but they appear to be confined to eukaryotes. The earliest known examples are in the algae (around 2.0 to 1.5 Ga) where there are different sulfated sugars in brown, red and green algae. They can form extensive gels of polymers with molecular weights often exceeding a million and form a part of the polysaccharide surfaces of the algae but they are not mineralised, *e.g.* carrageen in seaweed. In a similar but loosely structured manner sulfated polysaccharides are found in the extracellular matrices of higher animals but again they have not formed part of biomineralised structures. They are to be contrasted with the polysaccharides of coccoliths which are in part sulfated but it is the non-sulfated highly carboxylated sugars in their vesicles which aid calcium carbonate deposits (Section 4.18). It would appear that the common sulfate esters have very little affinity for cations and in saccharides give rise to necessary open meshes, because as anions they repel one another. They are extremely important then in allowing facile diffusion of many molecules inside connective tissues of multicellular animals.

To summarise: sulfur is an essential element in all living systems. Its abundance and then availability made it from the earliest times a very useful base and the most essential redox buffer inside cells by utilisation of its redox properties in –SH and –S–S–. Note the value of the redox potential close to that of iron (see Figures 5.1 and 1.4). The affinity of $R{-}S^-$ and S^{2-} heavy metals made it a selective centre for incorporation of selected metal ions in organic molecules as catalysts. No other element can replace it due to its availability, kinetics and oxidation–reduction properties. Once cells liberated oxygen sulfur was inevitably oxidised in the environment to –S–S– bridges and sulfate just before much of the oxidation of Fe^{2+} to Fe^{3+} was virtually complete. The appearance of sulfate has been followed using sulfur isotopes (Section 2.7). As sulfate it performed functions as an energy source and with HS^- and iron as a buffer of redox potential from 2.0 to 1.0 Ga (Sections 2.7 and 2.8). It is a strong acid monoanion, $-SO_4^-$, does not form strong H-bonds and has been the only such group available to maintain an open-mesh anionic organic, external network in higher organisms.

For much of the remainder of this chapter we shall be describing the presence and properties in cellular systems of the ions of the bulk inorganic metals. Before doing so we draw attention to the general difference between them and the non-metals in their chemistry. A major difference is that these metal ions exchange binding partners rapidly without catalysts. Consequently they come quickly to equilibrium between free and bound states in any given compartment including those in cells and in the environment. This allows us to appreciate their behaviour in terms of local binding constants and solubility products in compartments. Unlike the trace elements of Chapter 6 they do not change oxidation state. Generally they are relatively poor catalysts of acid/base reactions but they are excellent in controls of metabolism and homeostasis in the cytoplasm and later to some degree in extracellular fluids. They have strongly energised gradients across membranes so that members of the pairs, Na^+ and K^+ and Ca^{2+} and Mg^{2+}, occupy most strongly very different parts of space, K^+ and Mg^{2+} inside the cytoplasm and Na^+ and Ca^{2+} outside it. Hence they are all energised, pumped across membranes. Externally the divalent cations are involved with the above and other anions in many biominerals while internally the monovalent ions are concerned with osmotic and electrostatic charge balance. The use of the ions in message systems is a prominent feature developed from these gradients (see Sections 5.7.1 and 5.8). One striking exception to all these generalities is the appearance of non-exchanging magnesium in chlorophyll. To what extent were these changes unavoidable and are they all linked to the coming of oxygen?

5.4 An Introduction to Magnesium, Calcium and Silicon Chemistry in Organisms

These three are not trace elements – all are to be found in quantity in organisms, Mg inside cells as well as in the sea, but in the cases of Ca and Si outside the cell cytoplasm and extensively in biominerals of live organisms and in fossils and in the sea.[9–12] These solid structures of silica and calcium salts are easily recognised as shells and bones but there are also deposits of minerals such as carbonates, phosphates and silicates which are not of biological origin. They are described in Chapters 2 and 3. As emphasised before we are able to use the principles of equilibrium in Sections 1.2 to 1.4 in the discussion of Mg and Ca chemistries. The data from a number of more general studies indicate that whereas magnesium has been of great value internally to organisms in enzymes from the origins of life,[9,10] calcium was at first very largely rejected from cells and had only minor value[10] there, while silicon was apparently not used at all by prokaryotes and early single-cell eukaryotes.[11] While we describe calcium, magnesium and silicon in organisms we must remember that all three have been relatively plentiful in the sea at all times, though somewhat variably, with their available biomineral partners, carbonate and phosphate anions, which have been essential for all cells.[9–12] However the oceans have not been uniformly saturated at all times with any of the units of these minerals. We describe next the biochemistry of magnesium in more detail before we turn to that of calcium.

5.5 Magnesium in Cells

The major biochemical uses of magnesium[9] (Table 5.2) remain in the cytoplasmic reactions of organic phosphates in enzymes (Figure 5.2). As we stated above, the earliest Mg^{2+} phosphate chemistry even from the beginning of life extended from binding with several phosphate substrates to that of binding with the nucleotide triphosphates such as ATP and GTP, A = adenine and G = guanosine triphosphate, in energy-requiring steps, and in the corresponding diphosphates, *e.g.* ADP. Many of these pathways are common to all organisms (see Figure 1.10). The nucleotide tri- and diphosphates are often described in conventional textbooks without Mg^{2+} when they are in fact usually inactive. Mg^{2+} came also to have a role in the structures of RNA and DNA and it still controls their folded forms (and see Section 5.5 on signalling). There have always been controls over Mg^{2+} concentration in cells due to inward and outward pumps in order that these activities should be managed. In many cells Mg^{2+} and ATP are both held at around 3 mM in the cytoplasm but Mg^{2+} in the sea is around 20 mM so that Mg^{2+} is to some degree restricted in entry. We discuss why Mg^{2+} was selected and why it is uniquely suitable for its role with phosphates in cells in Section 6.3. As some Mg^{2+} phosphate compounds are commonly employed in so many pathways, including some transcription factors, Mg^{2+} ions at a fixed concentration have an equilibrated homeostatic balancing control in much of a cell's internal activity. Mg^{2+} is a weak catalyst in a few enzymes such as enolase. As very little of this chemistry has changed with time, and we treated it as an inevitable part of initial cell chemistry, we shall not describe it further here (see Chapter 4). These roles of Mg^{2+} in cells cannot be replaced by other metal ions (see Section 5.3).

Table 5.2 The Major Mg^{2+} Compounds in Cells

Compound	*Uses*
Mg^{2+} mononucleotides	Energy and phosphate transfer
RNA	Structure forming
DNA	Structure forming
Enzymes	Mild catalysis, especially of phosphate reactions
Chlorophyll	Light capture

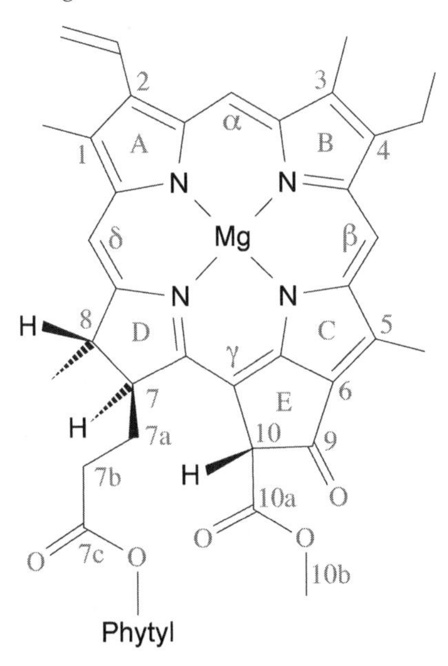

Figure 5.2 The formula of magnesium chlorophyll (green) from modern plants. Other chlorophylls have modified ring sites. The phytyl residue is a long aliphatic tail.

A very important step in evolution was the synthesis of the magnesium complex with chlorin to give chlorophyll in some cells which changed biology for all time in that as a light-capture molecule it separated future plant-like from animal- and fungal-like organisms. Again it is very rare for other elements to replace Mg in this complex and only one, very rare, case of a functional Zn chlorin is known. Mg^{2+} does not exchange at all easily from the chlorin complex and is not in equilibrium with free Mg^{2+}. It may be that Mg^{2+} became a partner of chlorin because it fits so well in the chlorine ring and the molecule does not fluoresce. Several different chlorophylls arose, separating groups of photosynthesising eubacteria and then red, brown and green algae and then especially those in the sea from those on land (see Section 4.13). The synthesis of porphyrins, from which chlorins are derived, is described in Section 6.2. Together with the synthesis of coenzymes these steps of evolution have an importance matched only by the original syntheses of the essential coenzymes and biopolymers. Chlorophyll is of unequalled importance since plants became the source of energy and food for all life.

5.6 Calcium in Organisms

The importance of calcium is much easier to follow as the evolution of its changing partner proteins and its minerals are readily recognised. The proteins have been analysed in detail by reference to DNA sequences in detail.[10] As stated the calcium ion is in equilibria related to its binding constants in a given compartment. In some of the calcium biominerals,[12] calcite not aragonite, magnesium is also present in considerable quantities. We must avoid confusion with dolomite, which is not a direct biological product but in which the magnesium content is highest. Much dolomite was formed by inorganic processing after burial of simple carbonates. Carbonates were not present when solid minerals formed first on Earth but were made slightly later in the sea from the products of weathering of the more temperature-stable silicates, giving Mg^{2+} and Ca^{2+} ions, accounting for the removal of much CO_2 and the formation of large stores of buried carbonates. Later CO_2 was removed in quite other ways, for example as buried organic matter in the Carboniferous Period. The use in biominerals was extended later to organisms in fresh water and on land.

An intriguing puzzle is that, as we have mentioned several times, despite the presence of a considerable variety of unicellular eukaryotes, the fossil record gives no evidence for mineralisation of any of their cells before about 0.75 Ga. After 0.54 Ga, however, such mineralisation appears commonly in both many single-cell and multicell eukaryotes of rather different families. We need to follow the way in which calcium functions evolved in eukaryotes before mineralisation and then ask what properties prevented and then stimulated mineralisation. It is clear that it is in part a consequence of oxidation of organic molecules (Section 4.17) which makes calcium binding different in eukaryotes from that in prokaryotes. We analyse this evolutionary development after we have introduced the essential nature of the function of signalling by calcium in all eukaryotes.[13]

The concentration of calcium in extracellular fluids, whether it is in the sea or the fluid between organised animal cells, has always been quite high in the variable range of around 1–5 mM, controlled by the solubility product of its salts, especially carbonate and phosphate, and limited by alkalinity and temperature. In the sea cell organic chemicals could then in principle bind calcium on outer cell surfaces provided the affinity of binding sites exceeded $K = 10^3\ M^{-1}$. This high calcium is clearly not found in fresh water, so it cannot bind easily to external molecules. Plants have a far lower level of calcium in extracellular body fluids than animals, which must pump it from fresh water into flowing extracellular fluids for the above general uses and to make bone. On the outside of some freshwater organisms there are higher affinity sites. Calcium in high steady-state concentrations has never been of great biochemical value inside cells. However it came to have a controlling role in the release of oxygen by photosynthesising prokaryote membranes associated with manganese, but in no case has calcium had a known activity in the enzymes of the cytoplasm of these prokaryote cells. Calcium is a weak catalyst in some external enzymes such as phosphatases. In single-cell eukaryotes the cation is found in vesicle stores and in assisting protein folding. It also developed a major novel role in cell signalling,[13] which is of such importance that we shall devote a special section to it (Section 5.7.1) after describing the evolution of signalling generally in Section 5.7. The calcium ion binding here is again always in fast exchange so that equilibrium considerations apply. These functions increased rapidly in multicellular eukaryotes when it came to act also as a morphogen. The two rises are coincidental with the two rapid rises of oxygen and the major changes of cell organisation which calcium signalling aids. We stress that its uses in signalling, like those of earlier elements such as iron and magnesium, but internally only, were a consequence of its chemistry but required its rejection and the creation of a strong gradient. Only Ca^{2+} ions have the necessary high concentration outside cells and fast exchange inside and outside them. In that sense it has an inevitable function in assisting the life of eukaryotes, compartmentalised cells.

It is interesting to observe how the binding of calcium to cell products is selected relative to that of magnesium.[3] The frequently observed higher affinity of calcium is usually by groups which have the negative ends of dipoles such as –OH, C=O and carboxylates (more oxidised carbon compounds) of proteins and saccharides on surfaces.[3] The binding is notably most selective in complexes of high coordination number, that is greater than six, in a cavity which fits the large calcium ion (Figure 5.3). These sites fail to bind Mg^{2+} as the binding groups of the Ca^{2+} sites cannot collapse to the small radius of this ion which remains preferentially hydrated, compare precipitation in aragonite and calcite. An interesting enquiry[10] is then into the evolution of both the extracellular protein and other oxidised structures which changed the value of calcium from that in bacteria, just in walls, to that in eukaryotes, in connective tissue, in biominerals and in message systems (see Section 5.7).

5.7 Introduction to Signalling

Any organism of no matter how many cells has to maintain a fixed pH, redox potential and a constant energy supply in the cytoplasm and body fluids for the maintenance of a steady state for reactions of chemicals. This is called a homeostatic condition. To that end there must be communication between the parts of the reaction network that require information transfer. The network

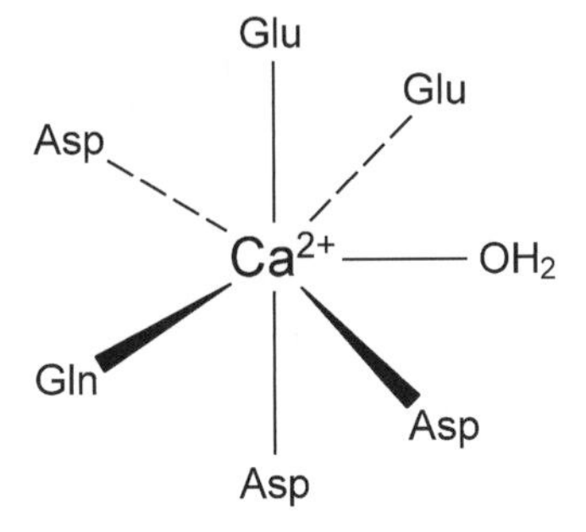

Figure 5.3 A typical coordination centre for Ca^{2+}. Each centre is made of Ca – O bonds greater than 2.3Å and each 7-coordination sphere is completed by water molecules.

includes the conformational control of the synthesis and activity of catalysts, enzymes, as illustrated in all cells, of contractile action (later in muscles) leading to motion in some cells and of degradative activities. Controls on the enzymes outside cells and coordination of activity are also a later necessary function, particularly in multicellular organisms. In the cytoplasm there have always been a number of mobile chemicals which exchange between different reaction pathways coordinating supplies of basic C, H, N and P units and energy. One example is the number of mobile coenzymes and cofactors, which bind in fast exchange to many gene targets or control units. One group of such outstanding small molecules, described in Chapter 4, are the cyclic monophosphate nucleotides, the nucleotide mono-, di- and triphosphates, often bound to Mg^{2+}. Another group is the glutathione redox potential control connected to Fe^{2+} ions. which also connects to gene expression, and a third is the nicotinamide coenzymes for H-, -P and substrate fragment transfer. Amongst simple ions, Mg^{2+} and Fe^{2+}, the only ones readily available in early seas and taken into cells formed many fast direct communication links within and between metabolic pathways from life's origin as we described in Section 4.3. To secure the condition of the interior of cells there must also be a coherent, cooperative set of reactions including uptake and rejection pumps. In Chapter 6 we shall explain how the controlled coordinated activity is achieved by the synthesis of pump proteins with binding strengths selective for each messenger. Hence any given part of this controlled homeostatic condition needs feedback control to the pumps from the build-up in concentration of any particular one of these messenger units so as to limit action, which is such that it prevents further concentration increase above a desired level. For example a cell must build and maintain a cytoplasmic concentration of free Fe^{2+}. It does so today by switching on the production of an uptake system (with pumps) to increase the not very available Fe^{3+} from the environment, and a vesicular reducing system for conversion of Fe^{3+} to Fe^{2+} (see Figure 6.5). The switch is a transcription factor, a protein, which is bound to DNA and in conditions of low Fe^{2+} drives translation of the Fe^{3+} uptake system. On reduction the increasing Fe^{2+} then activates the production of enzymes and iron-binding proteins and their balanced coordination. As the Fe^{2+} increases to an optimum level it slows increase of all these systems by reducing Fe^{2+} concentration through feedback inhibition, via the transcription factor, of translation and hence synthesis of input (pumps) so that there is eventually a steady state of Fe^{2+} and metabolic activities in the cell. In other cases build-up in concentration inside also directly inhibits uptake. Every cell must have had many of these feedback controls even very early in evolution, so that the concentrations of small organic molecules and inorganic ions were linked to the production and steady state of large molecular proteins, via transcription factors to genes and direct control of activity itself. This is the essence of feedback internal control signalling (Figure 5.4), linking genes, activity and the environment and gives rise to a characteristic of most, if not all, the free metallomes (Chapter 6). All cells have been made aware of many outside

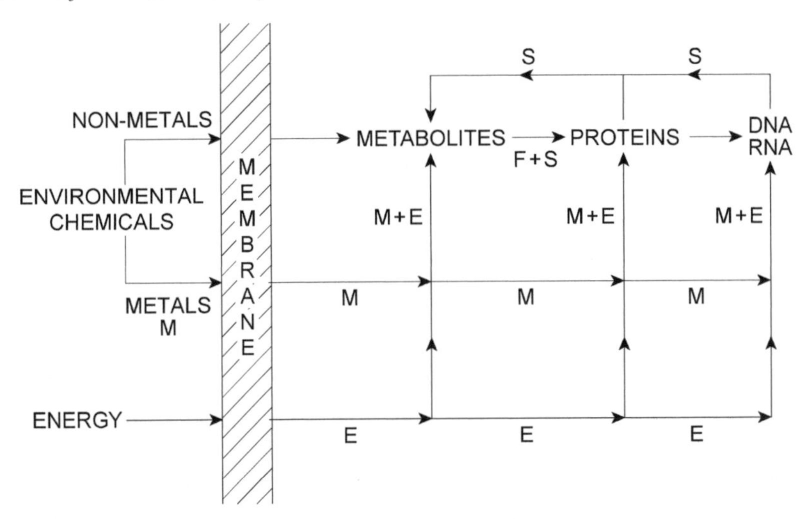

Figure 5.4 The essential feedback circuit for an ion such as Mg^{2+} or Ca^{2+}. For each ion there has to be a link between uptake/rejection pumps, metabolic circuits and the expression of proteins through exchange. Each ion has a free ion steady state. S is a substrate and F a control factor.

chemicals that can permeate the outer membranes and have response modes controlling their effect. Using such internal signalling any organism can adapt to long-term changes in external conditions by adjusting or introducing novel homeostasis, say, on going from an anaerobic to an aerobic environment.[13]

A prokaryote has a short growth period and its major needs are food for reproduction and energy. These cells have chemotaxis responses to gradients of desirable chemicals in addition to the ability to change homeostatic states slowly. They do not need a general rapid response to the environment as they are designed to survive by rapid reproduction and fast random mutation. They move as a whole cell and have an inflexible shape so that functions cannot be rearranged in space. By way of contrast a single-cell eukaryote cell has a long growth period and a flexible cell structure and gains as it can quickly adjust shape and cytoplasmic metabolism to advantage by being very aware of its environment while avoiding its dangers. It has vesicular structures including organelles too, which enclose several oxidation and glycosylation steps, and which need to be in communication with the cytoplasm. To this end there was introduced the remarkable evolutionary advance of a signalling system for conveying the information of the environment as fast as possible to a large number of receptor metabolism contraction targets, but largely not directly to the more slowly responding genes, in each cell. It was based on input of calcium and phosphate esters to the cytoplasm (Section 5.7.1). The signalling of the nature of the environment to the organism had to be still much more complex in the multicellular eukaryotes because the whole organisms needed to be aware of the external environment while the differentiated parts of the interior of the organism had to be aware of each other and yet able to act separately. The signalling devices which evolved had to be linked to larger and larger organic sensor devices, eventually to eyes and ears. As we have remarked above the carriers of the messages must be additional to Ca^{2+} as they have to be released from cell vesicles to stimulate other cells of the same organism. Calcium cannot act as a messenger between cells as it is in high concentration in extracellular fluids (contrast zinc in Chapter 6). In this chapter we shall be examining in particular the value of not only Ca^{2+} together with these many small organic molecules, but also later still with Na^+, K^+ and Cl^- acting in very fast messenger roles[3,4] (see Section 5.10). The value of these ions, like that of calcium, in exchanges or messenger functions arises from the gradients which arose since the beginning of cellular life when they, except K^+, were rejected as poisonous at their levels in the sea.

The use of gradients in exchangers in uptake of molecules or ions is observed in all cells but the messenger functions arose only when cells, eukaryotes, became flexible and needed long life. The calcium messenger role arose after the first somewhat rapid rise in oxygen in single-cell eukaryotes while the organic intercellular messengers arose only with the second rather rapid rise of oxygen. The sodium/potassium messenger role also arose a little later after there were multicellular organisms following the second rise in oxygen. Their different messenger activities directly reflect the chemical properties of the ions and molecules and their imposed energised gradients.

A different type of signalling from the above, which does not need to be very fast, is that which controls adaptation, or growth of particular cells in multicellular organisms.[13] It is linked to genes not to metabolism directly, via transcription factors which are proteins bound to the genetic code. The connection to metabolism is then through expressed proteins. We mentioned this in Section 4.20, when we described hormones. Changes in a multicellular eukaryote are not only due to effects of the outside environment but of the inside of the organism and it has an inbuilt timed self-assessment of its growth stage. The required hormonal control system has no direct calcium gradient connection because the switch on/off mechanism is not fast but very slow. Here zinc in transcription factors became particularly important (Chapter 6). There are hormones of this kind for controlling input of minerals including calcium. They are called glucocorticoids. Finally complex animal cells must recognise any infection and here a further new signalling reaction, the immune system, was devised in animals which has some dependence on the inorganic elements. We shall not describe this special activity here. The protective system of plants is very different (see Chapter 6). We return to the synthesis of organic messengers of all kinds in Chapter 6.

5.7.1 Detailed Calcium Protein Signalling and its Evolution in Eukaryotes

As we have indicated, from their origin, all cells, including the earliest of prokaryotes rejected calcium of necessity via pumps, ATPases, in their outer membrane (Section 5.3), because calcium ions coagulate DNA.[3,4,14] They had to limit cytoplasmic calcium concentration to below 10^{-6} M, thus creating a gradient of at least 10^3 to 10^4 across the outer membrane with the sea. It is this gradient which became the source of the evolution of a major messenger system in eukaryotes (Figure 5.5). Here we see the realisation of the possibilities stemming from the unavoidable rejection of calcium from the earliest of cells. Later it allowed controlled in-cell transfer to storage vesicles at 10^{-3} M. These stores could also be released by a mechanism we describe below. The high external calcium concentration is maintained in all multicellular animals in its body fluids but it is lower in plants in fresh water. The permitted entry of calcium into the cytoplasm of both kinds of cell, via the opening of a calcium channel directly on a cell's contact with external change is followed by binding of it to receptor proteins. The calcium concentration rise in the cells activates or deactivates certain enzymes, especially those for phosphorylation or phosphate ester hydrolysis and those which activate contraction. This requires a protein which acts to cause fast, millisecond, internal activity in response to the external change (Figure 5.6). These components of calcium signalling were known first in the single-cell eukaryotes such as yeast, but have been found in all single-cell plants and animals and all multicellular organisms (Table 5.3). Subsequently the calcium must be

$R{-}C(=O){-}O{-}CH_2$
$R'{-}C(=O){-}O{-}CH{-}CH_2{-}O{-}PO_2^-{-}O{-}CH_2{-}CH_2{-}N^+(CH_3)_3$

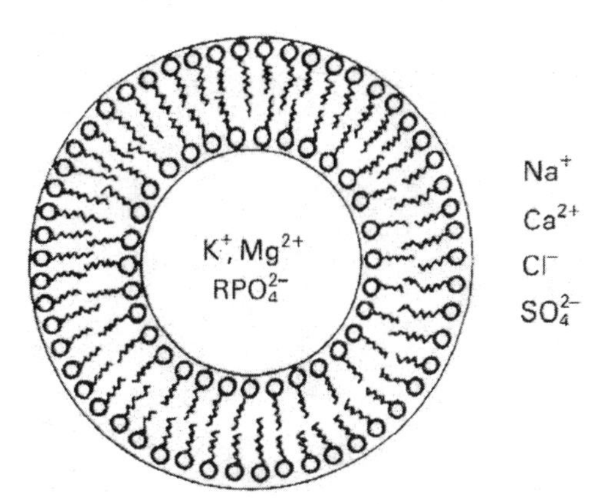

Figure 5.5 The gross distribution of seven ions of a eubacteria cell in the sea. Three free ions are in high concentration internally > 10^{-3} M while four others are below that level and in large excess in the sea.

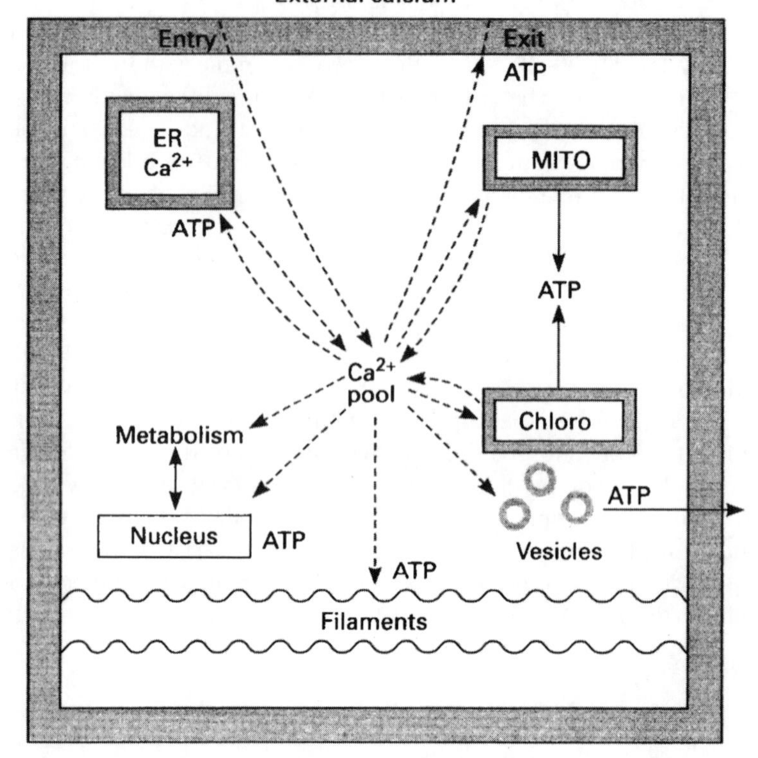

Figure 5.6 The basics of signalling by the calcium ion with stimulated input channels and exit pumps. Mitos are mitochondria; chloro, chloroplasts. The calcium store is in vesicles of the endoplasmic reticulum, see Fig. 5.8.

Table 5.3 Some Classes of Calcium Proteins

Protein	*Location and Function*
Calmodulin*	Cytoplasm, trigger of kinases etc.
Calcineurin*	Cytoplasm, trigger of phosphatases
Annexins	Internal associated with lipids, trigger
C-2 domains	Part of several membrane-link enzymes
S-100*	Internal and External: buffer, messenger, trigger
EGF-domains	External growth factor but general protein assembly control e.g. fibrillin
GLA-domains	External, associated with bone
Cadherins	Cell-cell adhesion
Calsequestin	Calcium store in reticula
ATP-ases	Calcium pumps

*EF-hand proteins

removed rapidly from a cell so that the cell is ready for a new message but phosphorylation has a longer retention time. For this purpose there are pumps out of the cell cytoplasm into the endoplasmic reticulum (ER) (see Sections 4.8 and 4.11), and into mitochondria as well as to the outside. The pumps must return the standing calcium concentration to 10^{-6} M. There was then required receptor proteins with a binding constant of 10^{-6} to 10^{-7} M together with the outward pumps which were already present from the evolution of prokaryotes. Now however they must have sites which stop outward pumping as the calcium falls to the intracellular level of 10^{-6} M. There were then several steps in evolution of the calcium signal, those of the channels, the pumps, stores in vesicles, and the receptors, but also of external carrier proteins for fast transfer and buffers to maintain fixed resting states. The expression of the proteins involved in these internal messenger systems in both single-cell and multicell eukaryotes has been studied directly knowing their present-day DNA[15] (Table 5.4).

The single eukaryote cells of fungi, which are thought to have evolved well before the Cambrian Explosion, have today all these essentials of signalling which are absent in bacteria and Archaea. Taking yeast cells as an example they have at least one calcium entry channel and one receiver activating calcium-binding protein in the cytoplasm for turning on enzymes for phosphorylation and one for activating the reverse step in a calcium-dependent phosphatase. Both the enzymes have been shown to have two domains of so-called EF-hand construction in a protein for binding Ca ions (Figure 5.7), called calmodulins,[16] with a binding constant of $K = 10^6\ M^{-1}$. All later eukaryote cells have these proteins. The separate domains of EF hands are of much lower affinity and the prokaryotes have only a few such binding sites of one EF-hand domain (of low affinity). Moreover the calcium sites do not bind ions of smaller size, including Mg^{2+}, for the reasons given in Section 5.4. Harold[17] observes in his book The Way of the Cell: "One of the great moments of evolution was the appearance of key molecules including calmodulin (the

Table 5.4 Distribution of Different Ca^{2-} Binding Protein Motifs in Organisms

			Binding Proteins			
	Excalibur	*EF-hand*	*C-2*	*Annexins*	*Calrecticulum*	*S-100*
Archaea	-	6*	-	-	-	-
Bacteria	17	68*	-	-	-	-
Yeasts	-	38	27	1	4	-
Fungi	-	116	51	4	6	-
Plants	-	499	242	45	40	-
Animals	-	2540	762	160	69	107

See Ref [15] *Single hands only see Chapters 3 and 5 **Note** The Table is based on DNA sequences available in 2004. The activities of the proteins are unknown in most cases

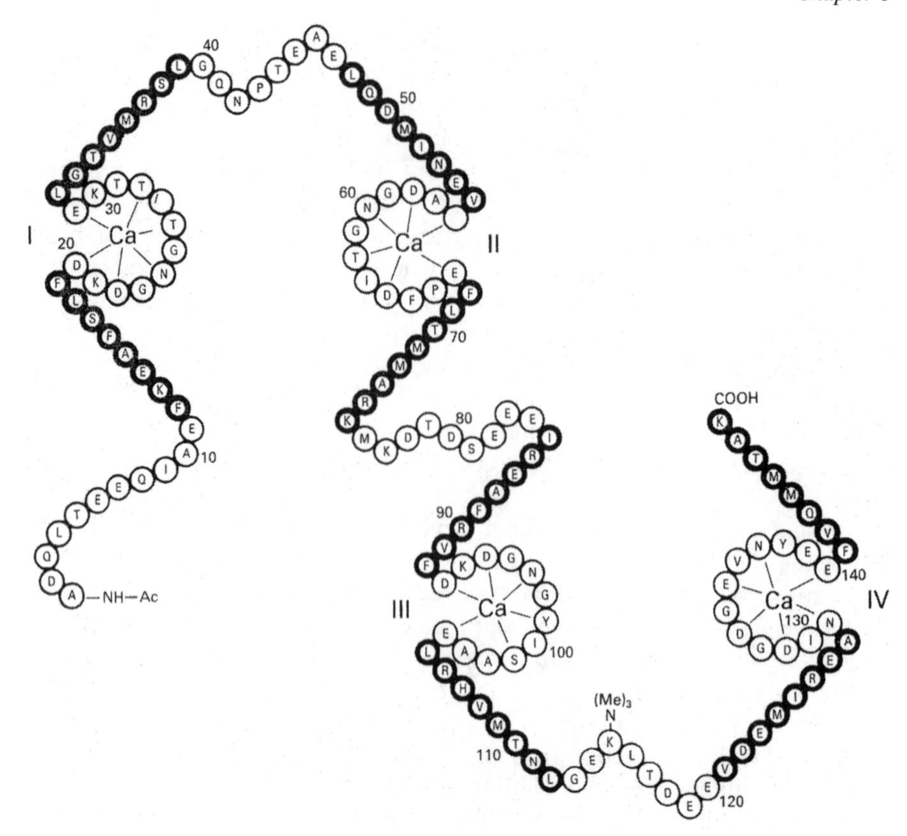

Figure 5.7 The sequence of cadmodulin showing the four cooperative binding sites, 7-coordination with added water ligands.

EF hand)".[15–17] Table 5.4 shows that multicellular organisms have a greatly increased number of such proteins as cell differentiation increased.[17]

As stated the calcium pump is found in even the earliest prokaryotes. It too must bind internally with a $K = 10^6\ M^{-1}$ but externally with $K = 10^3$ M only to stop further lowering of Ca^{2+} concentration. Now while the common possession of these pumps implies general gene inheritance from the earliest cells, the common possession of all the different signalling proteins in single-cell fungi, plants and animals, which all arose close to 1.5 Ga, is difficult to explain. It arose also in all their different subclasses of eukaryotes at very similar times. In Chapter 7 we shall examine gene duplication as it is a major evolutionary step, potentially providing a mechanism for the advent of new genes. See also the similar problem of the origin of biomineralisation almost simultaneously in many types of organism (Chapter 3 and Section 4.18).

Simultaneously with entry of calcium into the cytoplasm as a signal by the membrane proteins the calcium released into the cytoplasm was greatly enhanced by release from vesicular stores.[18] This is an amplification device caused by an increase in sugar phosphates, inositol triphosphate IP_3, in particular[4] (Section

5.2), which are released from the outer membrane when activated by selected events at their receptor to the ER vesicles protein for Ca^{2+} release (Figure 5.8). The stores in the ER vesicles are of free calcium at about 10^{-3} M but there is much more weakly bound to calcium proteins, such as calrectinins, which is rapidly releasable. The number of these proteins also increased with organism complexity (see Table 5.4). Once again the system of vesicle storage and release is common to all eukaryote branches of the tree of evolution. Are the branches truly independent in 'trees' as they are pictured by many biologists or are they connected by some means and are quasi-simultaneous?

The number of calcium signal proteins today includes several which are not strictly calmodulins such as the S-100, C-2 and annexins which have evolved relatively recently in many groups of cells (see Table 5.4).[17] Some of these proteins are also calcium transporters such as calbindin or buffers such as parvalbumins.[15] Such metal ion carriers are called 'chaperones' in a later discussion of different metal ion transfer in Chapter 6. They all have a modified EF-hand structure including those involved in mitochondrial/ER interactions.

Calcium ions also have the function of acting as morphogens in multicellular organisms. Here there is a steady-state calcium gradient in extracellular fluids which is thought to aid location of all cells as they migrate. It is proposed for example that such a calcium gradient aids the repositioning of cells when the gills of the earlier grub-like form changes to the wings of a flying insect.

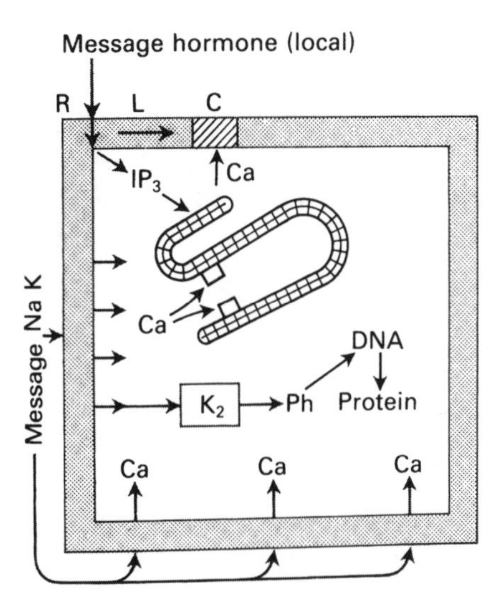

Figure 5.8 The details of calcium activation by either a hormonal or transmitter message at a receptor R, (top) which releases the inositol phosphate, IP_3, to the calcium store in a vesicle. The calcium may activate (3) a kinase K_2 which via phosphorylation, Ph, affects DNA so as to generate proteins. Calcium may also open a channel in concert with other membrane events. The various calcium stimulated activities are shown in Fig. 5.6.

In conclusion to this part of this chapter, signalling and its evolution and development (Table 5.5) are extremely important in evolution. Initially it was and it remains the essential process by which the internal homeostasis is managed. The ability of coenzymes, phosphates and certain ions to act as messengers internally could be matched in communication between the inside and the outside generally by only one simple available chemical outside the cell in the sea, the calcium ion, as has been shown repeatedly above. Only calcium could act in this way as it had a large outside/inside gradient and had very fast on/off binding. The system required special internal proteins which are specific for calcium, including both receptors and outward pumps. We have shown earlier the principles of this specificity and the new proteins which have the required binding sites. We do not know how they arose as no prokaryote has these special proteins and there is today a great variety of them performing many functions.

Calcium could not act also as the source of information between cells in differentiated cell organs. Each differentiated cell needed to be able to handle, *i.e.* to release as well as receive, selective messengers which must be stored internally in high concentration in vesicles of donor cells[19] and released to low outside concentrations. This is the reverse of the messenger action of calcium but it is the entry of calcium to donor cells that releases the stored organic molecules from vesicles and it is calcium that has brought about the binding of these organic compounds in receptor cells. The messengers became the synthesised organic compounds described in Section 4.21 and Figure 5.8, one class of which stimulates the calcium system (see Figure 5.4). These organic compounds are largely synthesised by oxidation steps that required the evolution of selective (copper) enzymes (Chapter 6). They evolved again in the same period around 0.75 to 0.54 Ga, as mineralisation which we describe next. Moreover the number of them increased rapidly together with extra uses of calcium signalling and the number of differentiated cells (Figure 5.9). Note again that the use of the same or similar organic messengers is common to

Table 5.5 The Devices Introduced for Calcium Signalling

Change	*Location*
1. Compartmental concentration of Ca^{2+}	Endoplasmic reticulum (chloroplast and Mitochondria). Other vesicles, Golgi
2. Calcium channels	Membranes (cell membranes and vesicle membranes)
3. Calcium receptors	Membranes of vesicles. Many enzymes (kinases). Filaments for contraction (calmodulins, annexins, troponins)
4. Calcium pumps	Membranes (driven by ATP)
5. Links to phosphate signals	Kinases and phosphatases with attached Ca^{2+} trigger proteins
6. Calcium stores	Calsequestrin in reticula
7. Nucleus calcium concentration	Calcium signals to the nucleus from the Surrounding endoplasmic reticulum

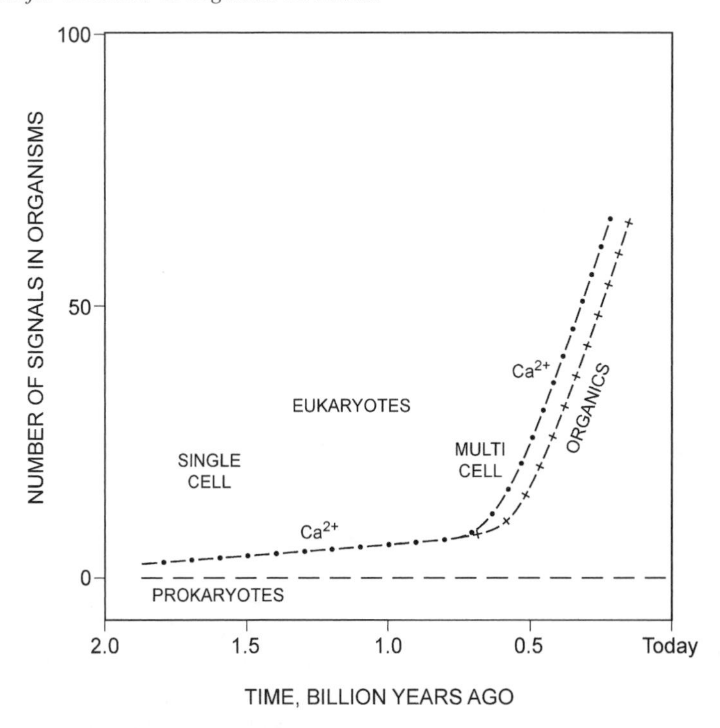

Figure 5.9 The increase in organic messengers with the increase in use of calcium sensitive responses in multicellular organisms.

many types of different organisms. Finally yet faster signalling connected to Ca^{2+} messages evolved in the nerve system of animals (Section 5.7). In this system Na^+/K^+ messenger signals do very little except pass electrolytic currents along membranes and opening Ca^{2+} channels at the ends of all nerves, synapses. The outstanding and very extensive use of the inorganic ion of calcium illustrates the strong interaction of environmental inorganic and cellular organic chemistry in organisms. The puzzle remains as to how this chemistry arose but it appears to be an unavoidable feature of eukaryotes.

5.7.2 Weaker Binding Sites in Vesicles

In the course of this discussion of calcium signalling we have referred to the proteins in vesicles which bind calcium. These proteins have binding constants around 10^3 M^{-1}, similar to those of the proteins which bind this ion on the outside of cells in contact externally with Ca^{2+} concentration of 10^{-2} M, for example in extracellular fluids of animals. These binding sites do not bind other ions. They are an essential part of signalling but similar proteins are present in other parts of the endoplasmic reticulum which act as chaperones – proteins guiding folding in the vesicles – and yet others are found in connective tissues, lectins. These are all weak Ca^{2+} binding sites. The question arises as to how their selectivity is achieved. One important point is that they do not bind Mg^{2+} which is often present in equal concentration. The design of binding sites is based on the same principle as discussed for calmodulin (and see Figure 5.7), except that the donor groups bind more weakly. The binding constant for Mg^{2+} can be as low as 10 M^{-1}. Very importantly all the binding sites are in equilibrium with free calcium ions. This point is stressed as it marks the big difference between binding of inorganic ions, both in the environment and in cells and that internal to organic molecules (Chapter 4). It gives one reason for the equal imperative presence of inorganic, fast reacting, elements as opposed to that of the organic slow-reacting elements. Both, together, provide the essence of life as we see again in the next sections and Chapter 6.

5.8 Sodium/Potassium Messages

The final development of message systems is the evolution of the nervous system.[3,20] This required the growth of long cylindrical tubular extensions of cells, neurons, so that messages could be delivered from a specified donor to a particular receptor cell along a wire-like construct (Figure 5.10a). A requirement for the greatest possible speed in message transmission arose after 0.54 Ga from the increased size of animals, with corresponding increase in the distances from senses for awareness to the muscle cells before action. Fast movement itself was also required by an animal as it had to capture prey and avoid capture itself. It is a scavenger, not a primary synthesiser of metabolites. Plants do not need such mobility as they take in energy from the Sun and primary material from solutions (earth) and the atmosphere which flow to them and growth is the obvious objective not movement. They do not have nerves or muscles. The fast message in animal neurons utilises the transient rejection of sodium from and the uptake of potassium into cells which arose from the essential stabilising gradients across all membranes from the beginnings of prokaryotes (Figure 5.10). The outward sodium pump is in the earliest known cell and quickly restores the stationary state. The novelty required for nerve action was a channel opening and closing unit in the membrane of the nerve cell tubes. Such channel opening was stimulated by physical contact or by organic transmitters released by calcium at the tubular cell termini, synapses (Section 5.4). The synapse construction can be traced back to the evolution of some of its proteins in unicellular yeast.[21] The transmitters are the well-known agents such as adrenaline and amidated peptides but there are very many more, perhaps close to a hundred, in the brain, many more than elsewhere (Table 5.6). To achieve a satisfactory electrical charge migration along a nerve requires that the gradients of Na^+ and K^+ across the cell membrane at rest are held very strictly constant. (One of the simplest first checks on patients' health by a hospital is a measurement of the concentration of these ions in their circulating fluids.) The long term constancy of the two ions in the extracellular fluids is controlled at intake by mineral corticosteroids linked to zinc finger transcription factors generating suitable

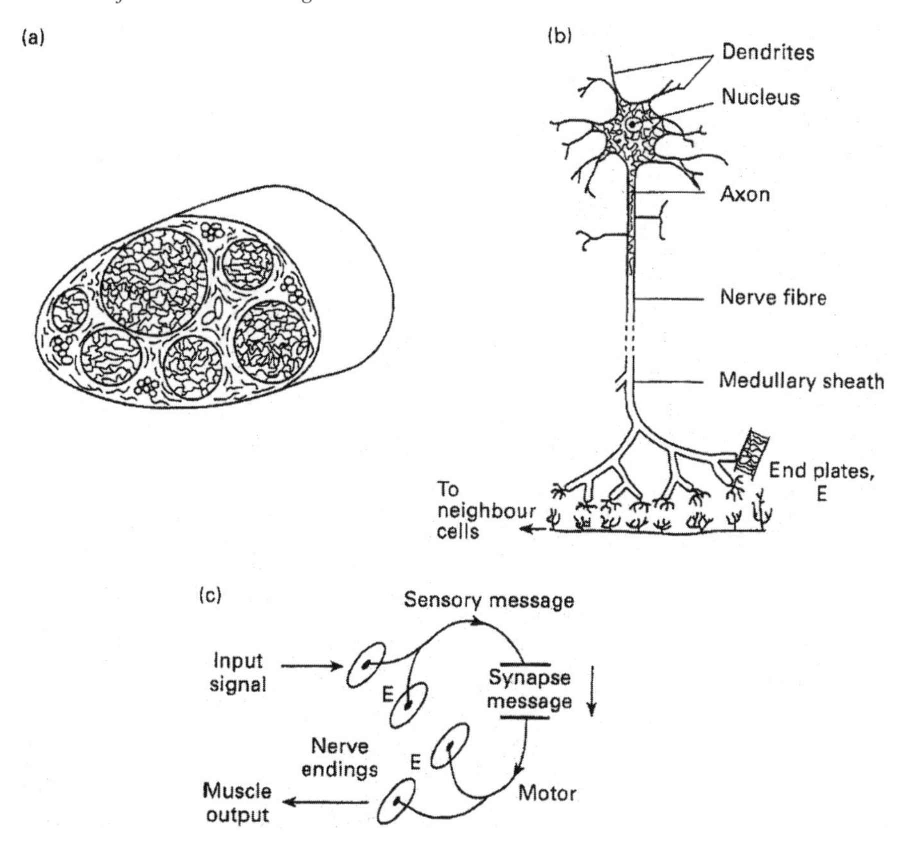

Fig. 5.10(a) General impressions of a nerve cell (a) cross-section of a trunk, (b) a neuron, (c) nerve connections via synapses.

proteins as is the case for calcium. The corticosteroids too are the product of oxidation of the basic sterol molecule. The link with the environment is again strong. No other ions can replace Na^+ or K^+ in cells due to their availability and chemistry, or rather lack of chemistry. The final development of this extremely complex combination of different ions, Na^+, K^+ and Ca^{2+} and organic molecules, transmitters and hormones, together with the advances in cell construction and cell growth is found in the brain of animals. All of the development can be traced to the potential provided by the available inorganic elements in the diet together with the necessary energisation of their concentration in cells.

The gradients and their uses depend on the ability of the cell to synthesise proteins or other molecules which, when placed in a membrane, can discriminate between Na^+ and K^+ by equilibrium binding strength.[3,20,21] Na^+ can be bound the more strongly due to its smaller size provided a cavity is made which can collapse negative partial charges of polar groups around this ion. A discrimination in favour of K^+ is achievable if the cavity collapse is prevented once it has reached a size to fit K^+ optimally. Na^+ will not bind as

(a)

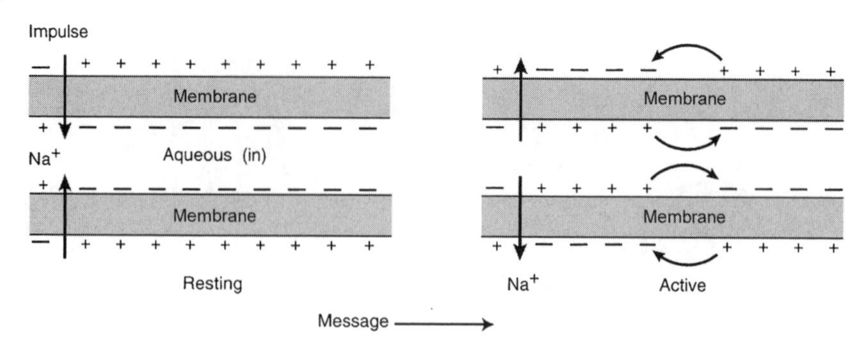

(b)

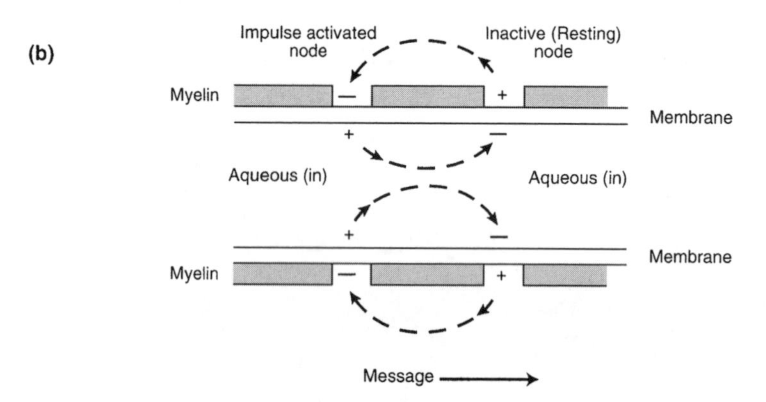

Fig. 5.10(b) The propagation of an impulse via Na^+ ion movement which is compensated by K^+ fluxes. The spread of the depolarisation wave is increased in advanced nerves with limiting input/output of ions at nodes of Ranvier. Note the myelin coat.

well to such a site because its hydration energy is greater than any binding energy for it in such an ill-fitting cavity. This discrimination based upon controlled size of holes is due to protein-folding constraints. For a flow of potassium there must be several such sites in a linear series in the protein which crosses the membrane.[21] The reverse flow of sodium has similar requirements but its rejection requires energised pumping by the Mg-dependent Na/K, ATPases. Essential selection between Na^+ and K^+ is seen to be under the same principles of local equilibria based on simple electrostatics as that between Mg^{2+} and Ca^{2+}. (The divalent ions cannot bind in these holes because water binds to them more strongly than the protein's binding groups.)

One of the fascinating features of evolution was that the use of ion rejection, especially of Na^+, Cl^- and Ca^{2+}, was necessary before the coded system could be stabilised. Subsequently the handling of Na^+, K^+, Mg^{2+}, Ca^{2+} and Cl^- by protein pumps and channels had to be controlled (Figure 5.11). They were

Table 5.6 Some Messenger Compounds Between Cells in the Brain

Excitatory (+)/inhibitory (−)
Glutamine (+)
Glycine (−)
GABA (γ-aminobutyric acid) (−)
Acetylcholine
ATP (+)
Noradrenaline (norepinephrine)
Dopamine
Serotonin
Inorganic signals
Na^+, K^+, Ca^{2+}, (Zn^{2+}), Cl^-)
Nitric oxide
Carbon monoxide
Peptides
Neuropeptides (large number > 20)
Substance P
Cholecystokinin
Corticortrophin-releasing factor
Melatonin

P.S. There may be many more which are used in small quantities.

then connected to a great variety of functions, for example controls of heavy metal ions from the sea. How did it all originate and develop using proteins at different periods? In fact the puzzle is present from the origin of life as DNA genes and gene products could only have a stable kinetic existence in cells of controlled ionic media. As mentioned before the sequence of evolution is often recognition of a poison, followed by rejection and then use. Maybe much development is the result of stress due to changes in external or internal

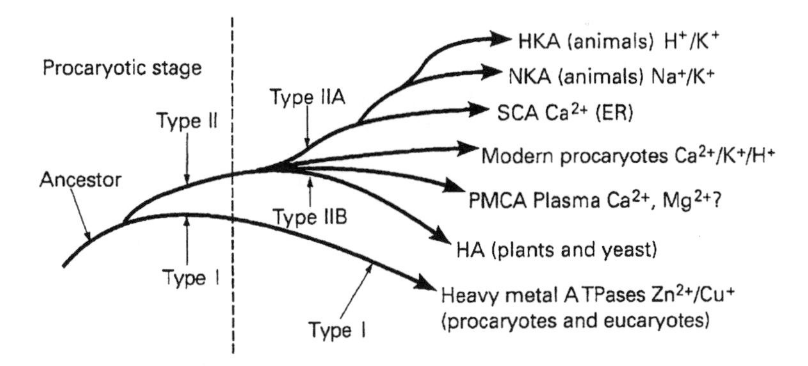

Figure 5.11 A possible evolutionary relationship of the ATP-ase pumps for metal ions. Type II acted very early on Ca^{2+} and Na^+ expulsion while Type I acted on trace transition metal ions. There is good reason to believe that the membrane part of the pumps has a common origin but the aqueous appendages are very different over time (After J.V. Moelles *et al.* (1996) *Biochem. Biophys. Acta*, 1286, 1-51).

chemistry. We return to the problem of the original inorganic and organic chemistry changes coupled to the arrival of new coded information and proteins in Chapter 7.

The most remarkable achievement of evolution is perhaps the rapid creation of the brain (Figure 5.12). It is composed of a vast interconnection of nerve cells, neurons and glia,[19,22,23] the neurons leading to and from processed senses to operating organs such as muscles but beyond that the brain is undoubtedly active all the time and is able to create 'pictures and sounds *etc.*' which are not real and are said to be 'in the mind'. We are a long way from disentangling the real and the imagined in the brain. The reference list has some published work of the involvement of inorganic ions in the brain processes but we are far from providing a full description of the processes of the brain,[23] which we know evolved soon after the Cambrian Period because a brain is seen with the nerves in invertebrates, *e.g.* in the octopus and nematode worms. The development of the brain which allowed animals to control their environment to some degree is the final way organism evolution took place. We discuss what this evolutionary development could mean to today's environment/organism ecological system in Chapter 7. Here we observe only that many advances in chemistry took place between, say, 0.6 and 0.4 Ga in so many different types of organism. We know that a common factor is the second rise in oxygen throughout the ocean. How was this genetically possible? Further major chemical change had to await the impact of meteorites or the coming of man.

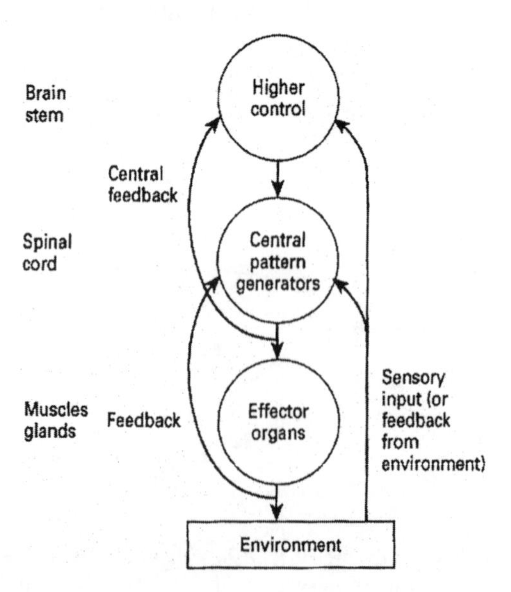

Figure 5.12 Hierarchical organisation of the neuronal circuits. One can consider the direct effect of the environment through nerves and the central response generators to muscles evolving before the higher control with more extensive memory and ability to look ahead.

The description of the brain brings us to the end of the examination of the evolution of message systems. Unlike the major developments of organic chemistry (Chapter 4), and of trace elements in Chapter 6 the evolution of the message systems based on Ca, Na, K and Cl has no obvious immediate connection with the rises of oxidation potential of the environment. This is in contrast with the evolution of later organic chemical messengers and of that of biominerals seen in the next sections. The inorganic message systems did develop, however, selectively, with different degrees of cellular complexity and compartmentalisation. Messengers are clearly a part of the essential nature of complexity. What we wish to stress is that although this evolution of use of all these bulk elements of the different messenger systems does not arise directly from the rises of oxygen it coincided with them.[1] Moreover they did not arise from a selection of possibilities but were an inevitable accompaniment of stages of complexity, each element playing a unique role due to its properties, including availability and chemistry, or lack of it, in water compartments, separated by lipid membranes. There must have been some connection between the oxygen level changes and these changes but what is it? We maintain that the evolution of compartments and the chemistry of them was itself inevitable due to the need for separate compartments for different oxidation reactions and oxidation potentials of the environment, and that this is inseparable from the rise of messengers. A similar problem arises with the evolution of biominerals.

5.9 The Evolution of Biominerals

We summarised the accounts of these minerals in Chapters 3 and 4. Here we stress the value of understanding the evolution of biominerals as they gave rise to fossils which are the strongest direct evidence of evolution (Chapter 3).[12,24,25] They are of two major kinds based on calcium and silicon. We followed the progression of the deliberate employment of calcium in biominerals using the fossil record, through the part of the gene complement expressed as calcium-binding proteins, and by isotope composition of deposits starting with the mineral systems of the simplest single-cell eukaryotes. We then passed through the stages of coral, other algae, to the minerals of the Cambrian Explosion fossils, 0.54 Ga, to the somewhat later appearance of silica in plants, 0.4 Ga. Finally we observed that there were later periods of mineralisation, especially of brown algae, 0.3 Ga, and last of all of grasses, 0.05 Ga. Some mineral formation was noted, for example of the Sr and Ba sulfates. We describe how the minerals were supported in nucleation and growth by organic matrices.[26] These matrices were crosslinked organic polymers laid down by oxidation. The whole process is described under the terms induced and controlled mineralisation, Chapter 3.

Now it is known that both induced and controlled mineralisation rely for the organic matrices upon internal synthesis. The question arises as to the extent the inorganic mineralisation is also under genetic control. Now we have noticed that inorganic calcium carbonate may be precipitated in one of two allotropic forms and the balance between the formation of calcite and aragonite depends on the relative concentrations of calcium and magnesium in the sea,[12,27] as described in Section 2.10. The growth of calcite is inhibited by magnesium so that when the ratio of Ca:Mg is below 1:3 aragonite forms but calcite (which may be so Mg-rich that it is included under dolomite) forms when the ratio approaches 1:1. The switch of allotrope to aragonite in organisms is seen at such times as in the Permian/Jurassic Periods both in shells built directly from seawater such as corals and in minerals from inorganic sources (Figure 5.13).[27] These are close to periods of extinction and burial of much carbon with rises and falls of weathering and varying Mg^{2+} and Ca^{2-} (Figure 5.13). The suggestion is clearly that environmental inorganic chemistry, not genes alone, is partly controlling of organism chemistry. Those organisms which nucleate $CaCO_3$ in vesicles, that is in the absence of seawater magnesium, only give rise to pure calcite, not magnesium calcite or aragonite (Figure 5.13). This final sophistication of calcium carbonate shells, with the aid of calcium and carbonate uptake in vesicles, is seen in coccolithophores. Here large sections of pure calcite shell are formed completely in vesicles before they are exported to sites on the outside of the single cells. It is advantageous because it allows building of a three-

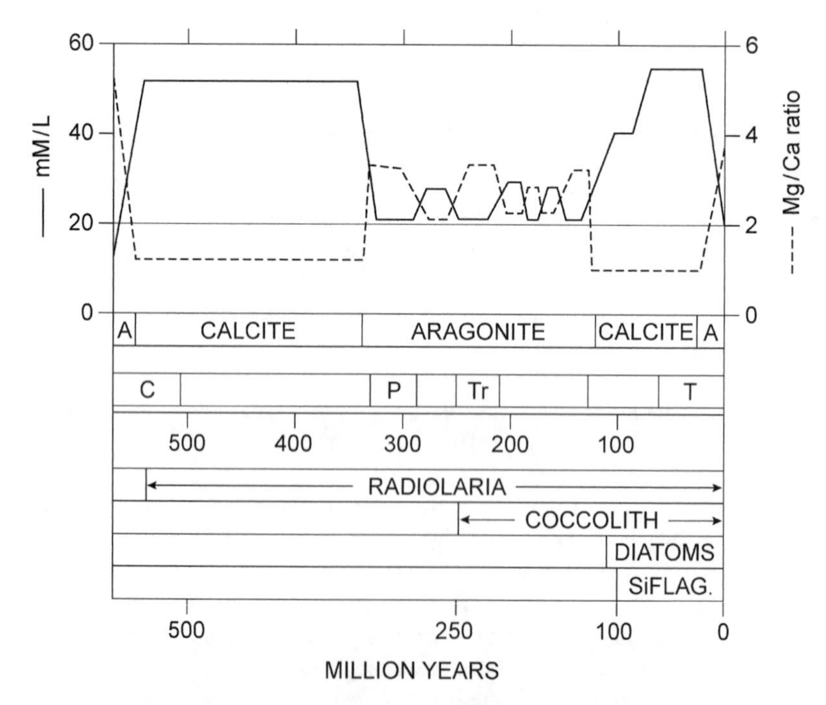

Figure 5.13 The fluctuations of calcium concentration and of the Mg/Ca ratio in the sea with time, millions of years ago and the corresponding switches between calcite and aragonite precipitation in the geological periods, C to T, showing also the evolution of four groups of single-cell euykaryotes with shells. SiFLAG is silica of flagellae.

dimensional shell enlarging it by adding pieces. The formation of the crystallites in the vesicles is guided by a special highly oxidised polysaccharide (see Section 4.2). The whole process of mineral formation must be under gene control here. This type of biomineralisation is also found in the formation of silica shells of single-cell eukaryotes.[12] The link between the controlled formation of biominerals following the rise in oxygen around 0.75 Ga is discussed later.

5.10 Calcium and Phosphates: Apatite

Here we repeat a part of Chapter 3 to stress the importance of calcium in various aspects of living forms. A problem with the biomineral carbonate is that once calcite is formed externally with a given shape it is very difficult to manipulate it with growth. Growth is then related to a single theme with expansion of a shape. Expansion around the rim of an opening is one common possibility, forming the shape of a cone. In Section 3.6 we described the formation of other shapes which, from their fixed beginning in shells, were related to a cone. Calcium hydroxyphosphate, apatite of bone is very different.[28–30] It has the great advantage of modifiable shape in three dimensions within an organism during growth. It is found in vertebrates which have an internal jointed skeleton. The vertebrates are multicellular organisms which evolved some 400 million years ago, that is somewhat later than the invertebrates. Controlled growth of their skeletons is made possible through the piezoelectric properties of bone.[30] Hydrogen ions can migrate in the structure so as to change surface charge where the structure is under stress. In certain regions the bone surface charge increases as the hydrogen ion migrates and this encourages bone dissolution while on the oppositely charged part of the structure the bone grows. This remarkable property makes bone into a living structural component of parts of an organism differently shaped during growth unlike the biominerals of invertebrates. External stresses in growth cause to some degree the exact shape of specific animals. The bone grows in the enclosed extracellular liquid and is linked to the homeostasis of calcium, hydrogen (H^+) and phosphate ions in the extracellular fluids. As is well known the calcium and phosphate ions are near the precipitation solubility limits in the extracellular fluids of vertebrates. There is then a required further connection of the mineral to uptake of the two component ions and the pH of the fluids. There are cells, osteoblasts, which attach themselves to bone and are actively engaged in controlling its shape by adjusting local pH. Special proteins in the vertebrate fluids are involved in the binding of calcium, many of which are phosphorylated and readily generate catalysed nucleation. The uptake of calcium and bone growth, like that of other living organs, is controlled by a hormone, vitamin D, the synthesis of which requires oxidation of the basic sterol structure. The hormone acts via zinc finger transcription of DNA expression. The synthesis of this hormone is quite curiously dependent on sunlight too so that here again the environment with genes, not genes alone, controls a feature of organisms. The evolution of the required osteoproteins is found in many species, probably first in fish. Proteins common to all the calcium minerals and bone are the collagens, described in Section 4.17, and which are dependent on copper enzymes for crosslinking synthesis and zinc proteases for degradation necessary for growth. The complexity of bone synthesis is then strongly linked to inorganic element availability and genes. It does contain considerable quantities of adventitious metal ions and there are variations in the fluoride content. Note that certain shrimps actually make calcium fluoride and sharks' teeth are heavily fluoridated.

Great advantages accrue from this adjustable structure with rigid sections relative to external shells. Although it provides reduced external protection (but note the strong skull protecting the brain), it permits valuable, jointed structures which have a variety of flexibility in motions of the animal's different parts. In this flexibility there is some parallel with certain internal silica structures in plants. Notice that the 'mineral' formed has a shape not related to a symmetry but to a use.

Before going to the next topic we must stress again the remarkable evolution in the handling and uses of calcium linked to the presence of this ion in the environment and to oxidation. This is quite separate from the origin of its gradient at the beginning of cells. It is seen in signalling, adhesion and biominerals. The whole apparatus for handling calcium seems to rise almost simultaneously in many diverse eukaryotes over two relatively short periods of time, around 2.0 to 1.5 Ga, and before 0.5 Ga, respectively. We shall show that a requirement for the development related to its functions is the presence of novel proteins. Its increased value around 0.5 Ga followed from the advantageous uses of copper to make, and zinc to break extracellular structures externally, which evolved with the second rise of oxygen (Chapter 6). The earlier rise in value associated with the first single-cell eukaryotes and the first rise of oxygen connects mainly to the changed mobility of the cell membrane due to oxidative synthesis of cholesterol and surface proteins.

5.11 Silica

This section too contains some material from Chapter 3 but we wish to give greater weight to the chemistry of silica here.[25,30,31] Silica precipitation is a simpler problem than that of calcium salts because silicon is not part of extensive metabolic networks. Even so it has to be precipitated locally in isolated spaces and can form either an external shell or an internal structure. It, like calcium, requires uptake channels and the precipitation requires nucleation but as the silica is amorphous the precipitation is easily induced so long as the free $Si(OH)_4$ is in high enough concentration. It would appear that in many cases silica is formed in vesicles and, once precipitated, transported as small spheroids (Figure 5.14)[32] to specialised spaces outside the cells, that is internally in extracellular space in multicellular organisms, so as to form a patterned network, *e.g.* in diatoms or sponges and inside of a framework of grasses, respectively. Most broad-leaved plants and trees do not have silica deposits but they are found in the hairs of the stinging nettle. The structure of the plants is of varying degrees of mobility, dependent on whether the microspheres are packed more or less densely. Internal packing allows considerable mobility in plants but where external packing is dense, it gives a rigid shell or edge of extreme sophistication, for example diatoms (Chapter 3). Note that no pH control is required because at pH = 7 the silicon in solution is $Si(OH)_4$ which by condensation would give eventually SiO_2 though the actual product is $SiO_n(OH)_{4\text{-}2n}$. It is possible however that pH may control the surface condition of the opal material, $–Si–OH \rightarrow –Si–O^- + H^+$ and that these anionic species may be involved in nucleation.[11,32] Note that calcium minerals are thought to be nucleated by anionic groups. We know that there are protein pumps for silicon uptake and an enzyme for the hydration of external silica to $SiO_n(OH)_{4\text{-}2n}$. The enzyme is very like carbonic anhydrase indicating its probable evolutionary origin (see Section 6.8), and clearly its formation is then related to that of calcium carbonate. Both biominerals evolved at close to the same time, 0.75 to 0.54 Ga.

Figure 5.14 Electron microscopy of some silica in diatoms.

Finally note that silica seems to have evolved first as a biomineral in multicellular sponges.[31] Curiously it developed in unicellular eukaryotes only later and in photosynthesising organisms later than in animals. Bursts of its formation at different times in many groups of organisms, plants and animals, are not easily explained by mutagenesis but may be incidents related to changes in the sea following periods of extinction due to natural disasters, when an explanation relating to the environment (with gene changes) may be more direct that any chance searching.

5.12 The Nature of the Matrices Supporting Mineralisation: Summary

Three points guided us to conclusions about mineralisation (see Section 4.9). We stress them again as they make such an important change in evolution, they undoubtedly date very clearly both more oxidising conditions and the appearance of multicellular organisms as seen in fossils.[12,25] This period is one of a very great development in the chemistry in organisms while little cytoplasmic chemistry has changed.

1. The calcium and silica controlled biominerals appeared at first at the same periods of evolution very close to the time of the appearance of multicellular organisms and perhaps first closely around or soon after 0.75 Ga. This extensive mineralisation occurred close to the second rise of atmospheric oxygen.
2. Some examples of controlled mineralisation are also found for single-cell eukaryotes at almost the same time, shortly before 0.54 Ga (Chapter 3). Evidence for single-cell controlled biomineralisation is now firm from around 0.60 Ga.[12,32]
3. Those minerals with controlled form require nucleation and growth control and we know that both are managed by organic molecule surfaces which must have special properties and arise at a given time of development.[31]

The control of multicellular growth is related in time to development of external organic messenger systems used in signalling and is also closely related to that of synaptic activity in nerves (see Section 5.7). The clear factor in common is the need for a novel extracellular matrix. How did it arise?

The general impression of the new proteins of the eukaryotes and especially those of the multicellular eukaryotes which bind calcium, including those required for external matrices and binding systems, as we described in Section 4.17, is that they use chemistry much of which is due to oxidation, and often glycosylation, in special compartments, requiring special trace elements (see Section 6.8). Glycosylation is often a basic part of the structures for silica minerals too. Such protein sites act as a crosslink as well as nucleating sites. They are usually moderately mobile so that they can deform somewhat to allow crystals to grow in the spaces around them. The common properties of these external proteins and the receptor calcium-binding internal proteins in the cytoplasm are that they all contain a large number of charged side chains which can bind to calcium through oxygen centres such as $-COO^-$, with $-OH$ and $C{=}O$. While the origin of the external proteins is clear we do not know how the internal proteins arose with eukaryote evolution.

5.13 Conclusions

We commented in the early sections of this chapter on the essential nature of the chemistry of phosphorus and sulfur as partners for organic elements in compounds because they are required to increase the kinetics of transfer of material and energy and in the case of sulfur to maintain a steady-state redox potential in the cytoplasm. Essential in this context implies an inevitable use even before life began. Sulfur became very valuable also in stabilising crosslinking extracellular matrices and, later, in sulfate esters, forming open-mesh structures. Any requirement here is in the evolution of structure. Quite similarly we have to recognise that certain bulk inorganic metal element ions are essential for different aspects of all life. For example the stability of cells depends on the control of osmotic pressure and electrolytic charge. Thus, given the nature of the sea in which life began, control over the high concentrations of Na^+, K^+ and Cl^- was required before any reproductive life could begin. Here binding of the ions is not of great importance except in membranes. It is this compulsory rejection from the earliest cells which generated the possibility of its future functional values seen in nerves and the brain messages. Both Na^+ and Cl^- were largely rejected. The expulsion of Na^+ ions, giving a gradient across the membrane, became used as an exchanger for input of required material. The sea has also always been high in Mg^{2+} and Ca^{2+} ions which bind quite well at equilibrium to organic molecules and which they then either crosslink or activate mildly. Consequently control over Mg^{2+} and Ca^{2+} inside cells was necessary again to protect the organic molecules from damage (precipitation) and from too much association before reproduction. The Mg^{2+} ions are the less effective in causing crosslinking and the more effective of the two in affecting rates of change, catalysis, and it has a controlled, only somewhat lowered, concentration in cells relative to the sea. Its concentration was and remains such as to allow it to be employed essentially in part as a mild catalyst while causing crosslinking in stabilising RNA and DNA folds, and in part as a communicating control between cooperative pathways in the cytoplasm. In fact Mg^{2+} had a major role, from the origin of life with nucleotides, in energy transfer especially for synthesis, directly in some enzymes and generally as a mobile control between many activities, homeostasis. It is the only cation which can be in cells to bind and activate pyrophosphates and in chlorophyll to capture light energy. Ca^{2+} ions are more dangerous as they more readily bring about association of molecules, even precipitation, and are then much less useful in catalysis in cells. They are reduced to a much lower level inside all cells to avoid these problems. The calcium ions remained of value outside cells and we have seen that Ca^{2+}, together with magnesium ions, became responsible for stabilising walls and membranes and outer structures of prokaryotes. There is a curious function of Ca^{2+} in the active site in the water-splitting enzyme which releases oxygen with magnesium.[34] The function of the two ions, Mg^{2+} and Ca^{2+}, was then a requirement of the very first cells in very different functions. The value of calcium, not so much magnesium, changed greatly during evolution with the coming of oxygen leading to cells of longer lifetime and greater flexibility, the eukaryotes. Their longer lifetime required the ability to recognise their environment, knowledge of which came from use of the steep Ca^{2+} gradient, outside to inside, of cells. External events were relayed by fast impulses of Ca^{2+} to the inside of the cell.[16] This gradient was formed initially in the earliest cells because cellular organic chemistry is not possible in its presence. On entering the cell, Ca^{2+} bound at equilibrium to many receptor centres, so modifying cooperative cell activity. It was the only ion or molecule which could function in this way and this use of its gradient evolved as an inevitable required component of eukaryote evolution. Note that equilibrium applies equally in different compartments due to energisation of ion concentrations and organic synthesis and trafficking of selected proteins with calcium binding centres. Later in multicellular eukaryotes the calcium messenger system became linked to organic molecule long-range messengers between cells in multicellular organisms. At the time the oxidation of the outside surfaces of cells gave rise to controlled precipitation and growth of calcium biominerals. We also noted that together with the creation of biominerals from calcium salts, others made from silica appeared.[12,20,27] Later the activity of the calcium messenger system was extended to synaptic transmission using oxidised small organic molecules when nerve tissue evolved leading to the development of the brain. The message transmission along the nerve itself turned the Na^+ and K^+ gradients into functional value in nerves which only evolved once oxidation of extracellular tissue gave rise to advanced animals. In many cells the Ca^{2+} ion also acts in transmission of synapses. It is clear that some essential functions of these bulk ions were present from the beginning of life but that others evolved with the stages of oxidation in the environment. All of the functionings of five cations, Na^+, K^+, Cl^-, Mg^{2+} and Ca^{2+} are dependent in large part on equilibria, binding with fast exchange in individual compartments or in membranes. No other elements can replace them. The equilibration is important for response to the environment unlike the slower exchange of P and S compounds so uniquely useful in control of internal metabolic cell activity and their redox potential, and the non-equilibrate exchange of energy between non-metal 'organic' elements H, C, N, O in their compounds (Chapter 4). The properties of these ions therefore generate kinetic coordination of activities. It may well be true that only by the deployment of the seven elements of this chapter in the observed ways by single-cell and then multicellular organisms was it possible for systems of sufficient kinetic survival strength to arise and develop cooperative, coordinated chemistry both at the initial stages of life's origins and through all evolution. But we must note that these ions and elements except sulfur are only capable of acid/base functions and cell chemistry required additional redox functions reliant on transition metal ions (Chapter 6). To see why their functions changed too we shall show in Chapter 7 how their biological chemistry fits with the organic chemistry of Chapter 4 and the catalyst chemistry of Chapter 6, into what may well be the best of all, perhaps the only, total chemical system of great kinetic survival strength able to exist. The whole system of elements, inorganic and organic, in compounds in the interactive environment/organism combination, is the essence of evolution. In the next two chapters we indicate how the realisation of the possibilities presented by the sodium and calcium gradients came about through the development of oxidative chemistry. Quite outstandingly the changing organisms' chemistry in numerous ways changed cooperativity. All its controls are major features in the inevitability of the fast reactions of inorganic ions toward quantitatively understood equilibria which are governed by the constants, solubility products, stability constants and redox potentials (Chapter 1). The system environment/organisms is driven by energised use of chemicals in cells with the slow oxidation of the environment due to oxygen release. The exact kinds and numbers of so-called species (see Chapter 7), may well be due to chance exploitation of environmental change but the sum of the species changes in chemical terms is directional and inevitable. This will be apparent again in the next chapter on the part played by the trace inorganic elements in relatively fast evolution between, say, 2.5 and 2.0 Ga and again between 0.60 and 0.40 Ga.

References

1. D. Voet and G. J Voet, *Biochemistry*, 3rd edn, Wiley, New York, 2004.
2. L. Stryer, J. Berg and J. Tymoczko, *Biochemistry*, 5th edn, W. H. Freeman, New York, 2002.
3. R. J. P. Williams and J. J. R. Fraústo da Silva, *The Chemistry of Evolution*, Elsevier, Amsterdam, 2006.
4. M. J. Berridge, *Annu. Rev. Physiol.*, 2005, **67**, 1.
5. A. Lepland, G. Arrhenius and D. Cornell, *Precambrian Res.*, 2002, **118**, 221.
6. C. de Duve, *Life Evolving: Molecules, Mind and Meaning*, Oxford University Press, New York, 2002.
7. J. R. Wilcock, C. C. Perry, R. J. P. Williams and R. F. C. Mantoura, *Proc. R. Soc. B*, 1988, **233**, 393.
8. J. R. Wilcock, C. C. Perry, R. J. P. Williams and A. J. Brook, *Proc. R. Soc. B*, 1989, **238**, 203.
9. J. A. Cowan, *Introduction to the Biological Chemistry of Magnesium*, VCH, New York, 1995.
10. R. J. P. Williams, *Biochim. Biophys. Acta*, 2006, **1763**, 1139.
11. C. C. Perry, see Silica in ref. 12, 2003, chapter 10, 291.
12. P. M. Dove, J. J. De Yoreo and S. Weiner (eds) *Reviews in Mineralogy and Geochemistry, Volume 54, Biomineralization*, Mineralogical Society of America and the Geochemical Society, Washington, DC, 2003.

13. B. Gomperts, I. Kramer and P. Tatham, *Signal Transduction*, 2nd edn, Academic Press, New York, 2009.
14. J. J. Fehrer, *Am. J. Physiol.*, 1983, **244**, 368.
15. R. Morgan, S. Martin-Almedina, M. I. Gonzalez and M. P. Fernandez, *Biochem. Biophys. Acta*, 2002, **1742**, 133.
16. E. Carafoli and C. Klee (eds), *Calcium as a Cell Regulator*, Oxford University Press, New York, 2008.
17. F. M. Harold, *The Way of the Cell*, Oxford University Press, Oxford, 2001.
18. H. Coe and M. Michelak, *Gen. Physiol. Biophys.*, 2009, **28**, F96.
19. S. G. N. Grant, *The Biochemist*, 2010, 32, April.
20. T. Shinoda, H. Ogawa, F. Cornelius and C. Toyoshima, *Nature*, 2008, **459**, 446.
21. Y. Jiang, A. Lee, J. Chen, V. Ruta, M. Cadane, B. T. Chaity and R. MacKinnon, *Nature*, 2003, **423**, 33.
22. E. R. Kandel, *Science*, 2001, **294**, 1030.
23. A. Sigal, H. Sigal and G. M. Shepherd, *The Synaptic Organisation of the Brain*, 4th edn, Oxford University Press, New York, 2004.
24. A. H. Knoll in P. M. Dove, J. J. De Yoreo and S. Weiner (eds) *Reviews in Mineralogy and Geochemistry*, *Volume 54*, *Biomineralization*, Mineralogical Society of America and the Geochemical Society, Washington, DC, 2003, p. 329.
25. A. Sigel, H. Sigel and R. K. Sigel (eds), *Biomineralization: From Nature to Application*, Wiley, Chichester, 2006.
26. J. Erez in P. M. Dove, J. J. De Yoreo and S. Weiner (eds) *Reviews in Mineralogy and Geochemistry*, *Volume 54*, *Biomineralization*, Mineralogical Society of America and the Geochemical Society, Washington, DC, 2003, p. 115.
27. T. K. Lowenstein, M. N. Timofeeff, S. T. Brennan, L. A. Hardie, and R. V. Denice, *Oscillations in Phenerozoic Seawater Chemistry: Evidence from Fluid Inclusions*, 2001.
28. K. Hoehn and E. N. Marieb, *Human Anatomy and Physiology*, 7th edn, Benjamin Cummings, San Francisco, 2007.
29. J. D. Currey, *Bones – Structure and Mechanics*, Princeton University Press, Princeton, 2002.
30. A. K. Davis and M. Hildebrand in A. Sigel, H. Sigel and R. K. Sigel (eds), *Biomineralization: From Nature to Application*, Wiley, Chichester, 2006, p. 255.
31. K. E. Richmond and M. Sussman, *Curr. Opin. Plant Biol.*, 2003, **6**, 268.
32. H. Hutter, B. E. Vogel, J. D. Plenofisch, C. R. Norris, R. B. Proenea, J. Spieth, C. Guo, S. Mastural, X. Zha, J. Schole and E. M. Hedgecock, *Science*, 2000, **287**, 989.

CHAPTER 6

*Trace Elements in the Evolution of Organisms and the Ecosystem**

6.1 Introduction

In previous chapters we have shown the reasons for the specific value and uses of the major elements H, C, N, O in forming kinetically stable energised organic molecules (Chapter 4). Associated with these molecules in cells are the kinetically more labile S- and P-linked groups needed for many, relatively slow, exchanges of molecular fragments of C/H/N/O and, in the case of S, redox homeostasis (Chapter 5). In Chapter 5 we showed also that it was necessary to manage the cellular chemistry of the fast exchanging common elements, Na, K, Cl, Mg and Ca. They perform essential functions in stabilising the cell while being maintained in energised gradients, acting in fast exchange in some weak catalytic functions and as messengers of external events. We also discussed the roles of Ca and Si in minerals. The description of the functional significances of all these bulk elements was examined in the analysis of the nature of the presumed earliest cells and then in the whole of organism evolution. The changes in cell chemistry were shown to be closely linked to the changes in the environment to oxidising conditions. There is missing from this account the absolute requirement for extra elements in well-controlled concentrations in order to catalyse the difficult reactions of organic molecules in the selected paths which we observe in both the inside and outside of all cells, as mentioned in Chapter 4. They also act as controls over major

* The reader unfamiliar with inorganic chemistry may well wish to read Sections 6.1, 6.3, 6.6 and 6.13 to follow the extent of the evolution of trace elements in cells and in the environment in Chapter 2.

Evolution's Destiny: Co-evolving Chemistry of the Environment and Life
R. J. P. Williams and R. E. M. Rickaby

Published by the Royal Society of Chemistry, www.rsc.org

activities by exchange internal to cells. In this chapter we shall describe the role of these, the trace elements, in organic chemical pathways, stressing our belief[1,2] that several trace elements, some available to the earliest life and others increasing in presence at different times, were critical for both the origin of life and its evolution.[3–5] While those elements available in the original biological chemistry catalysed many essential reductive chemical steps the consequences of this chemistry were that the rejected oxidised waste, O_2, from these reactions changed the environment. The resulting combination of the oxidation of the non-metals, H, O, N, C and S and of the metals in the environment was the material driving force of cellular evolution. Evolution depends upon antecedent organisms through modified reproduction as a mechanism but upon working within the limitations and changes in the environment as a cause. Here the environment is not only the sea but also any non-living source from which the community of organisms can feed and obtain the essential trace elements, *e.g.* soils which evolved by weathering. Virtually all life was in the sea from 3.5 to about 0.5 Ga, by which time chemical evolution was virtually complete and it is therefore the contents of the sea which are our major concern. Now it is not essential for all organisms to have exactly the same trace element complement, although these elements generate necessary products for all cells, so long as some cells provide others with the necessities they fail to make. Immediately note that animals cannot obtain C/H compounds from CO_2 but plants and algae do and they have very different complements of trace elements from animals, for example plants need more manganese. The community of organisms then works cooperatively using the environment optimally. This is not to say that there is no competition in the whole community, because we know that competition, especially between very similar species, is fierce and organisms feed on one another. There is no real competition, however, between organisms which differ in trace element requirements as they depend on one another, giving rise to symbiosis between the different chemotypes.[1,2]

There are several textbooks on the value of the trace elements in different organisms.[1,6–8] They stress in a detailed way that without these trace elements there could be no life because even the basic units from the original environment needed for life, H_2, $-CH_2-$, CO, CO_2, N_2 and H_2O, cannot be transformed via organic chemical intermediates to the necessary nucleotides, proteins, saccharides and lipids in the cytoplasm of cells, without them (Table 6.1). The metal ions are frequently in special coordination states, called entatic states, which protein binding has conferred on them (Section 1.2). The power of some of these trace elements is augmented by their ability to change oxidation state (Figure 6.1), and this is a major factor in evolution. The total content, of bulk and trace metal elements, of each organism is referred to as its total metallome[1,2] (Figure 6.2), and major changes in the total metallome reflect major evolutionary advances of groups of organisms, that is of chemotypes. Once we have described the availability and the principles and practices of uptake and internal binding in cells, Part 6A, we shall turn for their biological

Table 6.1 Selective Enzymic Catalysis of Small Units[2–5]

Metal Ion	*Selective Reaction*
V*	Halide activation
Mn	O_2 production
	(Glycosylation)
Fe*	e transfer
	OH transfer
	O_2 transfer
	H_2/CH_4 activation
Co	H_2 and CO activation
Cu	C-H oxidation e.g. CH_4
	Aromatic radical production
	Dinitrification
Zn	Peptide hydrolysis
	CO_2 hydration
Mo*	O-atom transfer
	N_2-Fixation (+ Fe)
Se	H^--transfer
	O-atom transfer
W	Reduction reactions (very early)

Notes* All three metal ions, Mo, V, Fe can activate N_2/H_2 reaction. At least six of these elements are essential in all organisms

significance to Part B. There we analyse the changes of the metalloprotein content in prokaryotes and in eukaryotes separately, noting also the linkage to the changes of available non-metal compounds. In Section 6.9 it will be apparent that organisms have eventually taken full advantage of the different properties of the whole range of the elements of the Periodic Table available at a certain time (see Figure 1.1). At least one element from each group of the table

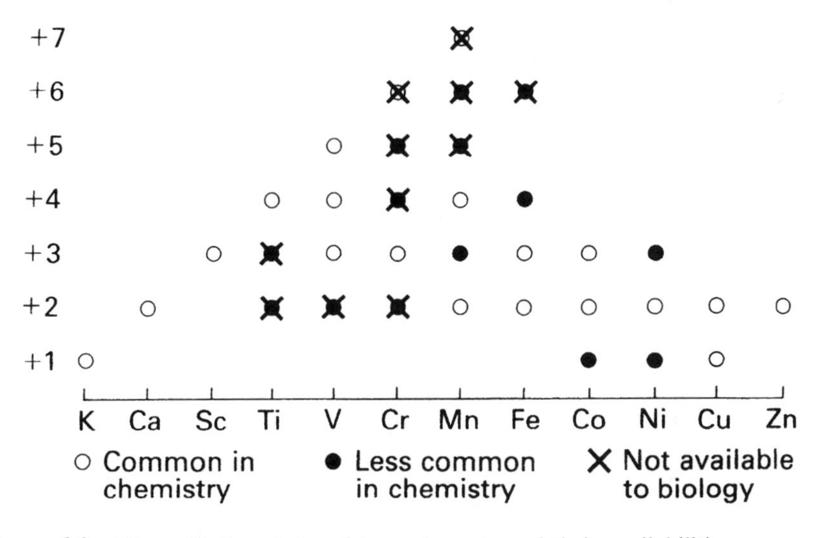

Figure 6.1 The oxidation states of trace elements and their availabilities.

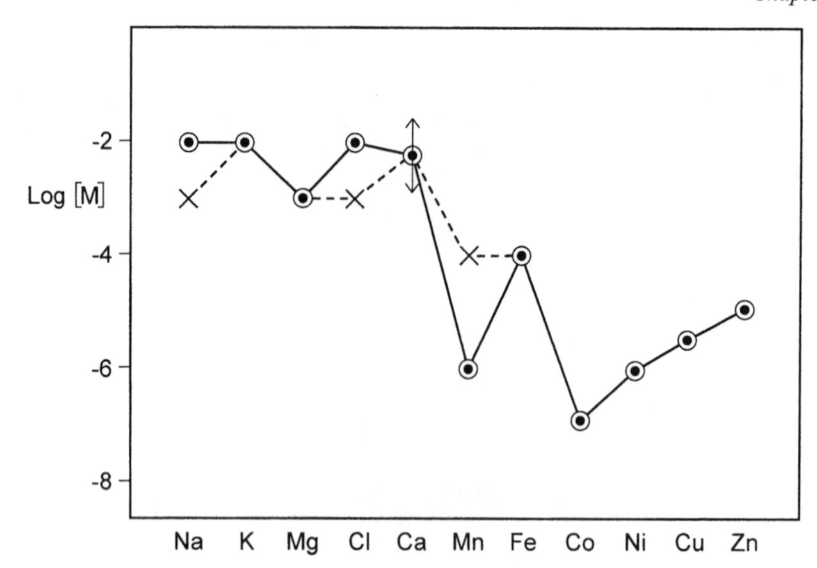

Figure 6.2 The average total concentration of free and bound metal ions in animals, o, and in plants, x. Other groups of species have different metallomes.

is used in organisms today and it will be seen to be the best, within availability, for its observed role, which for trace elements is mainly in catalysis, broadly classified in Table 6.2. The ability to substitute one metal ion for another is quite difficult for acid-base catalysts but extremely hard for redox catalysts as their redox potentials are very specific (Table 6.3; Figure 1.6). Substitution is likely only if a preferred metal ion becomes rate-limiting due to its low environmental concentration (see Sections 6.10 and 6.11). Such substitutions are also very rare for bulk elements. In Section 6.12 we look at the chemical

Table 6.2 New element biochemistry after the advent of dioxygen

Element	*Biochemistry*
Copper	Most oxidases outside higher cells, connective tissue finalization, production of some hormones, dioxygen carrier, N/O metabolism
Molybdenum	Two electron reaction outside cells, NO_3^-, SO_4^{2-}, aldehyde metabolism
Manganese	Higher oxidation state reactions in vesicles, organelles and outside cells, lignin oxidation (note especially plants); O_2 production
Nickel	Vitrually disappears from higher organisms
Vanadium	New haloperoxidases outside cells
Calcium	Calmodulin systems, signalling; general value outside cells
Zinc	Zinc fingers connect to hormones produced by oxidative metabolism
Selenium	Detoxification from peroxides, de-iodination?
Halogens	New carbon-halogen chemistry, poisons, hormones
Iron	Vast range of especially membrane bound and novel oxidases; peroxidises for the production of hydroxylated and halogenated secondary metabolites; dioxygen carrier and store

Table 6.3 The Classified Catalytic Value of Different Trace Metal Ions*

Acid/base Reactions	*Reductive Reactions*	*Oxidative Reactions*
Mg^{2+}, Zn^{2+}, Fe^{3+}	Ni^{2+}, Fe^{2+}, Co^{2+}	Cu^{2+}, Fe^{3+}, Mo^{6+}
Ca^{2+}, Mn^{2+}, (H^+)	(RSe^-)	V^{5+}, Mn^{4+}, (R_2Se)

value of certain non-metal trace elements. We consider that each element is used optimally. Section 6.13 draws together all this chemistry in a survey.

Now the long-term inevitable change from the initial reducing organic chemistry in cells in a reducing external medium was to that of reducing cells in an oxidising external medium, brought about by the release of oxygen increasing over time, as described earlier. We showed the effects of this unavoidable increase on the evolution of organic chemistry in Chapter 4. The oxygen had little influence on the chemistry of the bulk elements except for sulfur (Chapter 5). In this chapter we show how the incremental increase in oxidation of inorganic minerals and elements in solution in the sea brought about changes of inorganic chemistry in addition to that of the organic chemistry in cells. Evolution could only be brought about if the inorganic elements acting as catalysts, as they became available, were employed not only for oxidation, linked to organic chemical change, but also as novel controls. They then had to be used separately from the primitive reductive chemistry.

The essential trace elements to which we refer[2–5] are mainly the transition metals V, Mn, Fe, Co, Ni, Cu and Mo (W) to which we add Zn, with Se and the halogens, the last group being non-metals (see Figure 1.1). Notice that they cover nearly all chemical groups of the first row in the Periodic Table which includes the transition metal elements except those of Ga, Ge, As, none of which is essential. Some of the other elements have always been essential in all cells, Fe and Mn and small quantities of Zn for example; others have been most extensively used in the early reductive organisms, W, Co and Ni. Two, Cu and Zn, became employed mostly in later organisms, Cu especially in oxidising extracytoplasmic fluids, became required relatively recently. Many trace elements found in minerals have no biological function but several are of value in research into evolution of the environment, because their oxidation states in sediments changed following conditions throughout geological times. We have mentioned for example the relative concentration of certain REEs as a guide to oxidation conditions (see Table 2.1), while trace elements such as Pb and U and others and their isotopes are of great value in dating minerals generally. In this chapter we are only concerned with the value of those trace elements active in cells as their chemical functions are introduced or modified on going from prokaryotes to the more and more advanced cells of eukaryotes. (We do not refer to the use of further trace elements in medicine by man[9] in this book.)

Before we begin an examination of the essential value of trace elements in organisms we need to describe the general chemistry of these elements in the environment and their uptake into cells. Section 6.2 will outline the problems of their changing but restricted availability, which immediately sets them apart from the readily available bulk non-metals of Chapter 4 and the bulk metal ions of Chapter 5. Due to their mineral chemistry the trace elements have always been in low concentrations in the sea and the concentrations vary with the oxidising power of the environment (Chapters 1 and 2). Thus uptake and rejection of trace metals are dependent on selective energised pumps which also drive their distribution into compartments (compare calcium distribution). The condition in which they are held is of great importance because some functions of their ions need them to be labile, in fast exchange as for Mg^{2+}, or fixed in compounds (compare non-metals), or in selective exchange (Section 6.3). The extreme exchange difference is exemplified in man's chemistry in that between aqueous complex ion chemistry, fast exchange, and organometallic chemistry in all solvents, no exchange. The condition of limited exchange of metal ions is very uncommon in the environment. Further complications arise from the redox conditions, which affect the oxidation state of the trace metal ion, and pH, which affects binding in a compartment. Rates of change of redox states are usually rapid. Most of the properties of functional interest are dependent upon equilibria, quantitatively analysable by stability (binding) constants, solubility products and standard redox potentials (Chapter 1), as we show in Section 6.4. We then give some examples of how increases in availability have affected systems in some cases while in others reduced availability has precluded evolution of one kind of organism dependent on a particular element (Section 6.5). Once the chemistry of these trace elements, including their availability and binding and exchange properties in cells has been described, we can turn in Part B to their extreme importance in participation and functions in both the earliest forms of life and during evolution. This second part of the chapter will be divided into sections on the trace element contents and functions of the cells at differently evolved stages of life. Function is dependent upon the metalloprotein produced in primitive and later organisms, their proteomes and metallomes (Sections 6.6 and 6.7, respectively), which we analyse before we attempt an evolutionary survey (Section 6.13). Section 6.12 covers the essential trace non-metals, such as selenium.

Part A. The Chemistry of the Trace Elements

6.2 The Availability of the Trace Elements

Availability at a given time can be assessed from data from non-biological sources as described in Chapter 2, but it is useful to confirm it from the ability of organisms to take in and use elements as seen in organisms today (see Figure 6.2) and known to live in environments considered to be like those of a given earlier period.[1–5,10] Most of the section will be a reminder of the contents of Chapter 2. All uses of trace elements have been affected by availability but we take it that only Mg^{2+} and Fe^{2+}, with possibly some Ni^{2+}, Co^{2+} and Mn^{2+} were in sufficiently high concentration in certain locations, to be readily taken up initially in reducing conditions from 3.5 to 3.0 Ga (Chapter 2). Iron is a special case as it is very abundant in the universe and in reducing conditions when it was around 10^{-5} to 10^{-6} M as free Fe^{2+} in the sea. In these conditions it was in amount between a bulk element, 10^{-3} M (Chapter 5), and other true trace elements, 10^{-8} M. At this early period simple cells had much greater contents of iron than any other trace element. We have seen in Chapter 2 that Fe^{2+} iron together with sulfur were barely changed over a period of more than a billion years, 2.5 to 1.0 Ga. After this period Fe^{2+} became insoluble Fe^{3+} and sulfide became soluble sulfate. By about 1.0 Ga, the required Fe^{2+} iron became present in the sea at less than 10^{-9} M. Some Zn^{2+} was obtainable in the early reducing conditions but apparently with considerable difficulty and Cu hardly at all. Many of the limitations were due to the insolubility of compounds with sulfur in a reduced state (see Figure 2.13). As described in Chapter 2 it is generally accepted that the availability of trace transition metal elements was then greatly altered by the increase in oxygen as many, trapped initially in sulfides, were oxidised but, as stated, Fe^{2+} was oxidised to Fe^{3+} and it was largely precipitated and reduced in availability. Oxygen in the atmosphere rose generally from some extremely low level at 3.5 Ga largely in two steps, around 2.5 to 2.0 Ga and close to 0.75 to 0.40 Ga, with buffering between these two dates due, as explained, to the oxidation in the sea of the relatively large amounts of ferrous iron and sulfide (Section 2.8), and then with some smaller variations to today. There could have been only slow individual changes of the availability of the trace elements during the buffered times, that is, say, mostly until 0.75 to 0.54 Ga, when the changes become more rapid, as seen in sediments (see Table 2.11). The elements more easily removed from sulfides could have been considerably increased during and just after the first oxygen step while the most difficult to remove would only have increased considerably in availability approaching and in the second step. Inorganic redox equilibria indicate that molybdenum would have appeared relatively readily while copper and cadmium would have been most difficult to bring into solution (see Figure 1.6). Cobalt, nickel and then zinc, most like copper, had a more intermediate release pattern. Any availability would follow the general pattern of a redox curve, selective in mid-point potential for each sulfide, with oxygen increase (see Figures 1.5 and 2.2). However we must remember that we use 'buffering' in the period 2.0 to 0.75 Ga to describe the environment as if it were homogeneous. Before we do so we draw attention again to the close parallels between environmental change and organism evolution, outlined in Chapter 4.

Before proceeding further we need to consider again this non-uniform character of and hence availability in the sea over long periods (Section 2.9). The content of bulk water-soluble elements such as Na, K, Cl, Mg, Ca may not have varied greatly throughout all of evolution (Chapter 5), and may well have been only somewhat affected by life. There would then have been little variety in their concentrations from the top to the bottom of the sea. However, particularly before there were large numbers of organisms, that is before 0.75 Ga and especially in the period 2.0 to 0.75 Ga, there was considerable nonuniformity of other elements, mainly the trace metals. A very important point is that the organisms at the surface, often related to algae (plants) as they require light responsible for oxygen increase, would have been the first to feel the oxidative changes of the trace elements including the loss of iron. The gain of Cu and Zn was supplied in fair part from the surface of the land by oxidative weathering. We must not expect therefore that evolution of chemistry in organisms would have been the same over these 1 billion years, that is simultaneously equivalent at the top to especially plant-like and to the bottom of the sea to metazoan-like animals. The estimated time of ocean mixing from about 0.75 Ga to today has decreased greatly to around 2,000 years, which is fast on the evolutionary scale of our concern. Difficulty in describing the exact environment of organisms is increased when it is recognised that much of the original life evolved at the bottom of shallow water or in sediments with little exposure to light and no oxygen and has remained there. Even today there are quite considerable changes in available elements with depth, especially where there is strong surface organism productivity. An examination of deep static 'anoxic' regions has shown very recently that what may be quite exceptional organisms can live in very unusual environments in the extremes of deep sea.[10–12] In some ways the best we can do is to represent the availability and oxidising changes of the trace elements in the whole sea over time in a single drawn-out redox curve connecting two steeper steps over time, as described in Figure 2.3.

It has been proposed that the progress of cellular evolution should be recognisable both through element deposition in sediments, reflecting their availability (Chapter 2), and also through the introduction of new uses of these trace elements in cells.[1] The assertion is that organism evolution proceeds deterministically as a consequence of the primary environmental redox change, which greatly affected trace element availability, and possibly not just by improving adaptation by chance mutation in the altered environment. We return to these analyses in Chapter 7.

Finishing this section on natural changes of availability we stress that the equilibrium thermodynamics of redox chemistry, together with that of binding constants described in the previous section, in the environment due to fast exchange has resulted in giving evolution of organisms a predictable character through changes dependent on element availability. In this respect they differ from the other elements, C/H/N, which, when they became less available remained in conditions of virtually no exchange but of course all the changes were due to increased availability of oxygen. Oxygen has fast reactions in the environment but the cytoplasm of cells always has been protected from it.

Finally we draw attention to man's industrial activity in which 'availability' has changed through his application of quite novel chemistry. Perhaps the most striking activity is the use of reduction not oxidation to obtain elements as such, for example elemental aluminium and silicon. We return to the possible problems in the long-term future at the end of Chapter 7 because many of man's products are not compatible with biological chemistry.

6.3 The Principles of Binding and Transfer of Trace Elements in Cells

In the above description of the bulk metal elements in the environment in both Chapters 4 and 5, and of the cellular chemistry in Chapter 5 we were able to assume that as binding was in rapid exchange in all compartments the chemistry in each one could be treated using equilibrium constants. There was in addition the pumping of ions between compartments to give valuable ion gradients. Somewhat differently the chemistry of most of the trace elements can be related to their absolute equilibrium-binding constants to anions and organic molecules in a particular compartment. A novel consideration is due to the increased strength of binding of some of these trace metal ions to proteins and small molecules, which makes exchange slower. All these ions, like the bulk metal ions, are also out of equilibrium between compartments because they are also pumped across membranes, but far from equally so far as different trace metal ions are concerned. However their resultant gradients are rarely of functional value. In this section we give a *qualitative* view of the part of the system in which simple absolute equilibrium considerations can be used and of other parts where they cannot be easily used but can be replaced by equilibrated exchange between bound forms in some, but not all, cases.

Because the bindings of the active trace elements are largely in complexes with proteins we devote most of the section to these complexes.[1–3] The binding strengths are such that in some extreme cases there is virtually no exchange and the number of molecules of a metalloprotein is then controlled by the limited synthesis of the protein and by almost equal numbers of metal ions allowed in the cell. The free metal ion here may have no function and is kept at an extremely low concentration but the bound forms can now be catalysts and exert controls. In other cases of considerably weaker binding the free ion concentration, which is considerably higher, is in rapid exchange with binding to proteins and is used in internal communication and management of cooperative metabolic action through exchanges, much like magnesium in the previous chapter. Some of these free metal ions are also used in contact with genetic material through rapid exchange binding to transcriptive factors, that is they act as general controls over synthesis of certain of the own proteins in all cells.[10–12] The metal ions in the first slow-exchanging group, for example copper, behave like relatively slow exchanging non-metal elements, such as sulfur and phosphorus. While those in the second rapid-exchange group, such as Mn^{2+} and Fe^{2+}, behave like acidic exchangeable protons (see the end of Section 5.3). We must also consider, however, the intermediate case of exchange of strongly bound metal ions directly between proteins, maybe avoiding free ion involvement, for example Zn^{2+}. Finally the properties of some elements change dramatically with their oxidation state and/or spin state, both in binding and in exchange, for example iron.

The problem of controlled concentration is aggravated by the poisonous nature of several of these ions in excess so that there has to be selective mechanisms for homeostatic control of the free ions. The level of some is decided in part by input and exit pumps and buffers, much as we have seen for calcium ions in Chapter 5. The pumps work by binding on one side and then using energy to increase (input from the environment) or to decrease (output to the environment) binding by conformations of the proteins (see Figure 6.3). Equilibrium is closely established in the different bindings on opposite sides of the membrane. We saw in the previous chapter that the calcium ions are pumped *out* of the cytoplasm down to below 10^{-6} M in *all* cells but eukaryotes also pump calcium ions *into* the reticula. Both concentrations are in local equilibrium between free and bound states. The ions Mn^{2+} and Zn^{2+} are also pumped out of cells and into certain reticula to an optimum level and are often found free and weakly bound in vesicles at equilibrium in heightened concentration. The concentration of free Cu ions, largely pumped out of cells, is very low everywhere and requires separate description of its binding, as does zinc in the cytoplasm. Free iron-like magnesium is generally retained in the cytoplasm and rarely enters vesicles but it is pumped into some cells and out in others. The pumps in all cases are ATPases, linked to Mg, where Mg.ATP is equilibrated magnesium adenosine triphosphate and its hydrolysis gives the energy to mobile protein groups in the pumps. Many of these inward and outward pumps have feedback control from those elements with relatively high cytoplasmic free ion concentration and smaller binding constants, *e.g.* Mg^{2+}, Mn^{2+} and Fe^{2+} when sufficiently high cell concentrations are reached and free ion binding to pumps prevents further ion entry. The free ions diffuse rapidly in cells and so equilibrium is established throughout them. Further control of metal ions can also be by the binding of the metal to transcription factors which control the expression of uptake proteins and enzymes which bind the metal ions.

The mechanism of transfer in the cytoplasm of strongly bound ions with little exchange with free ions is not by free diffusion but by carrier proteins. The first recognised was for some cases of calcium flow through cells[13] but subsequently there have been found selective carriers for a number of other ions.[14–17] The carriers are small proteins with considerable binding constants which are present in high enough concentration so that their diffusion is faster than the very low concentration of free ions. They have been called chaperones.[14–17] For these ions it is not the free ion which interacts and exchanges with other proteins of the cell, giving enzyme activity and also acting directly in transcription factors (Figure 6.3), but the carriers or chaperones. It may also be that it is these bound carriers which feed back to the pumps to maintain homeostasis (for an example see the case of molybdenum, Figure 6.3).

The movement of some ions is more complicated in that they are taken into cells as complexes and in the case of iron, for example, the intake today is generally of bound Fe^{3+} (ferric). This scavenging for iron is so important (Figure 6.4), iron is the major trace element, and different from uptake of other elements, that we describe it in Section 6.5. In Chapter 4 the free iron control of oxidative chemistry was shown to parallel that of hydration/

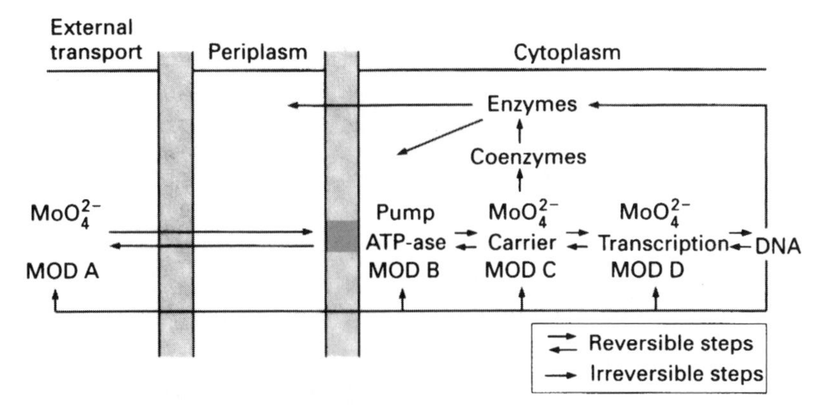

Figure 6.3 The uptake into a cell and incorporation of molybdate as a functional catalyst. Many tightly bound metal ions are involved by similar but very selective schemes. MOD refers transcription factors. Internally there may be fast exchange between proteins, even equilibration, but no effective exchange with free ions.

dehydration (condensation and hydrolysis) chemistry aided by magnesium from the earliest times. It is also the case that Mg has strong binding in porphyrin, chlorophyll.

6.4 The Importance of Quantitative Binding Strengths and Exchange in Cells

The previous section has described qualitatively uptake, binding and exchange of trace metal ions for the general reader. The value of the different metals can only be fully appreciated by a more quantitative approach. (This section is not essential for a general appreciation of trace element binding and exchange. It illustrates that the approach to evolution through quantitative inorganic chemistry is quite different from that through genes. It is a thermodynamic treatment.) Because several of the ions of the trace and other metal elements are often bound in relatively rapid exchange, selective thermodynamic equilibrium constants play a dominant part in the quantitative way they are combined in precipitates or complex ions with organic and inorganic units. This allows a detailed approach to the function of metal ions which it is necessary to understand because it is of great consequence to the non-equilibrium chemistry of non-metals.[1,2,6–8] The constants give a selected metal ion a particular function in a given compartment just as we saw for Mg and Ca in Chapter 5. Binding depends on the binding constant, K, and the concentrations of the ion [M], and its binding group [L] such that $[ML] = K[M][L]$ (Section 1.4) in a compartment. Consider again as an example the selection for Mg^{2+} where cytoplasmic $[Mg^{2+}] = 10^{-3}$ M. The best biological Mg^{2+} binding centres are of an octahedron of O-atom donors in L, give a binding constant of about $K = 10^3\ M^{-1}$. Mg^{2+} then distributes equally to this

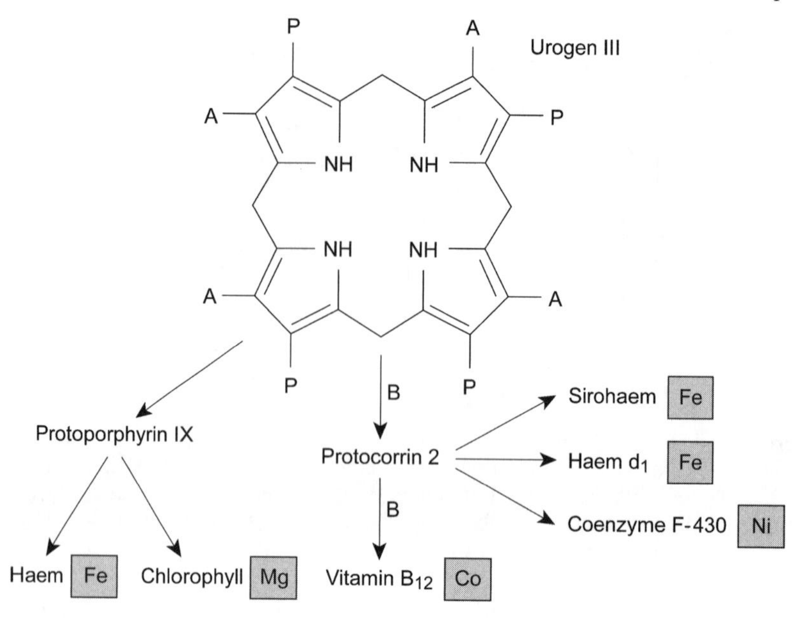

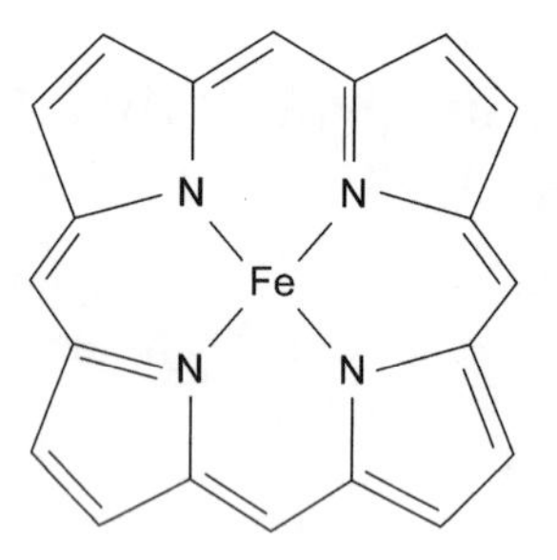

Figure 6.4 The relationship of the four porphyrin derivations of uroporphyrin. Fe, Co, Ni can be low-spin complexes in proteins. All four metal ions do not exchange. Beneath is an outline of heme. Note that Co, Ni, Mg and some iron porphyrins are not found in all organisms.

type of site in many small molecules, enzymes, in its pumps and for feedback to them to prevent excess uptake, and in its transcription factors, all in the cytoplasm or on a membrane to it. As there are so many sites the free [Mg^{2+}] at about 10^{-3} M is a cooperative characteristic of the cytoplasm of all cells. There is no competition for these sites from any other ions because their concentrations are kept too low to be able to bind these weak binding sites, *e.g.* pyrophosphate (adenosine bound with pyrophosphate, ATP, is the actual compound most used today), despite greater binding strength, *e.g.* of Cu^{2+} and Zn^{2+}. 10^{-3} M Mg^{2+} ions are then an essential component in the homeostasis of much of acid/base metabolism.

In the cases of stronger binding, often to N- or S-donor centres (see Figure 6.5 for examples), to which Mg^{2+} and Ca^{2+} do not bind even at high concentration, there is a general rule that stability constants of binding to organic molecules follows a fixed series of M^{2+} ions, the Irving–Williams order (see Figure 1.4):[3]

$$Mn^{2+} < Fe^{2+} < Co^{2+} < Ni^{2+} < Cu^{2+} > Zn^{2+}$$

Their binding constants usually rise from 10^{-6} M^{-1} (Mn^{2+}) to 10^{-15} M^{-1} (Cu^{2+}). All their free ions that are at low concentration in the environment are also low in concentration in all cells, usually below 10^{-6} M and often below 10^{-10} M (Figure 6.6), so that they cannot compete for the weakest binding sites and do not compete with Mg^{2+} nor with Ca^{2+} in cells or in the environment. The figure is the inverse of Figure 1.4, showing that the metal ion concentrations in cells are generally a direct reflection of the strength of inorganic complex ion equilibria. The transition metal ions are bound either by different ligands in proteins or special selective small chelating agents. From the above equation the ion that binds to a given protein in a given compartment depends on the properties of the protein's selective binding groups, the concentration of M, decided by the action of selective cell in-and-out surface and vesicle pumps, and the concentration of the protein, [L], decided by synthesis, and the extent of protein pumping into a given vesicle compartment. Protein synthesis itself is linked to the concentration of M by the transcription factor, so that for strong binding there is little free L or M (see Section 5.5). The degree of discrimination

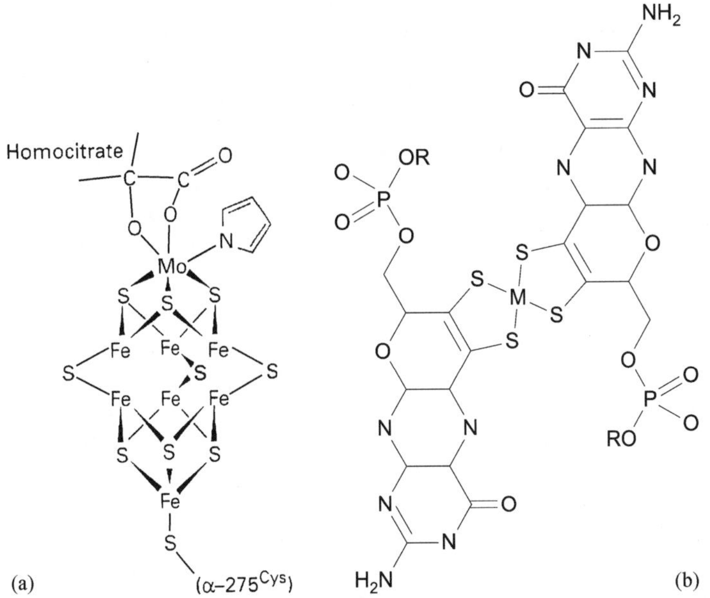

Figure 6.5 (a) The Fe, Mo sulfide complex, Fe_7MoS_9, of nitrogenase; (b) the Mo thiolate complex. Exchange with free ions is slow in both.

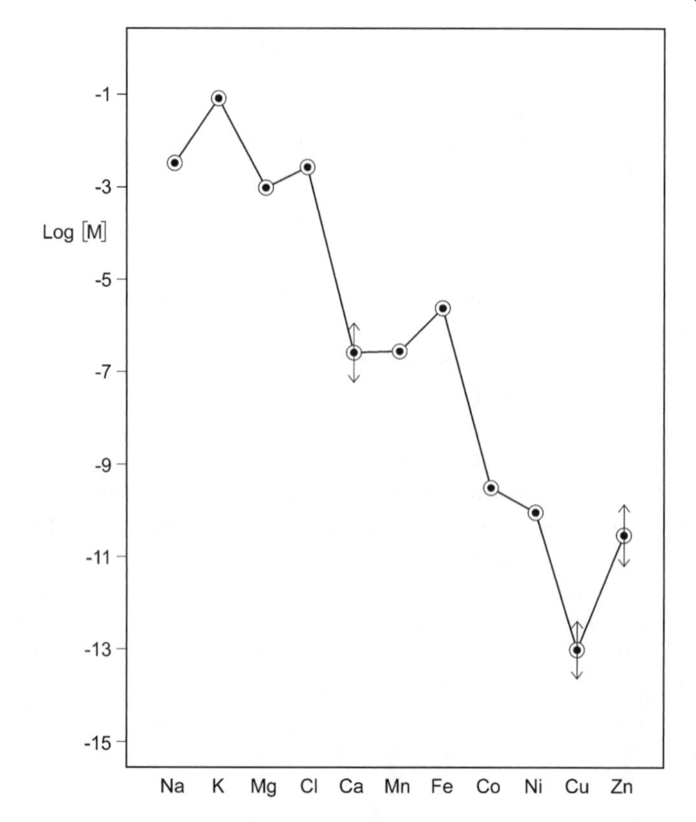

Figure 6.6 The concentrations of free metal ions in the cytoplasm. The values of the free metallomes are fairly constant in all cells in keeping with the dominant reductive chemistry in the cytoplasm, chapter 4.

in the above series is then in part dependent on a number of chemical factors of the binding groups, that is the number of thiolate, amine or carboxylate groups (see Figure 1.4), their stereochemical properties and the resultant spin-state conditions of M^{2+}, and the compartment into which they are placed. To limit binding of the strongest binding metal ion, copper, to one selective group of proteins, all of its binding constants fall in a range close to $K = 10^{15}\ M^{-1}$. The concentration of free Cu ions in all compartments is effectively close to the inverse of this value 10^{-15} M controlled by pumps, probably by linked feedback from ML. To avoid production of excess binding protein the synthesis of the protein must stop so as to give the desired level of the copper metalloproteins. This again requires feedback, in principle from copper complexes, to its synthesis via transcription factors on DNA. Copper is distributed amongst its active sites by carrier proteins, chaperones, selective to each receptor target. All copper exchange is between proteins of similar high binding constant. In this way the whole metabolism of copper and its proteins is arranged in a homeostatic condition such that its proteins are not free to bind less strongly binding groups which bind selectively to metal ions in the above series. By way of contrast, the lowest binding strength within the Irving–Williams series is of Mn^{2+} and Fe^{2+}, so that especially Fe^{2+} at a free concentration of 10^{-6} M connects together many coordinated activities of different enzymes as already described (Section 6.4), and relates directly to transcription factors which control synthesis of its own proteins and much of oxidative metabolism, all with $K = 10^6\ M^{-1}$ (see Figure 6.3). The Fe^{2+} ion can also be carried by a special protein in equilibrium with free Fe^{2+} to a selected site of very different binding character (see porphyrins below). While free Co^{2+} and Ni^{2+} have no known role, free Zn^{2+} in all probability coordinates many gene products in eukaryotes, as it binds equally to many transcription factors, zinc fingers, with a binding constant of around $10^{10}\ M^{-1}$.[17] However this ion binding does not exchange sufficiently rapidly to act like Fe^{2+}. The zinc ion, like the copper ions, has several selective carriers (chaperones). They are found to link together the zinc binding sites of zinc fingers, the receptors of hormones and growth factors, directly because zinc exchange is known to be fast between these proteins,[14–16] but not with free ions. This gives directed homeostatic distribution of zinc to different cellular activities. All the chaperones of zinc have similar binding constants so that distribution is balanced. When bound it behaves therefore in a similar manner in bound form to a free ion as an internal control but at a much lower exchange rate. In contrast it may well be that the earlier zinc binding sites, for example in carbonic anhydrase, did not exchange as they have binding constants of $10^{12}\ M^{-1}$ or greater.[2,17] There is a similar quantitative chemical as well as a genetic contribution that affects the intake, binding and functions of the other trace elements.

There is a separate quantitative influence of binding strength which is on the redox potential of ions which have more than one oxidation state, as is the case particularly for iron, copper and manganese in biochemical resting states and in active sites of enzymes. Most copper proteins inside cells are in the Cu^+, cuprous, state as Cu^+ is often bound more strongly than Cu^{2+} and the standard redox potential of Cu is then well above its standard potential +0.1 V.[18] Neither oxidation state exchanges with free copper ions rapidly. Iron has a lower standard redox potential at pH = 7 due to hydroxide binding to Fe^{3+}, but the reducing conditions of the cytoplasm hold it as free Fe^{2+}. Mixed oxidation states of iron are also observed in Fe_nS_n complexes of electron transfer proteins. Outside cells both copper and iron in normal (today's) ambient conditions are in their higher oxidation states. In the higher oxidation state bound Fe^{3+} is not so available as much is lost as ferric hydroxide, but Cu^{2+} is not so affected. While iron was and remains responsible for redox activity in the cytoplasm of cells copper enzymes are usually found outside cells or in vesicles. Manganese is occasionally bound in cells in the oxidised Mn^{3+} state, which is important in some enzymes, *e.g.* the oxygen-producing one, but free manganese ions circulate in the Mn^{2+} state in the cytoplasm and vesicles, for which exchange is fast, like Fe^{2+}. In redox enzymes the rates of change of oxidation states, electron transfers, are usually fast for both these two metals often due to the constraints of structure.[18] While the control of oxidation state is managed by protein binding strengths in cells, in the environment it is controlled by the lattice constraints of oxides, sulfides and silicates and solubility, that is solubility products. Examples are the precipitation of BIFs, of high-spin Fe^{3+} in Fe_2O_3, of low-spin Fe^{2+} in pyrite, FeS_2, and of the mixed valence state copper sulfides. The fast equilibrium electron exchange of all oxidation states of ions in minerals allows the balance of these states in the known oxidising and reducing conditions of the environment to be dated (see Chapter 2). In organisms it has a very important role in redox homeostasis and energy capture as well as in many enzymes. The equilibrium of metal ion complexes in redox exchanges are essential quantitative features of trace metal chemistry in organisms in contrast to the non-equilibrated redox states of non-metal chemistry (Chapter 4).

Each of the trace metal ions can be seen now to have its own complement of proteins in a particular spatial division free from confusion from other metal ion binding. Given the control over intake to a compartment of metal ion and protein concentration by feedback from the individual free ion with the magnitude of the different binding constants from above 10^{15} M^{-1} for copper, to 10^{11} M^{-1} for zinc and to around 10^{6} M^{-1} for Fe^{2+} and Mn^{2+} based on protein chemical structure and thermodynamic properties of the ions, the metalloprotein complement of each compartment of cells is decided. It is this that lends a degree of predictability to cell biochemistry in that, as already mentioned, these binding constants are then close to the inverse of the limited concentration of the ions defining a fixed free metallome in the cytoplasm of all cells (see Figure 6.6). Binding constants of metal ions via free ion concentrations make direct environmental connections also through availability and pumps. Availability changes (in the sea) based on redox potentials and solubility products in the environment together with redox potentials and stability-binding constants to organise molecules in cells are then major predictable determinants of evolution. The implication is clear. The metal ions which connect pathways of all kind coordinate acid/base (Mg^{2+} particularly), redox (Fe^{2+} particularly) and growth (Zn^{2+} particularly) pathways. They are then of the very essence of biological chemistry.

Now as already noted some of these free concentrations are so low that control and distribution of such metal ions lies with carrier and buffer proteins. A totally different, kinetic way of gaining selective ML complexing is then possible. M is selectively bound at first to a protein, using the above thermodynamic selection, but then M is transferred in a specific manner from the particular ML_1 (for example by employing weak selective binding for Mg^{2+}, relatively moderate for Fe^{2+}, strong for Cu^{2+}, Zn^{2+} or Ni^{2+}, and intermediate for Co^{2+}) to its final binding destination by insertion into another site ML_2. This kinetic selection is due to the recognition of the shape of L_2 by ML_1, where L_1 is called a chelatase but it could again be called a chaperone, giving a unique pair transfer reaction from a given ML_1 to L_2 for each metal ion.

$$ML_1 + L_2 \rightarrow ML_2 + L_1$$

The ML_2 complex may not have an absolute binding constant which is useful in discussion as there is no reversible exchange with free M or ML_1, but the relative reversible binding constants for different ML_1 remain vital. Examples of ML_2 are found especially amongst the porphyrins Mg chlorophyll, Fe porphyrin, Co corrin, and Ni(F430) (see Figure 6.4), and show exclusive ML_2 formation and functions. The further action of each of these complexes is separate from their simple ions and they are separately controlled.[3] It is as if a second kind of metal ion had been made from Mg, Fe, Co and Ni, especially as Fe, Co and Ni in porphyrins are often in low-spin states while the free ions are all high-spin. Note the parallel with non-metal, non-equilibrium organometallic chemistry. The chelatases are early examples of carrier proteins or chaperones which may also interact with storage proteins and transcription factors but, in the case of iron, such exchange is more often through free Fe^{2+}. The distribution in compartments of ML_2, like that of different M, can be quite idiosyncratic.

The special synthesis of the four or more of the separate porphyrins is known to branch out from uroporphyrin. One branch leads to haem and diverges earlier to chlorophyll, another branch leads to vitamin B_{12}, others to cytochrome d_1, sirohaem and F-430 (see Figure 6.4). We note that neither Cu nor Zn ions have porphyrin complexes and these ions became very important only later in evolution.

A very intriguing feature of the Fe, Co, Ni and Mg porphyrins and their variations is that they were all present very early in evolution, usually before 3.0 Ga, and that examples of three of them, Fe, Co and Ni, are found in the earliest cells, prokaryotes, of which we have knowledge. They then represent a very early case of evolution of catalysts to take on substrates such as CH_4 and H_2 and also to provide excellent fast states for electron transfer and reduction. Earlier still we believe that metal/sulfide complexes were the original catalysts for these reactions. Porphyrin must have been synthesised very early in evolution, possibly from cyanide. The basic uroporphyrin synthesis is simpler than that of some nucleotide bases. This paragraph illustrates how little we understand about the origin of the combination of inorganic with organic chemistry which gave rise to life.

While examining the trace element concentrations we note the peculiarity of the location in cells of some of the metal binding reactions of ions of the porphyrin-related compounds, giving haem and chlorophyll, and of the Fe/S cofactors (see Figures 6.4 and 6.5). They are all synthesised in the organelles of eukaryotes, haem and Fe/S in the mitochondria by proteins, many of which are shipped there from the cytoplasm where their synthesis is coded in the parent DNA though the original genes were in the organelle bacteria. After synthesis the small units are distributed to all parts of the organism, and in the case of vitamin B_{12}, Co porphyrin distribution is also to organisms which do not make it. This is again a clear example of mutual dependence of organisms, symbiosis. Also note that much like other factors some are synthesised in all cells, haem and Fe/S, some in plant cells only, chlorophyll (Mg^{2+}), and others are synthesised in special symbiotic lower organisms, *e.g.* vitamin B_{12} (Co) and donated to higher animals. This vitamin is not required by higher plants. To our knowledge F.430 (Ni) is only found in Archaea. Sirohaem and cytochrome d_1 are absent in higher organisms such as man, which cannot metabolise NO_3^- or SO_4^{2-}.

Much of the above discussion of metal ion compound formation relates to the cytoplasm and its free metallome but in the other compartments different selection applies. In them the concentration of metals and proteins is different, as is the redox potential. For example Mg^{2+} and Fe^{2+} ions are not often found in vesicles while Ca^{2+} and Mn^{2+} are frequently and Zn^{2+} occasionally there.

We summarise the approximate generalised concentrations of free ions, the free metallome, in the cytoplasm of aerobic cells today in Figure 6.6. Due to the restrictions necessarily applied through quantitative equilibria and selectivity of binding the free metallome in the cytoplasm is virtually invariant in all cell types (see Section 6.5), and is then a characteristic of life. How different are these metal ions from the organic chemicals of cells much though they are mutually interactive. The total metallomes are variable from one very different group of organisms to another (see Figure 6.2), while this is not so between organisms of a group (see Section 4.8 for parallels with general cytoplasmic chemistry). The free metallome will be compared with the concentration of trace elements generally in the environment in Section 6.5. A very similar total metallome (see Figure 6.2), found in a particular separate large group of organisms led us to call the group collectively a chemotype. Metallomics then describes the special study of the metalloproteins in organisms and is an extremely valuable quantitative way in which to follow evolution. Frequently the metallome sum is about 30% of the total proteome. Once again note that changes in metal ions in both the environment and organism metal ion systems are the most valuable of all evolutionary chemical markers until about 0.75 to 0.4 Ga.

The above account of preferential thermodynamic binding of trace and bulk metal ions to protein side chains has been fully confirmed by structural studies of extracted proteins.[18–20] The studies show the frequency of selective, relatively weak, binding of Mg^{2+} and Ca^{2+} to specific oxygen donor centres, of Mn^{2+}, Fe^{2+} and Zn^{2+} frequently to combinations of stronger oxygen (carboxylate) and nitrogen sites, in some cases in moderately fast or slow exchange. They also show Zn^{2+} and Fe^{2+} binding to thiolate groups, while copper and molybdenum are bound largely and very strongly by thiolate and nitrogen donors, all in slow exchanges. As indicated earlier, in these structures the selectivity is enhanced by the matching of preferred stereochemistry,[18] of spin states and of sizes of the metal ions to the structure of the binding cavity.[1,6–8]

There are also complexes with nucleotides, usually weak and in rapid exchange where Mg^{2+} and K^+ are the main ions involved (see Chapter 5). Finally binding to saccharides, often in membrane structures or externally, is also common and we have described the example of calcium complexes in Chapter 5.

In summary selectivity is aided by the quantitative control of metal ion concentration and its location in particular compartments of cells, the controlled synthesis of proteins and their transfer to compartments and their sequences. The ions are often closely at equilibrium in their combinations either directly with the ions or by exchange between bound complexes. The particular description applies to the specific metal ion binding to transcription factors, carriers, pumps, enzymes and structural proteins[14,16] and applies in forward and feedback reactions. Interestingly each time a new metal ion activity is found in an organism it is accompanied by the synthesis of a large number of equilibrating proteins to control its uptake and distribution apart from its participation in many enzymes in special compartments. In this way homeostasis is advanced. The example of zinc was given in Section 6.5. How did this complexity arise and evolve? The function of trace elements differs greatly from that of non-metals and bulk elements in that the second two groups of elements are in plentiful supply, much though they require energy to manage their transformations and concentrations. The non-metals are in slow exchange although some of their components, *e.g.* coenzymes, in fact exchange and are also engaged in both pathway coordination and catalysis. The use of trace elements also depends on energy but their availability changes grossly. This is not obviously linked to the DNA code but they must influence it to lead to the evolution of organisms with reproduction and heritable characteristics (see Chapter 7).

It is extremely important to realise, as in Chapter 5, that unlike the analysis in Chapter 4 the evolution of inorganic element chemistry is dependent directly on the environment and then on selection by cells of the new environment factors at the protein level with changing equilibrated internal concentrations and not just at the level of random change of the code. It is this that requires the code to be an evolutionary mediator of the changes of the environment and of cells. But how are they coupled? The next paragraph draws attention to some particular changes of availability, before 0.54 Ga, which affected evolution.

6.5 Examples of the Thermodynamic and Kinetic Limitations on Uptake of Metal Ions

The presence of oxygen brought great accidental advantages to the evolution of organisms but there were many disadvantages. We have mentioned the release of superoxide and oxygen which menaced especially Fe/S proteins in cells. Further disadvantages sprang from the oxidation in the environment which we have also described in Section 4.11. There were two dangers: (i) the rise in 'poisonous' metal ions, such as copper, but this was turned to advantage in special compartments, and (ii) the precipitous drop in available iron as oxidised Fe^{3+} which formed insoluble $Fe(OH)_3$. Under (i) we have discussed the production of special proteins in the cytoplasm and of membrane pumps to remove the possible excess of (poisonous) metal ions. It is equally important to recognise the difficulty of obtaining a metal ion from a very low availability in the environment.

We can make a quantitative estimate of the involvement of protein binding of the trace elements at different times in evolution relative to their availabilities as seen, for example, in cyanobacteria.[10] Although, for example, we do not know for certain what is the limiting concentration of zinc which could be scavenged to a low concentration from the environment by organisms at very early times we do know that all organisms retain some zinc. This indicates that the binding of zinc at 3.0 Ga, very primitive times, must have been stronger than that limited by its solubility product in an outside medium of, say, $10^{-3}HS^{-}$. Free zinc was then at or a little below 10^{-10} M which suffices for zinc retention as the stability constants of some ancient non-exchanging zinc proteins are known[2,16,17] to be about 10^{12} M^{-1}. In general these proteins do not exchange zinc, for example carbonic anhydrase, where protein production limits the role of zinc in the earliest cells. However weaker binding zinc proteins of eukaryotes, which are used in signalling or carrier exchange functions, have a binding constant at or below 10^{10}. They could only function in this way after sulfide was removed and zinc availability increased. This reflects the low content of zinc in unicellular eukaryotes and then the rapidly increasing zinc in multicellular organisms that is after, say, 0.75 Ga in zinc enzymes and zinc fingers. We can relate the increased possibility of obtaining zinc after H_2S was removed as following a redox titration curve (Section 2.11.2). At somewhat similar environmental redox potentials to those affecting free zinc we can follow the availability of copper and other trace elements. As given above the corresponding binding constants for copper are around 10^{15} M^{-1} while in sulfide media free copper ions are at a concentration much less than 10^{-15} M. This explains the low values of zinc and the virtual absence of copper in organisms in a sulfide medium (Section 2.11). After 0.75 Ga copper became more available with the removal of sulfide. Copper with zinc then provided catalysts to make possible multicellular organisms.

In Section 5.5 we discussed the importance of Fe^{2+} in the controls of prokaryote metabolism. To compensate for the loss of easy access to Fe^{2+}, (ii) above, cells synthesised scavenger molecules of high binding strength for Fe^{3+} which were exported to the environment.[21,22] The scavenger molecules, siderophores, were synthesised by oxidation reactions. These molecules, bound to external Fe^{3+}, were then imported via special pumps and reduced. Generally the scavengers are phenols or hydroxamic acids and must be selective such that they do not introduce aluminium into cells. The multicellular eukaryotes have a more complicated solution.[21] They carry the iron, Fe^{3+} taken in their diet, on a protein in extracellular fluids, transferrin, and deliver it to specific cells. The iron protein is taken into a vesicle when the iron is reduced to Fe^{2+} and passed from the vesicle into the eukaryote cell cytoplasm. In both prokaryotes and eukaryotes there is also a storage protein, ferritin, which holds considerable amounts of what is effectively a biomineral ferric hydroxide/oxide so as to buffer the concentration of iron. Notice that both solutions of the iron scavenging problem require considerable synthesis of new proteins and hence several novel genes. All the uptake proteins and siderophores are expressed through equilibrium controls including Fe^{2+}-binding transcription factors such as Fur, as shown later in Figure 6.10. The two different solutions for prokaryotes and multicellular eukaryotes must have arisen at the times of the two rises in oxygen, respectively at about 2.5 and 0.75 Ga, which caused decreasing iron availability for all cells. The multicellular set of uptake proteins must have arisen quickly and generally because each of the required types of protein appears to have arisen at a closely fixed time in very different groups of organisms. This is an expansion of the control of Fe^{2+} described already.

An excellent, different and earlier specific example of the influence of a catalytic trace element availability coupling to protein supply in prokaryote evolution is provided by the work of Konhauser *et al.*[23] on the levels of nickel in BIFs, which is taken to reflect its changing solution concentration throughout the first 1.5 billion years (Section 2.11). Notice in Figure 1.4 the much smaller difference of nickel affinity for sulfide relative to oxide than for other metal ions. Nickel was apparently relatively plentiful, at least in some locations, in the period from 3.5 to about 2.7 Ga, but approaching and around the time of the first great oxidation event, 2.5 to 2.0 Ga, nickel in the iron deposits went into very considerable decline (Section 2.11.2).[23] The estimated values in solution fell from 400 to 120 nM by 2.5 Ga and then declined further to the second oxidation event at 0.5 Ga to the value around 10 nM today. The history of nickel-utilising organisms, especially Archaea, parallels this availability (Section 2.11). While in the earliest period of life from 3.5 Ga nickel supported widely what is believed to have been a large number of Archaea, the Archaea are now found only in niches. The correlation has been extended to the effect of siliceous acid in the seas on nickel from the earliest times to today as it is considered that the nickel is limited by precipitation of its silicate.[24]

Part B. The Evolution of the Metalloproteins, the Metallomes and their Functional Value

6.6 Introduction

We regard the evolution of oxygen with the changes of the trace elements in the sea (Sections 2.11 and 6.2) as the fundamental material drive of the organic/inorganic chemistry evolution coupled to the capture of light to reduce carbon chemicals. We have noted in Part A of this chapter quantitative considerations of the principles of the chemistry of the ions of these elements which allow them to be selectively taken into cells and combined selectively, often with novel organic chemicals, as the inorganic elements changed concentration in the environment. We shall now continue quantitative analysis by looking at the numbers of different metallo-proteins from the earliest to the most recent cells. The number of combinations increasingly depended in large part on the development and separation of chemistry, mostly oxidative from reductive, in separate compartments. We can follow evolution of cellular chemistry therefore by a study of the proteomes of the cells using DNA sequences. This is a quantitative study of protein type, not concentration. We shall divide the proteomes into those of prokaryotes from those of the eukaryotes in the following two sections before turning to the uses of the metal ions in life.

6.7 The Evolution of the Metalloproteins of Prokaryotes

The examination of the analysis of the metal ion proteins in organisms began some years ago has become much more sophisticated with the introduction of bioinformatics.[4,5,25–31] The importance of proteins with metal ions can hardly be overestimated as they represent 30% of the organism's proteins. We have already mentioned the changes in cyanobacteria[10] with the coming of oxygen. The conclusions of the study are in agreement with those of a wider analysis by C. Dupont *et al.*[4] (also see references 30 and 31). They studied the changes in trace element-binding protein domains with increase of genome size, not just amongst Eubacteria and Archaea, which we describe first, but in eukaryotes, as a good indication of trace element history. Types and numbers of domains are known to increase generally in evolution (Figure 6.7). We shall discuss the functional gains of the increase in domains in Section 6.9. Amongst prokaryotes changes in types and numbers of domains of metalloproteins in Archaea are very different from those in Eubacteria in the earliest period of life and in later periods both differ from eukaryotes. Anaerobic Archaea have an almost three to five-fold greater variety of nickel proteins, and fewer, if any, copper proteins. They have a different distribution of iron proteins, all increasing to an optimum value with size of the genome quite quickly, from those of anaerobic Eubacteria which also have very few copper proteins. In both these prokaryotes the number of zinc proteins only increased slowly,[2,4,5] while Fe, Co, Ni and Mn domains increased proportionately to genome size. The prokaryotes do not appear to have been capable of further increase in genome size after 2.5 Ga. During the period before 3.0 Ga it is generally agreed that all cells were more or less anaerobic, implying that prokaryotes generally had few oxidative enzymes within an optimal domain size no larger than, say, 3,000–4,000 genes or 1 million bases. The nature of the proteins of iron which began with the Fe/S proteins was soon accompanied by haem proteins. They were added to in aerobes by the non-Fe/S non-haem proteins, the α-β glutamate requiring high-spin iron proteins in relatively small numbers (Table 6.4). By contrast the nickel, cobalt and manganese protein remained constant but the copper proteins were novel to the aerobes.

We observe that as oxygen increased[32] many of the new catalytic metalloproteins were to be found in the periplasm or on the surface of the outer membrane facing this space (see Figure 4.12). They include not just the reactions of O_2 but those of sulfate and nitrate in specialist prokaryotes. The

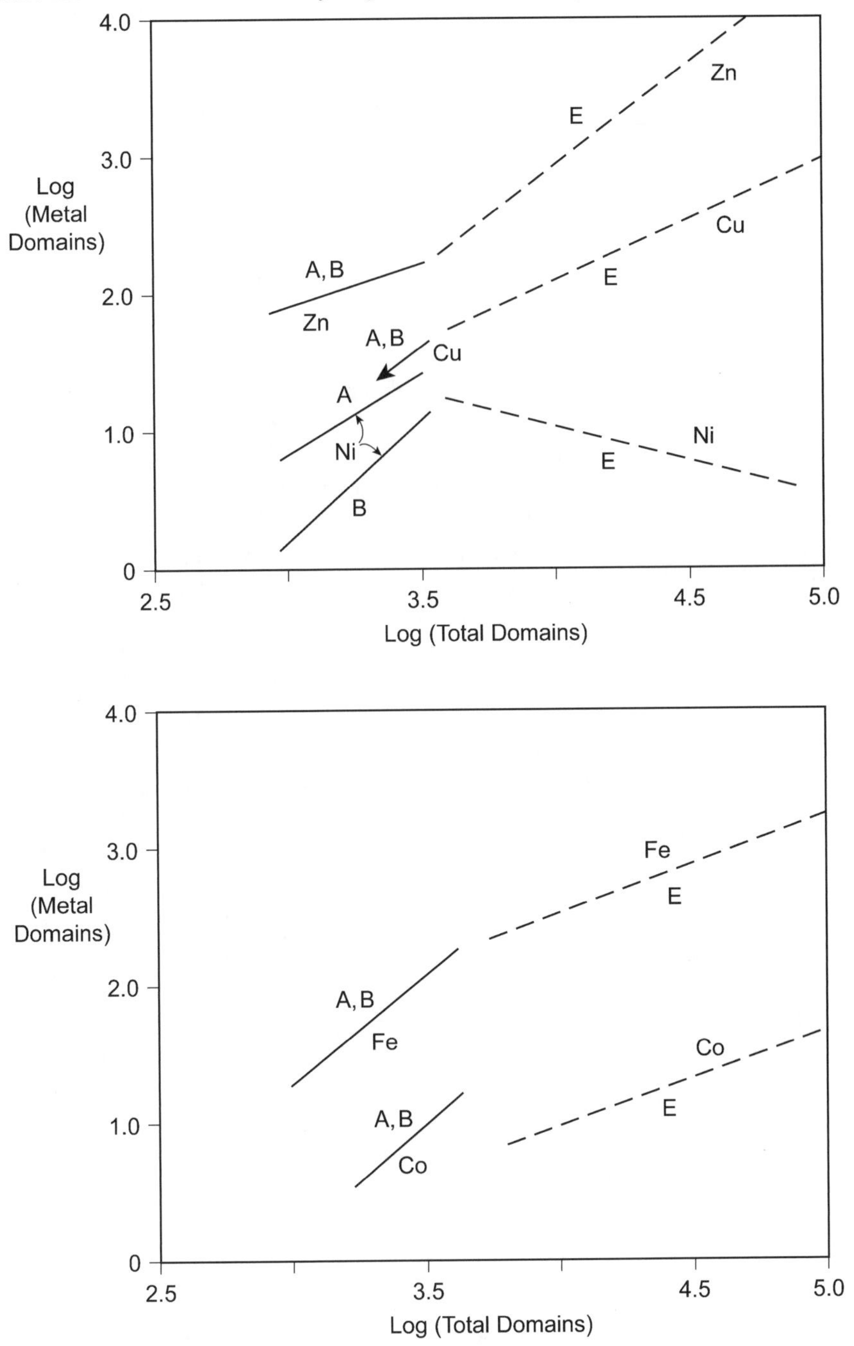

Figure 6.7 The number of domains of the proteins of various metal ions plotted against the total number of domains of all proteins for A, Archaea; B, eubacteria; and E, eukaryotes. After Dupont *et al.*[18] and personal communication from Dupont.

Table 6.4 Some Iron Hydroxylases

Organism	*Hydroxylation Involvement*
General	DNA-repair
Bacteria	Penicillin synthesis
Plant	Ethylene synthesis
Plant	Gibberellin synthesis
Animals	Procollagen synthesis
Animals	Factor of hypoxia induction (HIF)

See references 34 and 36 for details of these enzymes.

increased catalysts in the periplasm are copper and molybdenum enzymes with some haem-iron enzymes but no Fe/S proteins. None of these extra proteins seem to have increased genome size greatly.

An interesting example of substitution in evolution and of the protein domains is the change in metal ions found in 'cytochrome-c oxidase'.[33] The original protein may have been based on Fe only and have been for the reductive catalysis of NO but later it became the copper-dependent, O_2-reducing enzyme cytochrome oxidase seen today (Figure 6.8). A similar switch from Fe, now haem iron, to Cu is seen in some photosynthesisers in the replacement for electron transfer of a cytochrome (Fe) by azurin (Cu) protein.

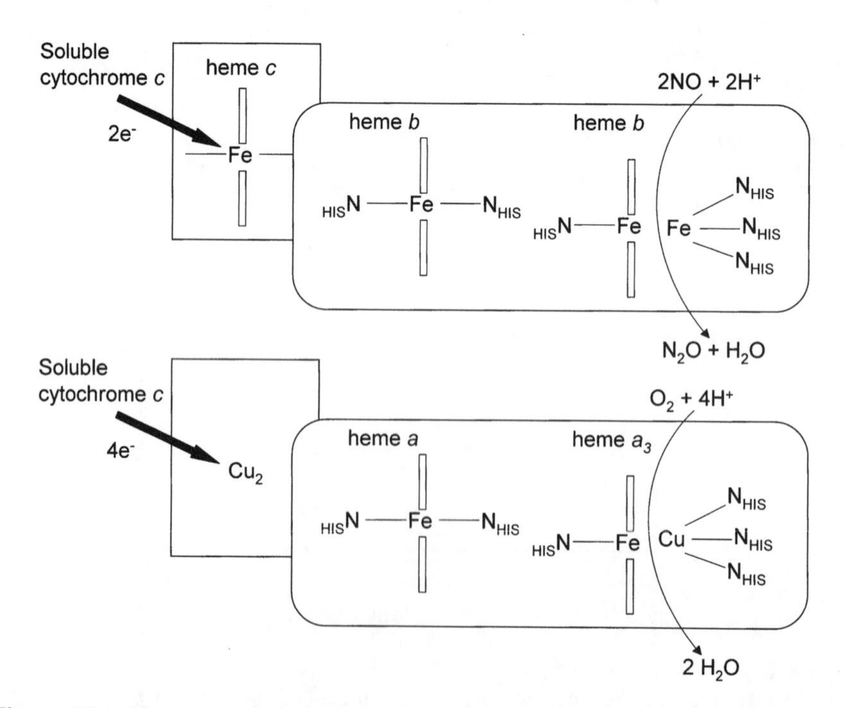

Figure 6.8 The relationship between nitric oxide reductase (top), an all iron enzyme, and cytochrome c oxidase (bottom) incorporating three copper ions.

Notice by contrast the increase, not substitution, in the trace element proteins of prokaryotes developed only with iron, not copper, oxidases in the cytoplasm for hydroxylation and oxidative removal of poisons, such as penicillin.[34]

In summary there are some changes in the types and contents of the prokaryote proteomes involving metal ions as oxygen increased introducing new proteins for handling oxidised nitrogen, sulfur and selenium substrates and metal ions, such as copper. Mostly these new chemicals acted externally as substrates in oxidases. The enzymes also catalysed oxidation of certain organic molecules to remove dangerous compounds internally.[34,35] There was not a great increase in proteome size, however, because the prokaryote cytoplasm survived with little or no change in basic structural and functional units. The major step was the probable evolution from them, at least in part and apparently due to oxygen, of eukaryotes with a different outer membrane, several new compartments and structural units and many different activities (Section 4.7). We do not understand this step and here we treat the metallome and proteins of eukaryotes separately from those of prokaryotes, as in Chapter 4. Before leaving the prokaryotes we see that much of their evolution is strongly coupled to feedback from oxygen and environmental oxidation of minerals related to the controlled concentration of free ions in the cells. If we consider it in terms of chemical systems as for the beginning of cell anaerobic chemistry, then we could take their new chemistry based on an aerobic environment to be a second chemical system added to the first. We put to one side any connection to genes or reproduction until Chapter 7, as we wish to investigate first the possible underlying causes of cell chemical change linked to the environment.

6.8 The Evolution of the Metalloproteins of Eukaryotes

The eukaryotes arose in two stages as single-cell organisms and as multicellular organisms, probably after the first and then with the second rise of oxygen. In these stages the cell content of proteins doubled from 5,000 in prokaryotes to around 5,000 to 10,000 in the first eukaryotes and then greater than 20,000 in later multicellular eukaryotes. In this section we shall sometimes treat single-cell and multicell eukaryotes together as it is not always easy to distinguish colonies of single cells from multicell organisms, especially from 1.0 to 0.54 Ga. Perhaps the greatest problem in understanding the effect of the trace metals in the environment on evolution is sure knowledge of the date of the first evolution of oxygen in some quantity, probably around 3.0 Ga, and that of the single-cell eukaryotes. The earliest date suggested for the evolution of single-cell eukaryotes is a little before 2.5 Ga, while dates much closer to 2.0 Ga, perhaps in surface water only, have also been proposed. A helpful feature is that, after they formed, many of these eukaryotes changed rather little as far as we know, even to today, and we can set them aside from the later period of well-established multicellular eukaryotes which did not evolve until after 0.75 Ga with the final rise in oxygen,[32] but have continued to change in complexity rather than in chemistry after 0.5 Ga until now. (The multicellular algae, bangiophyte red algae, dated at 1.2 Ga are exceptional and may not have arisen in surface water.)

In the following treatment of eukaryotes, as well as treating the total metalloproteome complement increases, we refer to the presence of different trace elements in different organisms as indicative of differences in their protein content. An example of trace element involvement in different paths of evolution of single-cell eukaryotes is illustrated by the evolution of the photosynthesising different forms of the algae.[5,36,37] We assume that the basic unicellular eukaryotic organisms are distinguishable here by their chlorophyll and carotenoids, originally in bacteria, which gave them different colours, *e.g.* brown, red, green (Section 4.16). Their development to a vast range of organisms in the sea and on land is confused by the presence of both colonial and truly multicellular forms, which all required an increase of manganese over the non-photosynthesising cells. We shall therefore use the title photosynthesising eukaryotes to cover both later single-cell, colonial and multicell 'plants', including all the algae. We describe the algae in some detail as they dominate the top layer of the sea and they are the major light absorbers and oxygen producers. It is thought that single-cell algae arose in a multitude of forms, perhaps four or five, around 2.0 Ga and that more complex definitely multicellular algae evolved generally some millions of years later, but note again the Bangiamorpha.[32] It has been known for at least 50 years that brown (red) algae have a greater calcium, strontium and barium content than their green partners.[38] An interesting more recent observation is that the green algae have a relatively higher Cu, Zn (and Fe) than brown (red) algae which have higher Co, Cd and Mn, relative in all cases to average P content.[36,37] The conclusion, in keeping with the suggestion from the observable fossil record, is that brown/red algae evolved in a less oxygen-rich environment than the green algae, before 1.5 Ga. The green algae evolved earlier at a time or place when there was a little more oxygen, copper and zinc, at close to 2.0 Ga. There is the possibility then that the green algae arose in shallow light-rich seas as a combination of bacterial cells, while the red/brown which are a combination of two such eukaryotes arose in deeper water. A probable supporting explanation of their different depths in the sea is that the sea allows shorter wavelength radiation to penetrate water to a greater depth than long-wave radiation, visible light. Hence the brown/red algae and their precursor bacteria were better adapted to life in deeper layers in the sea, and green algae thrived better, especially in shallow water and on land. It is also intriguing that generally green algae are the precursor of multicellular land plants while brown/red algae are the forebears of single-cell corals (colonies only), coccoliths, flagellates and diatoms, largely in the sea. The major 'multicellular' brown algae appear to be huge seaweeds but they are hardly distinguishable from colonies of cells. The difference in element content then relates to the difference in oxidation conditions lower in the sea. Further interest in brown algae arises from the unique use of cadmium 'carbonic anhydrase' in certain brown algae (see

Section 6.9). The increasing diversity of these organisms is one indication of the increase in types of trace element proteins available in evolution, that is of chemotypes. We comment more generally on the ratios of elements in organisms in Section 6.9.

More generally, just as we saw in the case of the Ca^{2+}-binding proteins[39] in Section 5.5 we observe that the single-cell eukaryotes already have a number of trace metal proteins which do not relate easily to the proteins of any prokaryote.[26] For example there are the zinc-finger type proteins (see Figure 6.13), and the new Cu/Zn metallothioneins[16] and Cu/Zn superoxide dismutases.[35] These differences are best seen by including the later multicellular eukaryotes in our further discussion of eukaryotes as the new proteins are often very much increased in them. This is shown in Figure 6.7 where the domains of metal binding are plotted against the total number of domains in the genomes.[4] The figure is somewhat misleading, as we noted earlier, because the Archaea (A) and the Eubacteria (B) have a maximum number of domains of below 5,000, independent of external conditions, while single-cell eukaryotes (E) have a much larger number of genes, say 5,000 to 10,000, and multicellular eukaryotes have between 15,000 and 40,000 domains, the latter dependent on higher oxygen. Taking all eukaryote metalloproteins together the number of zinc and copper domains relative to the total number in their genomes has increased most rapidly.[4,5] Man's DNA contains at least 100 copper domains and 2,000 zinc domains so that zinc is equal to or above iron proteins in importance in direct contrast to the proteins of prokaryotes which may have less than 10 copper and less than 100 zinc domains but at least 1,000 iron domains.[15] All the iron proteins do not increase proportionately in eukaryotes. Although there is a rise in haem and non-haem, non-Fe/S, α-glutamate-dependent proteins there is a relative fall in Fe/S proteins in eukaryotes. We turn to the formation of the new proteins in the next section. The number of nickel proteins falls considerably and that of cobalt rises little if at all, due to loss of genes in some organisms. Although these comparative falls do not follow those in the environment, these changes are all indicative of a generally fixed cytoplasmic reductive chemistry but with a decreased concern for substrates such as H_2, CO and CH_4, mostly lost from the environment by 2.0 Ga. In comparison there is a rapid increase in oxidative chemistry either in vesicular or extracellular fluids related largely to some rise in iron but a larger rise in copper proteins. In later sections of the book we show, however, that the reduced dependencies on Ni and Co in later eukaryotes are in part compensated by their dependence on symbiotic prokaryotes which remain requiring cobalt and nickel. Man is an example of dependence on symbionts and advanced plants do not use or require vitamin B_{12}. Note that increases and losses are generally systematic across phyla, reflecting the need for dependence on the environment evolution. Further changes include the role of manganese in glycosylation in eukaryote vesicles. A notable and seemingly quite peculiar environmental dependence of all organisms today is for molybdenum[40] but in the earliest prokaryotes this element may have been preceded in use by tungsten as molybdenum sulfide is insoluble. We have not included here the rapid increase in the number of non-metal selenium proteins to which we turn later (see Section 6.8). A review analyses the changes in trace elements within a wide variety of organisms, giving very extensive references.[4]

In passing we stress again that these observations concern the quantitative chemical element content of the proteomes of organisms.[2–5] They could be taken as evidence of the addition to the reductive chemistry of prokaryotes in an anaerobic environment of a system of largely oxidative reactions in a novel aerobic environment. The increase is more extensive in eukaryotes. Oxidation generally but not always was maintained separately from the cytoplasmic reductive chemistry of both prokaryotes and eukaryotes. Leaving on one side the need for changes of coding (DNA), notice that the most striking feature of the evolution of the eukaryotes is the changing protein chemistry with coalescence of at least two cells seen in all eukaryotes. In them there is a division of the proteome initially between the two of which the so-called organelles mitochondria and chloroplasts are the major centres of bioenergetics and oxidative metalloprotein pathways while the largest, the cytoplasm from a second proteome questionably maintains overwhelmingly the original reductive metabolic pathways and controls over both. Later the totality of the proteome of a plant or animal contains a greater number of participants from different organisms, evolving within a stepwise increase in the protein content of the eukaryote. It requires all the symbionts to donate some proteins, coenzymes, metal ions and substrates to their partner compartments. At the same time some of the metabolites are placed outside all of them in extracellular regions, an extra compartment included within an enclosing membrane or 'skin'. They are ideal places for parasites, not for useful symbionts. All these changes are related to concentration changes in oxygen[32] and trace elements in the environment (see Figure 6.2). In what follows we elaborate new uses of elements and where in the organisms of many compartments the uses arose.

6.9 Survey of the Evolving Uses of Trace Elements

As we have observed many times there is no possible way in which life could have begun without the trace elements. There are essential steps in the chemistry of all cells that need the ability of available metal ions such as iron to transfer electrons and of other ions such as magnesium to activate the chemistry of phosphate transfer (see Table 6.1). Some of these observations have already been made in Chapter 4 in the context of cell organic chemistry. We discuss the trace elements which act in redox reactions and their control first.[41–44]

At first presumably electrons were transferred via Fe/S centres in proteins in a membrane[44] but later this electron transfer process was aided by haem (iron) proteins (see Figure 6.4), and then much later by copper proteins. Their function in rapid electron transfer became the basis of both oxidative phosphorylation and photosynthesis, as described in Chapter 4. It is interesting that what are presumably some of the very early Fe/S centres and protein sequences of hydrogenase proteins are very similar to the first chain of iron catalysts before use of them in oxygen requiring oxidative phosphorylation.[42] Iron complexes, perhaps with nickel and cobalt centres, are also essential in the earliest reactions of the inert molecules in the cytoplasm, for example of H_2, CO, CH_4 and CN^-, to enable H, C, N incorporation into large molecules (see the second of the two above reactions). Notice the extremely primitive-looking iron centre of Fe/S hydrogenases with bound CO and CN[43] (Figure 6.9). Because Fe^{2+} and Mg^{2+} can exchange quickly from many enzymes and control proteins, they were also used to coordinate reactions. Thus the presence of iron and magnesium, associated with proteins, has always been an absolute necessity for the metabolome in its connection with their uses (Figure 6.10). Given iron's particular properties, considerable abundance, and several oxidation and spin states, it is unique, though not dominant alone in usefulness in cells. Hence the increasingly complicated energy-intense procedures used in its acquisition.

We have mentioned that the proteins of the iron-dependent nitrogenases have remained in symbiotic Eubacteria. Their function depends today mainly on iron together with vanadium or molybdenum centres (see Figure 6.5a). Note the extreme complexity of the protein structure based on the complex multiprotein unit (Figure 6.11a) for nitrogen metabolism (Figure 6.11b). A notable curiosity is the use of molybdenum[40] or vanadium as well as iron in these sulfide complexes in proteins. It was either extremely difficult or not desirable to transfer all the proteins and their use to one compartment. Is this to avoid complexity?

An immediate need for the earliest prokaryote metabolism, additional to the handling of electrons and one-electron oxidation reactions, was for the transfer of fragments such as O, H^- and CH_3^-. The transfer of O at low redox

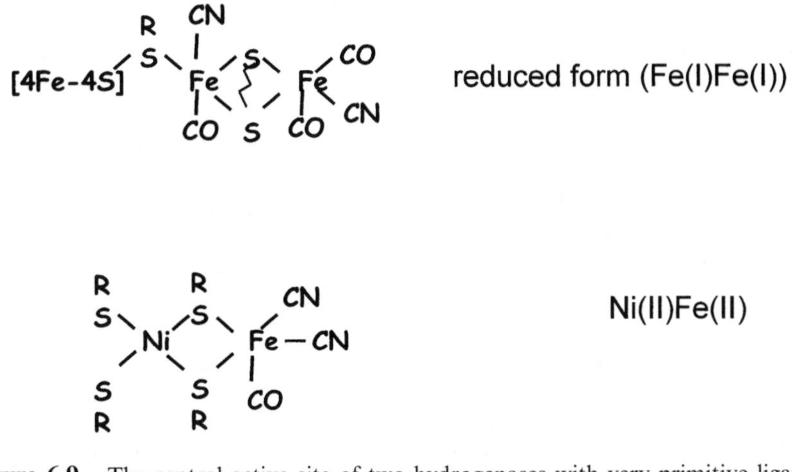

reduced form (Fe(I)Fe(I))

Ni(II)Fe(II)

Figure 6.9 The central active site of two hydrogenases with very primitive ligands. The Ni/Fe enzyme is the more primitive and sensitive to oxygen.

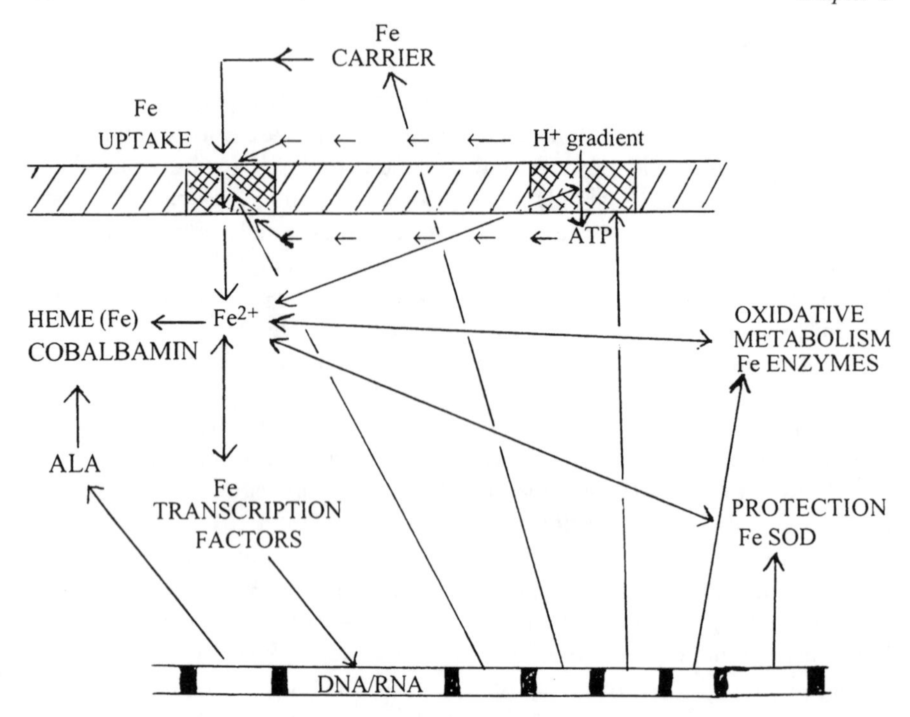

Figure 6.10 The interaction of free iron is with its uptake proteins, its carriers, its enzymes and transcription factors. ALA is α-laevulinic acid. The iron inside cells is Fe^{2+} except in the storage protein, ferritin (not shown).

potential requires a catalytic reaction centre which readily changes oxidation state by two at such a low redox potential. Metal ions that meet this condition easily are not found in the first transition metal series from Ti to Cu, with V and Mn as partial exceptions in certain structures of high redox potential chemistry, but are found in the chemistry of both later transition metal series,

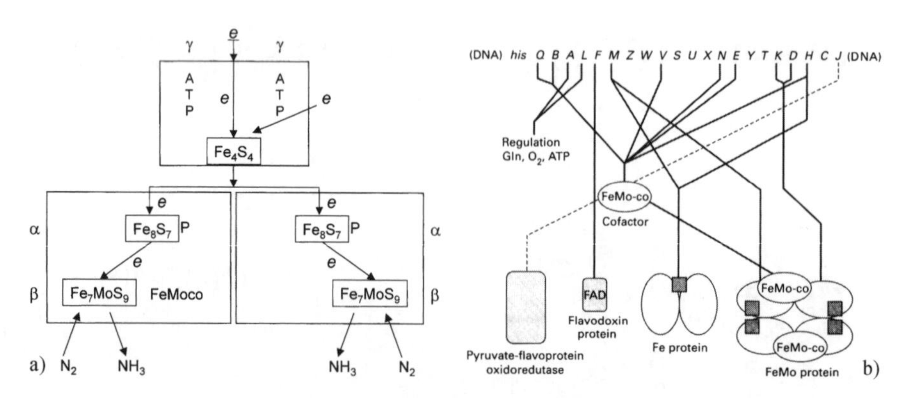

Figure 6.11 a) An outline sketch of nitrogenase. b) The complicated genetic connection for the expression of nitrogenase, see Fig. 6.5, including its regulation and its link to the flavoprotein. Nitrogenase has two separate proteins.

for example molybdenum and tungsten, and in non-metals such as selenium.[5] Today molybdenum catalyses the reaction

$$2RCHO + O_2 \rightarrow 2RCOOH$$

and perhaps much earlier it assisted the back-reduction of RCOOH to RCHO, both through the intermediate MoO.[40] Today Mo (Table 6.5) and selenium (see Section 6.8) are essential in all cells because the importance of molybdenum greatly increased in oxygen atom transfer in fully aerobic metabolism.

A different high redox potential O-transfer chemistry was needed for the production of oxygen from water in the later anaerobic cells which is linked to photosynthesis:

$$2H_2O \rightarrow O_2\uparrow + 4H \rightarrow \text{to reductive synthesis}$$

The catalyst, again seen only in prokaryotes or their related organelles in eukaryotes, is a manganese cluster with calcium and affected by chloride.[43] It evolved shortly after 3.0 Ga and certainly before 2.0 Ga. Manganese is a very high redox potential oxidising agent, much like copper which is active in the reverse reaction in oxidases of oxidative phosphorylation in high redox

Table 6.5 The Major Molybdenum Enzymes[40]

Enzyme Type	*Enzyme Families*	*Example*	*Reactions Catalysed*
Multinuclear M centre	Nitrogenases (Mo, V, Fe)	Mo-nitrogenase	Dinitrogen to ammonia
Mononuclear (pterin bonded)	Oxido-reductases	Xanthine oxidase	Purine or pyrimidine catabolism Aldehyde to acid
	I. Xanthine oxidase family (hydroxylases) (15-20 members)	Aldehyde oxidase Formate dehydroferase	Formate to CO_2
	II. Sulphite oxidase family (eukaryotic oxotransferases) (2-3 members)	Sulphite oxidase Plant nitrite reductase (assimilatory)	Sulphite to sulphate Nitrate to nitrite
	III. DMSO-reductase family (Bacterial oxotransferases) (2 pyranopterins bonded to Mo) (8-10 members)	DMSO reductase Nitrate reduction, dissimilatory; Terminal respiratory oxidase	DMSO to DMS Nitrate to nitrite
Others		Pyridoxal oxidase Xanthine dehydrogenase Pyrogallol transhydrolase	

Note The enzymes using oxygen evolved more recently

potential media. Manganese was of course available before copper. In principle it should be possible to use copper to produce oxygen from water but Mn has remained selected to the exclusion of any other element. This essential reaction, like those of reductive photosynthesis and oxidative phosphorylation, remained throughout evolution as prokaryote (organelle) coded, no matter that most of the genes were transferred from organelles to the eukaryote genome. We ask again what genetic information was in the original 'eukaryote' before the transfers and what was lost? Note in passing that there are no forms of life from its origin to today which are not heavily dependent on metallochemistry, yet little stress is placed on this inorganic chemistry in biochemical studies.

The coming of oxygen was met by slowly increasing use in oxidative chemistry. This is the time of the beginning of single-cell eukaryotes which may have been at first present in surface water from 2.5 Ga. Perhaps the first used in oxygen reactions evolved from the haem enzymes, maybe even just to protect against oxygen in the cytoplasm. The whole of organic chemistry is at risk from oxygen and especially the primitive Fe/S centres. We discuss the evolution of protection from this link in both prokaryotes and eukaryotes later in this section. The risk is unavoidably still with us and may well account for several disabling conditions in humans in old age. At the same time, from around 2.5 Ga, oxygen became used to give valuable products by oxygen addition to or hydroxylation of some organic compounds in the cytoplasm. The major enzymes were the P450 cytochromes utilising haem and novel oxidases based on the oxoglutamate-dependent simpler Fe^{2+} enzymes. We consider that both evolved before Cu enzymes.

The functions of iron, molybdenum and manganese evolved in the anaerobic and then the aerobic prokaryotes as did copper in the above reactions and denitrification. However they are most strongly associated with novel functions of eukaryotes.[1,2] The major value of copper (Table 6.6) developed most strongly, in fact very late, in multicellular eukaryotes mostly at the second of the two steps of oxygen increase in the atmosphere after 0.75 Ga. As an example of its use copper acts in the final oxidation steps in the production of many of the organic messengers required in multicellular eukaryotes, *e.g.* in adrenaline and amidated peptide synthesis (Table 6.7, Section 4.2). These enzymes are placed outside the cytoplasm in vesicles. Copper is also a major oxidative catalyst for extracellular reactions, especially in final crosslinking of proteins of some animal connective tissue[1,2] such as collagen (Table 6.8, Section 4.17). In plants it brings about phenol oxidations giving rise around 0.5 to 0.4 Ga to external lignin. Both lignin and collagen greatly aid stiffness of structure and hence allow the building of large structures, especially on land. There are two reasons for copper use in vesicles and outside the cytoplasm rather than iron: (i) In its strong complexes it has a higher ground state redox potential than iron. Protein binding is able to raise the redox potential, further favouring Cu^{+}, and therefore oxygen binding,[18] which enables copper protein to be used even as a cell-free oxygen carrier in extracellular fluids, in hemocyanin. The iron oxygen binding

Table 6.6 Some Copper Proteins and Enzymes: Localisation and Function[1,2]

Protein or Enzyme	*Localisation*	*Function*
Cytochrime oxidase	External face of mitochondrial membrane	Reduction of O_2 to H_2O
Laccase, tyrosinase	Extracellular	Oxidation of phenols (reduction of O_2 to H_2O)
Caeruloplasmin	Extracellular (blood plasma)	Oxidation of Fe(II) to Fe(III) (reduction of O_2 to H_2O)
Haemocyanin	Extracellular (blood plasma)	Transport of O_2
Lysine oxidase	Extracellular	'Cross-linking' of collagen (reduction of O_2)
Ascorbate oxidase	Extracellular	Oxidation of ascorbate (reduction of O_2 to H_2O)
Galactose oxidase	Extracellular	Oxidation of primary alcohols to aldehydes in sugars (reduction of O_2 to H_2O_2)
Amine oxidase	Extracellular	Removal of amines and diamines
Blue proteins	Membranes (high potential) thylakoid vesicles	Electron-transfer (many different kinds)
Superoxide dismutase	Cytosol	Superoxide dismutation (eukaryotes)
Nitrite reductase	Extracellular	Reduction of NO_2^- to NO
Nitrous oxide reductase	Extracellular	Reduction of N_2O to N_2
Metallothionein	Cytosol	Copper (I) storage
ACE-1 (MAC)	Cytosol	Transcription factor
Dopamine mono-oxygenase	Vesicular	Hydroxylation of Dopa
Coproporphyrin decarboxylase	Extracellular	Production of protoporphyrin IX
Ethylene receptor	Membrane	Hormone signalling
Methan oxidase	Membrane	Oxidation to methanol
Terminal glycine oxidases	Vesicular	Production of signal peptides
CP-x type ATP-ase	Membrane	Copper pump
ATx-1 (Lys 7)	Cytosol	Copper transfer

Table 6.7 Oxidative Synthesis of Small Signalling Molecule Messengers

Molecule	*Enzyme*
Sterols	Fe-oxidases, P-450 cytochromes
Adrenaline	Fe- and Cu-oxidases
Hydroxytryptophan	Fe-oxidases
Amidated Peptides	Cu-hydroxylases

Table 6.8 Some Cross-linking Processes

Reaction	*Catalyst*
Hydroxylation of proline	Iron proline hydroxylase (animals)
Hydroxylation of lysine	Iron lysine hydroxylase (animals)
Oxidative coupling of collagen	Copper lysine oxidase (animals)
Cross-linking of chitin	Copper tyrosinase (animals)
Incorporation of phenols (lignin)	Copper peroxidises (plants)
Incorporation of calcium	(plants)
Disulfide bridges	(general)

carrier proteins, haemoglobins, are only found in cells, erythrocytes, in the extracellular body fluid or internal to muscle cells, myoglobins. A different external iron 'oxygen' carrier neatly carries hydrogen peroxide, haemerythrin. (ii) The difference in compartment arises from the second advantageous feature as neither Cu^{+} nor Cu^{2+} exchange readily while simple external Fe^{2+}, not in haem, is easily lost from proteins, and removed as ferric hydroxide. Haem itself is also to some degree oxidisable. As we have mentioned earlier it is the late and increasing release of copper after, say, 0.75 Ga, from its sulfides that preceded the coming of multicellular eukaryotes (Section 4.22). We consider it to be essential for the arrival of multicellular organisms for the synthesis of both messengers and connective tissue.

Certain copper protein evolution can be traced back to its introduction in the prokaryotes,[5,26,43–45] for example certain oxidases and pumps. They are virtually all based on simple single β-barrel domains[18] (see Figure 4.10). The eukaryotes have many extracellular and vesicular copper proteins based on multiple copies of this domain. Was the gene for the domain carried into the eukaryote genome when the eukaryote organelles formed from oxidative bacteria? Tyrosinases, having a copper site like that of hemocyanin, not a β-barrel, and both are only found in eukaryotes.[43–45]

Before we leave the subject of oxidation/reduction reactions we remind the reader of the control over the network of iron proteins, enzymes, uptake, storage, transport of proteins by the free Fe^{2+} link to transcription factors in both the anaerobic and the aerobic cytoplasm.

One safeguard of the mechanism of the oxygen-using enzymes in the cytoplasm is that O_2 in them is not released as superoxide or hydroxide peroxide but reduced to H_2O or incorporated in a substrate. Additionally the cytoplasm is provided with enzymes that remove superoxide and hydrogen peroxide. Most of the early prokaryotes, as illustrated by mitochondria and chloroplasts, had either an Fe or an Mn superoxide dismutase and a haem or manganese catalase. Later some prokaryotes have a Cu/Zn enzyme but it may well have come about by gene transfer from eukaryotes. (All eukaryotes have a Cu/Zn superoxide dismutase in their cytoplasm.) Some, mainly Archaea, even have a Ni^{2+} superoxide dismutase in their cytoplasm.[35] It has been found in the earliest oxygen-producing prokaryotes, the cyanobacteria.[35] This form of the enzyme is undoubtedly earlier than the Cu/Zn enzymes which are believed to evolve with the eukaryotes but quite probably they do not predate either the iron or manganese enzymes. Unlike these two but like the Cu/Zn enzyme, the metal ion, Ni^{2+}, is very firmly bound in a planar hook of amino acids including these cysteines. This state is an expected geometry for rarely observed low-spin Ni^{3+}, a tetragonal field with a weak fifth ligand, generating the possibility of redox activity. Nickel superoxide dismutase is probably formed through Ni/Co permeases and a transporter, chaperone, also related to Co vitamin B_{12} synthesis. The suggestion is that both these metal ions were pressed into service before zinc and copper became available in any quantity and in such a way as to avoid confusion with functions of manganese and iron when damaging products from partial reduction of oxygen arose some 2.5 to 3.0 Ga. Like other nickel and cobalt and manganese enzymes they tend to be lost as zinc and copper enzymes became available, that is when eukaryotes arose.

Leaving transition metal ions aside, magnesium and zinc introduce quite different types of reaction because they do not act as a redox catalysts and are therefore of lower risk both outside and inside cells. Initially in prokaryotes magnesium was the outstanding catalyst for many relatively easy acid/base reactions and controls often with ATP, as described in Chapter 5. Mg became the centre of chlorophyll very early in evolution. Neither its use in acid/base catalysis or in light capture has changed, even to today. Possibly zinc was used in a few different tougher hydrolases and in ligases,[17,26,27] using water as a substrate (Table 6.9). The earliest known uses of zinc are in RNA and questionably in DNA synthetases and the equally important use in carbonic anhydrase, a few hydrolases, some proteases for digestion, and peptidases. The proteases and peptidases became of great value later in multicellular eukaryotes, externally and in the hydrolysis of connective tissue proteins to allow growth in multicellular organisms. They are known as metalloproteases. These proteases also aid removal after use of messenger peptides. We have shown the general rise in zinc proteins in eukaryotes compared with that in

Table 6.9 Some Zinc Protein Classes

Oxido-reducatse	*Dehydrogenases Cytochrome oxidase*	*All phyla*
Transferases	Aspartate Trans-carbomoylase	All phyla
	Kinases	All phyla
	RNA plymerases	Bacteria
	RNA transferases	Very Few
Hydrolyases	Metalloproteases	Mainly Eukaryote
	Endonucleases	All phyla
	Glycosylases	All Phyla
Ligases	t-RNA Synthetase	All phyla
	DNA Ligase	All phyla
	Carbonic Anhydrase	All phyla
Transcription Factors	Zinc fingers	Eukaryotes
	LIM Domains	Eukaryotes
Zinc Storage	Metallothioneins	Eukaryotes

prokaryotes in Figure 6.7.[4] Outstanding, apart from changes in use in hydrolysis, is the vast increase in zinc-finger transcription factors, of minor importance in prokaryotes (see Figure 6.8). The transcription factors are functional in the action of many hydrophobic hormones, sterols, thyroxins and long-chain aliphatic acids (Figure 6.12). These hormones control many features of growth. Here while free magnesium integrated many hydrolytic reaction steps, bound zinc probably did not exchange directly with free Zn^{2+} but underwent exchange between proteins.[30] With the coming of the first eukaryotes these indirectly exchanging zinc transcription factors controlled growth and were also linked to the corticosteroids which maintain the extracellular homeostasis of calcium, sodium and potassium. The first members of the family of zinc fingers may have evolved in prokaryotes (very few) and may have been introduced into eukaryotes by mitochondrial genes.

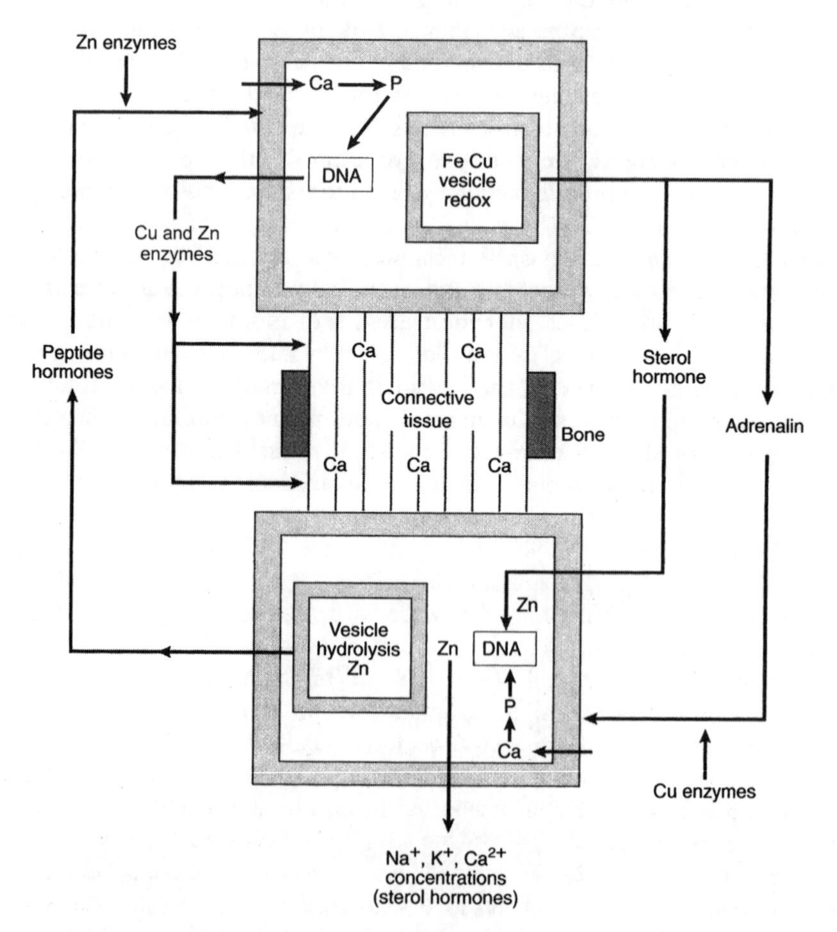

Figure 6.12 A schematic representation of the connections between multicellular eukaryotic cells showing requirements for the synthesis of connective tissue and of organic messengers and their respective receptor systems.

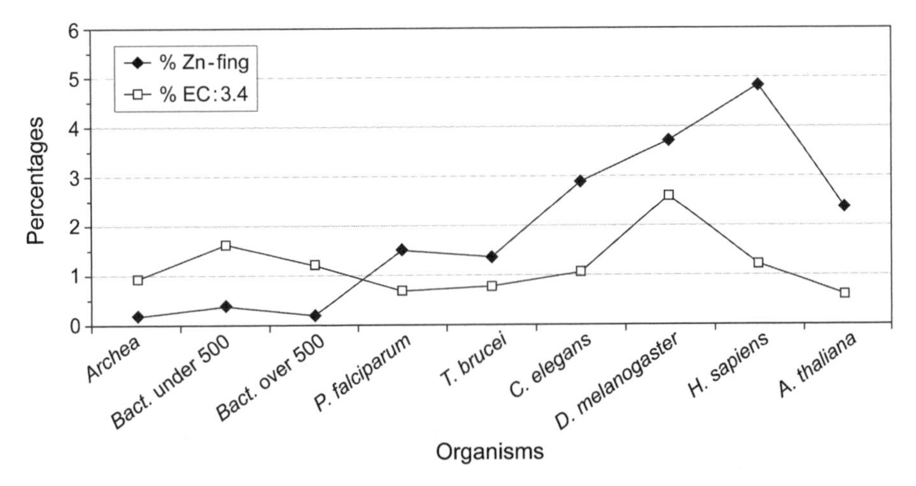

Figure 6.13 Timescale comparison of small zinc proteins in small genome prokarya, in larger ones, then in unicellular and finally in multicellular eukaryotes. It is notable that percentages of Zn-finger content rise in evolution. The percentage value of EC:3,4, a metallo-protease, in small bacteria is higher than in the large bacteria. Small bacteria are usually parasites, they need a bigger pool of proteases/peptidases to break down extracellular proteins for food.

The central value of iron, magnesium and zinc is remarkable. As described earlier, free iron by exchange controls the homeostatic balance of oxidative enzymes in all cells. Examples are control of the citric acid cycle and of RNA conversion to DNA. Fe^{2+} is also important in transcription factors using both Fe histidine/carboxylate, haem and Fe/S centres. They are linked to glutathione in maintaining redox homeostasis in all cells. From life's beginning in the cytoplasm magnesium has acted in controls of many hydrolytic reactions including condensation, and in enzymes for energy transfer phosphates and phosphorylation and is a major factor in their transcription. Many of these reactions are due to MgATP. Zinc is not involved in such fast exchange controls in prokaryotes. But in contrast with Mg and Fe it acts in slow controls in eukaryotes, especially multicellular species. It became increasingly involved in the controls of mineral balance and of many growth functions (see Figure 6.12). Its numbers increased greatly with complexity in eukaryotes (Figure 6.13).[2] As an example humans deficient in zinc have very stunted growth and do not go through puberty. Zinc has been described as a master growth hormone.[46] Free Mg and Fe and later Ca are controls in fast reactions of metabolism while zinc coordinates the activity of many processes which are very slow and gradually changed in significance over the lifetime of an advanced organism. It is then sufficient for Zn itself to exchange relatively slowly due to its strong binding, most likely through direct protein transfer in carrier chaperones.[8,9] One very striking different observation is that the shikimic acid pathway, for production of aromatic amino acids and some transmitters and hormones, which require zinc, has been lost in modern animals. This requirement for aromatic amino acids from outside is an example of the inability of certain living organisms to act independently and competitively with other chemotypes. The success of many driving processes of evolution depends on such simple cooperation. Finally free zinc ions are known to be important in certain vesicles of nervous tissue where they act as messengers, but this activity is not well described. (A further most interesting development is that in very recently evolved plants zinc has taken over the role of coenzyme B_{12} (Co) in some enzymes. We noted earlier that B_{12} is a vitamin in humans.)

The switch to the use of oxidative chemistry in vesicles or outside in eukaryote cells, employing especially copper together with growth requirements of multicellular eukaryotes, needing zinc proteases, presents us with the further problem, as explained in Chapters 3, 4 and 5. The changes in this chemistry are observed in many classes of organism at very close to the same time, from 0.54 to 0.40 Ga. (Note that we take a very short time to be a period of 100 million years for a large-scale chemical change in the total time period of life of 3.5 billion years.) Most obvious is the huge expansion of mineralised organisms or organisms with an outer skin or plastic coat, dependent on these metal ions (Chapter 3). The mineral and plastic developments apply to classes of organisms as widely separated as plants, with external lignin (bark) or internal silica structures, and to all groups of animals from flying insects to those on the bottom of the sea. How could mutation have brought about such common, almost simultaneous, changes requiring novel metalloenzymes with different functions amongst many groups of organisms at very similar times unless there is a connection by gene transfer or by the effect of the environment directly on genes? Remember that each introduction of a new element and often of a new process requires many proteins including pumps, carriers, transcription factors, enzymes and controls with messengers. We return to this question in the last chapter. As an example of information transfer we note the introduction of certain mineral nucleating proteins from bacteria into a modern sponge.[47] Is this a general mechanism? In a parallel development, in Chapter 5, we noted that the calcium messenger system arose possibly around 2.5 to 2.0 Ga in single-cell eukaryotes after calcium had been rejected from all cells. This messenger system is common to all eukaryotes. There is also the common use of calcium in biomineralisation in many different multicellular eukaryotes which seem to, or did, arise in contemporal fashion.

A further stage of evolution is that of the nervous system and then the brain (Section 5.10). Synapse-like protein structures have been discovered in protozoa[48] and their development can be followed to man. Nerves however are not known until their presence in the octopus and the nematode worm by 0.54 Ga with some degree of coordinate connection which appears to be the beginning of the brain. The central role of sodium and potassium ions has been mentioned in Chapter 5. Here we are more concerned with the evolution of nerve/nerve junctions and especially the molecules used as messengers in communication across them. We know that zinc itself is one of these messengers and that copper is instrumental in the production of many of the organic molecule messengers such as adrenaline. These messenger molecules and the enzymes responsible for their synthesis in the nervous system are common to many different groups of organisms and we have to wonder again how nerves appeared across such a wide spectrum of animals, apparently at close to the same time. There appears to be a huge general development from 0.54 to perhaps 0.40 Ga, immediately following a huge environmental change. A very interesting feature is that the trace elements are used differently. Each one has functions in which the element with the best properties for a function is employed and each case is seen in many classes of organism. While little major chemical change has taken place from 0.4 Ga almost to today the next two sections introduce the possibility that chemical change may be beginning again due to man's novel exploitation of the elements.

6.10 Effects of Metal Ion Limitations and Excesses on Growth

We have seen in Section 4.3 that there is a close parallel between the Redfield ratios C:N:P of 106:16:1 in very many, not quite all, algae and this ratio of elements in the sea.[50] Growth generally can be limited if there is an inadequate supply of N or P and it is common to add nitrogen, ammonium salts, and phosphorus compounds (K_2HPO_4) in agricultural practice. The pH is also controlled using lime, but sodium must be introduced only sparingly on land. Here we ask about limitations to growth based on trace elements.[49] We address the problem of life in the sea first (Section 6.5). It has been found that most required elements can be made growth-limiting when their concentration falls below some optimum value so that they are not sufficiently available. While iron appears to be close to this limit in the ocean zinc can also become growth-limiting today if for example growth of plankton is very rapid.[31] As different algae have different requirements for trace elements then the presence of one species rather than another depends differentially on adequate supply. However under stress, shortage of an element, an organism can respond in a few cases by substituting a similar element for that normally used. We give a possible case in the next section. Alternatively its genes could change to become independent of an element. For example higher plants use zinc in place of vitamin B_{12} as a catalytic centre in certain enzymes. This could be of concern in evolution generally as vitamin B_{12} is still required by higher animals which cannot synthesise it but use higher plants for food. Cobalt 'licks' are often put out for grazing animals. Limitations on the supply of mineral elements are quite common throughout the world, of Zn, Cu, Se and Mo for example, but there is also a new risk of excessive supply of other competing elements from our industries which are poisonous. We give an example in the next section.

Excess of a metal ion in an organism or its presence in an incorrect compartment can create serious problems. Alzheimer's disease in man is thought to be in part due to the binding of adventitious copper or zinc ions to the prion protein. On the other hand Parkinson's disease has been associated with difficulties linked to adrenaline which is produced by a copper enzyme. There are many different problems also with iron intake. Both adventitious Fe and Cu can liberate free O_2^- or H_2O_2 which are dangerous oxidising agents but this activity may be compensated by adequate removal of them by superoxide dismutase, described earlier, and peroxidase containing iron or selenium.

These points are made to illustrate possible effects of man's activities on the environment and then on organisms. A particular case is given next.

6.11 The Value of Zinc and Cadmium: 'Carbonic Anhydrases'

A special zinc series of enzymes (carbonic anhydrases) evolved very early in evolution to increase the rate of the reversible reaction:

$$CO_2 + H_2O \leftrightarrow HCO_3^- + H^+$$

The direct supply of CO_2 from bicarbonate of seawater, pH = 8.2, is apparently too small for an optimum function of ribulose phosphate enzymes of algae and plants and carbonic anhydrase assists conversion of HCO_3^- to CO_2 in their cells, pH = 7.3. The enzyme became of increased value in animals for the formation of bicarbonate for the calcium carbonate in biomineralisation. As bicarbonate the HCO_3^- is both transported and pumped, passing through membranes in a controlled manner. Here the reverse reaction from CO_2 to HCO_3^- is required. We draw attention to carbonic acid, drawn, as a possible form, as a four-coordinate tetrahedral intermediate, $C(OH)_4$, in the above reaction (Figure 6.14). Interestingly one step in the hydration and breakdown of SiO_2 (solid) is to give a similarly shaped, transportable tetrahedral, $Si(OH)_4$, on the outside of cells. The enzyme for this hydrolysis, silicase, is clearly related to one carbonic anhydrase (Section 5.8). Thus the formation of silica in vesicles has similarity with the formation of $CaCO_3$. Both biominerals appear in the fossil record at almost the same time and both may well require zinc in an enzyme.

This is a useful point at which to introduce a novel trace element, cadmium, of no known functional value until recently. Morel and his colleagues[51] have discovered that in diatoms a cadmium enzyme is present as an apparent carbonic anhydrase which normally contains zinc or sometimes cobalt in the organism. The appearance of cadmium in the enzyme is explained by Morel *et al.*[51] as being due to the considerable removal of zinc from the surface of the ocean by fast-growing organisms. This is an illustration of the way some cells can adapt to conditions for certain purposes although fast adaptation is of most use in prokaryotes and then unicellular eukaryotes. We would expect that prokaryotes will always lead evolution of changes but this substitution is not known generally in them. An interesting possibility is that, as it is found in diatoms, it acts not only as a carbonic anhydrase but as a silicase, especially as silica is not metabolised by prokaryotes. (Is it possible that Cd^{2+} in the environment only very slowly increased even after 0.5 Ga as its sulfide is very

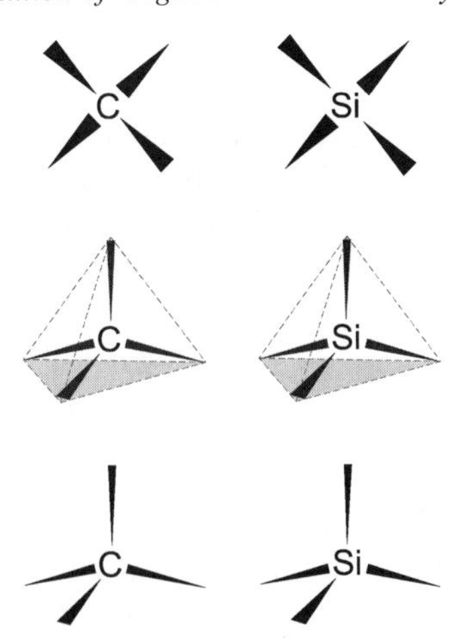

Figure 6.14 The proposed tetrahedral carbon intermediate in carbonic anhydrase and the structure of $Si(OH)_4$.

insoluble and so was not available to early prokaryotes?) If cadmium enters the food chain then higher animals will have to adapt to it because it is poisonous to them at present. We do not believe there is immediate risk but it is wise to be aware of potential risks. One other feature of cadmium biochemistry is the discovery of metallothioneins, the major proteins in the homeostasis of zinc and copper, as a protective protein against cadmium. It was discovered by Vallee in horses' kidneys as a Cd-binding protein.

6.12 The Special Case of Two Non-Metals: Selenium and Iodine

Selenium is the only non-metal trace element required by most, if not all, eukaryotes but not noticeably by most prokaryotes, excluding some anaerobes, and is the only heavy non-metal of general importance (Table 6.10).[52–54] The special value of selenium is in its two-electron reactions either in hydride or oxygen atom transfer. The hydride transfer function is usually to be found in anaerobic prokaryotes. In these two-electron reactions its special character is seen to be a general property of the elements of Group 16 of the Periodic Table, previously called Group VIB.

O – S – Se – (Te)

to which we could well add (Cr) – Mo – W of Group 6 (Section 6.6), which used to be called Group VIA. The later two, either Mo or W, are the only very heavy metals, *i.e.* heavier than zinc, where one or the other is of value, in all organisms.[1,39] Oxygen atom transfer, another two-electron change, is common to the elements of Group 16. The oxygen-transfer function of selenium in human health has been summarised by Gladychev[52] and Rayman[53] who are

Table 6.10 Some Selenium Proteins

Protein	*Function*
Glutathione Peroxidase	Antioxidant enzyme
Iodothyroxine Deiodinase	Enzyme for thyroxine T_3 synthesis
Thioredoxin Reductase	Enzyme for reduction of nucleotides
Formate Dehydrogenase	Enzyme for oxidation of formate
Xanthine Oxidase	Enzyme for xanthine oxidation
Some hydrogensases	Reversible oxidation of H_2 to protons

particularly concerned about its value in detoxification in humans. The selenium enzyme of great importance here is glutathione peroxidase, note the association of the three elements of Group 16, O – S – Se, in this enzyme, and which is only found in eukaryotes. This enzyme removes organic peroxides but not so much H_2O_2, which is removed by catalase, a haem enzyme. The haem enzyme is found in the cytoplasm of prokaryotes but catalase is generally in vesicles in eukaryotes. The great advantage of selenium over iron in these reactions is that Se=O does not readily undergo damaging one-electron reduction unlike Fe=O. Hence Se in an enzyme can be quite openly exposed in the cytoplasm with DNA and is easily accessible to large peroxides and other agents but Fe cannot be and, in its reactions of the small H_2O_2, the haem (Fe) centre is buried, obstructing access to all but this substrate.

Another recent function of selenium is in the enzyme for the removal of one iodine atom in going from the tetra-iodo to the tri-iodo derivative of a phenol which is the hormone thyroxine.[1,2] The use of iodide is mediated by a haem peroxidase. Its earliest use may well have been as a simple protective antioxidant with release of iodine in early algae but its sophisticated value in thyroxine is probably later than 600 million years ago. The evolution of thyroxine appears in fact to be as late as that of the vertebrates, later than the mineral corticosteroids, but it also uses binding to Zn-finger transcription factors to stimulate growth factors. Oxidation of other halides, bromide and chloride, does not occur in higher animals but lower organisms do activate bromination and chlorination with a vanadium haloperoxidase. The products are frequently very poisonous to man.

6.13 Conclusions

We have shown in this chapter that from the very beginning of life the essential requirement for trace elements is very largely in catalysis and in controls of internal homeostasis. Their functions are then unlike those of the more common organic elements examined in Chapter 4 and the bulk inorganic elements in Chapter 5. Their catalytic value in very many redox enzymes is due to their ability to change oxidation state rapidly. They are also strong acid/base catalysts. The use of these heavier elements was limited in the first instance by their availability, which itself is partly a function of abundance (see Figure 1.8).

The availability of these metal ions increased sequentially with the increase in oxygen as it oxidised minerals and elements in the sea. This increase arose from the demand for the reduction of oxides of carbon by cells. Note that certain relatively available elements were excluded as they have no property of clear selective functional value, *e.g.* Rb and Br. Although man, when looking for catalysts has explored the value of elements from all parts of the Periodic Table, organisms are generally restricted by the contents of the sea or soils to those elements in the first three rows of the Periodic Table (see Figure 1.1), which they can obtain most readily. Surprisingly molybdenum (tungsten) is an exception and is relatively available in the modern sea (see Figure 2.11). The major sources of the elements for life were generally more difficult to access on land. Organisms on land faced the difficulty that they had to obtain many of these elements, especially the trace elements from very dilute solutions. All cells in all environments had to bind them in proteins in order to create their special states and catalytic power and selected reactions. These properties are secured by the synthesis paths to the essential, particularly the larger, molecules and to degradations described in Chapter 4.

Unlike the elements of organic chemistry but like some of the major elements discussed in Chapter 5, the exchange in the environment and in cells of some free ions between different bound protein states is fast, but the exchange of a few free ions is slow. The second class is more tightly kinetically restrained. While the free ions are rarely, if ever, active as catalysts some are part of the control system of cells because they act as fast cooperative messengers and controls of metabolism and of their own concentration through fast exchangeable binding to pumps, catalysts and transcription factors. Elements such as free Fe and Mg, and perhaps later some free complexes such as haem, have then all played a very considerable role in evolution from the beginning of life, not just in their activities as catalysts but through fast exchange, where they have assisted homeostasis. The comparable organic molecules are exchangeable coenzymes. Moreover the elements iron, copper and manganese are able to undergo fast oxidation state changes so that they became an essential part of energy capture devices and, mainly later, in a large number of oxidative enzymes. Tightly bound ions, such as zinc, may not have been involved in exchanges at all from early enzymes in prokaryotes but later in zinc fingers at lower binding strength their exchange, maybe not have been with free ions but certainly is between proteins directly. Very early in evolution we also observed the increase in potency of the available trace metal ions by the incorporation of Fe, Co, Ni (remember Mg) in non-changing complexes such as porphyrins. The molybdenum was always in complexes with dithiolate ligands and was also not exchangeable. This tight-binding doubled the variety of strong catalysts. The exchange, direct or indirect, made it possible to use equilibria as deterministic guides with availability in a description of a directed evolution in certain stages of life.

It is important for us to know the concentrations of the trace elements as they play such an essential role in evolution. Unfortunately the timing of the changes of concentration was made somewhat difficult to analyse due to the failure of mixing of levels of the sea from 2.5 to 0.75 Ga. We drew special attention to rises in use of copper and zinc and to falls of that of nickel and cobalt, after 2.0 and 0.75 Ga. The eukaryotes evolved in two major steps, unicellular (around 2.0 to 1.5 Ga probably in surface water), then multicellular forms mainly about 0.75 to 0.5 Ga with the steps of oxygen and trace element increases. The oxidative organic chemistry in organisms was added to the existing reductive chemistry aided by this changed complement of trace elements in the same two periods as described above. A very important fact is that through Fe/S buffering in the period around 2.0 to 1.0 Ga the environment changed little, in accordance to a redox titration of reduced material with oxygen and there was little change in organisms.

However the essential anaerobic chemistry of the cytoplasm, of all cells, is vulnerable to oxygen and protective devices had to be developed but the risks remain to today as evidenced by diseases. Protection was aided in that much of the oxidative chemistry was confined to compartments outside the cytoplasm, especially the organelles, vesicles, and then extracellular space. The novel metal ions were mostly deployed as essential catalysts there. Some of the organic contents of the vesicles could then be exported to the outer cellular space as messengers and proteins to help connective tissue to form an outer skin. Meanwhile to the cytoplasm were added the extra controls needed for growth of eukaryotes which utilised the now more available space. Zinc and copper became essential growth factors linked to connective tissue and signalling between cells.

As we have seen in the cases of all the essential ions, including those of magnesium and calcium in Chapter 5, the ion levels either free or bound in cells have been tightly controlled by feedback at all times. The strictness of this control allowed description of characteristics of the metal content of cells and organisms under the two headings of the free metallome, very similar in all cells, and the bound metallome different in different chemotypes of organisms. These analyses led us to propose that there were, starting from the earliest known Archaea and Eubacteria and increasing up to man, large groups of organisms that have particular trace as well as bulk element dependencies. The number of chemotypes, differently dependent on trace elements, could only expand as the environment changed particularly through oxidation, especially of sulfides. The increased use of the metal ions, requiring pumps, carriers and transcription factors for them in cells allowed the whole cell apparatus to become more complex (Table 6.11). The increase in complexity of the organic chemistry was accommodated as described in Chapter 4 and to which we return in Chapter 7. In the next chapter we will return to the importance of the coupling of the evolution of organisms with that of inevitable previous trace element environmental change, mainly in the sea, which will lead to an analysis of how the environmental changes could have been coupled to the coded genetic molecules. This analysis, based at first on knowledge of proteins,[2,3] is being uncovered increasingly by the use of bioinformatics.[4,5,27] An intriguing point is that as the trace elements changed in the environment many groups of

Table 6.11 Metallo-protein Sets in Two Organisms

Protein set	*Yeast*	*Worm* (C. elegans)	*Metal*
Nuclear hormone receptor (Zn)*	0	270	Zn
Binuclear GAL cluster (Zn)*	54	0	Zn
Metalloproteases (Zn)	0	94	Zn
Na^+ channels*	0	28	Na
Mg^{2+} adhesion*	4	43	Mg
Calmodulin-like proteins (Ca) *	4	36	Ca
K^+-Channels (voltage gated)*	1	68	K
EGF, Ca^{2+}-binding cysteine-rich repeats*	0	135	Ca
Kinases (tyrosine)	15	63	Mg
Cytochrome P-450 (heme, Fe)	3	73	Fe

*Absent in bacteria. Note that values are also available for man and arabidopsis After S.A. Chervitz *et al., Science* (1998) **282**, 2022-7, see also K.N. Deotyarenako and T.A. Kullkova (2001), *Biochemical Society Transactions*, **29**, 139-47.

very different organisms appear to have evolved similar uses for these elements at much the same time. For example we have noted that the trace elements are distributed similarly in different compartments, especially in vesicular and in extracellular fluids where they were much concerned with extracellular matrices, biomineralisation and messenger synthesis in many different classes of organism around the same time. In concluding this chapter we ask, as we have done in previous sections, did these developments arise only through the pressures of survival demanding the best possible, chemical system of oxidation now largely external to the cytoplasm, to add to that of reductive chemistry in the cytoplasm? Was it that protection of the original anaerobic system led inevitably to the only possible incorporation of oxidative chemistry? Just as we ask if the kinetic strength of the compounds formed in reductive circumstances gave the optimal survival independent of any code, we ask now whether this was true of the newer oxidative chemistry. Moreover is it the case that both were dependent on the available bulk and trace elements in an overall cooperative chemical system?

We trust that the last three chapters have shown the fundamental importance to the chemistry in life including the involvement with the organic chemistry of the inorganic elements and in so doing have revealed how evolution was very largely knitted into the environmental changes in its first 3 billion years to 0.54 Ga. In particular we stress the importance of the changing availability and use of trace elements from 3.5 to 0.54 Ga. Since 0.54 to 0.4 Ga there has been little further evolution of chemistry but there have been many environmental fluctuations and further fluctuations will occur in the future. They have been so large at times that periods of extinction have occurred and will do so again. However fluctuations also involve changes of physical parameters, including temperature as well as of chemical availability which can cause different rate changes in living processes. All such stress could also lead to evolution of new organisms (Chapter 7). There is no guarantee that after an extinction period the progress of evolution will not have affected particular organisms strongly. This and other events could alter the distribution of trace elements when a new period of evolution is possible. A question then arises concerning man's activities, which we shall look at briefly in Chapter 7, Part C, for they undoubtedly are leading to a change in element availability.

In completing this with the previous chapter we note that inorganic chemistry in water moves quickly to equilibrium. This has allowed a quantitative approach to evolution very different from that of organic chemistry including genetics, and shows it to be inevitable not random in most respects. It could be called an equilibrium thermodynamic component of cellular chemistry.

References

1. J. J. R. Fraústo da Silva and R. J. P. Williams, *The Biological Chemistry of the Elements*, Oxford University Press, Oxford, 2001.
2. R. J. P. Williams and J. J. R. Fraústo da Silva, *The Chemistry of Evolution*, Elsevier, Amsterdam, 2006.
3. A. D. Anbar and A. H. Knoll, *Science*, 2002, **297**, 1137.
4. C. L. Dupont, S. Yang, B. Palenisk and P. E. Bourne, *Proc. Natl. Acad. Sci. USA*, 2006, **103**, 17822.
5. Y. Zhang and V. N. Gladychev, *Chem. Rev.*, 2009, **109**, 4828.
6. I. Bertini, H. B. Gray, S. J. Lippard and J. S. Valentine, *Biological Inorganic Structure and Reactivity*, University Science Books, Sausalito, 2007.
7. W. Kaim and B. Schwederski, *Bioinorganic Chemistry: Inorganic Elements in the Chemistry of Life*, John Wiley and Sons, Chichester, 1991.
8. E.-I. Ochiai, *Bio-Inorganic Chemistry: a Survey*, Academic Press, New York, 2005.
9. O. Sellinius, B. Alloway, J. A. Centeno, R. B. Finkelman, R. Fuge, U. Lindh and P. Smedley (eds), *Essentials of Medical Biology*, Elsevier, Amsterdam, 2004.
10. M. A. Saito, D. M. Sigma and J. M. M. Morel, *Inorg. Chim. Acta*, 2003, **356**, 308.
11. R. Danovaro, A. Dell'Armo, A. Pusceddu, C. Gambi, I. Heiner and R. H. Kristensen, *BMC Biol.*, 2010, **8**, 30.
12. H. J. Brumsack, *Palaeogeogr. Palaeoclimatol. Palaeoecol.*, 2006, **232**, 344.
13. J. J. Fehrer, *Am. J. Physiol.*, 1986, **244**, 363.
14. C. E. Outen and T. V. O'Halloran, *Science*, 2001, **292**, 2488.
15. L. Andreni, L. Barieu, I. Bertini and A. Rosato, *J. Proteome Res.*, 2008, **7**, 209.
16. W. Maret and Y. Li, *Chem. Rev.*, 2009, **109**, 4682.
17. W. Maret, *Metallomics*, 2010, **2**, 117.
18. H. Gray, B. Malmström and R. J. P. Williams, *J. Biol. Inorg. Chem.*, 2000, **5**, 551.

19. A. Messerschmidt, R. Huber, T. Poulos and K. Wieghardt (eds), *Handbook of Metalloproteins, Volumes 1–2*, John Wiley and Sons, Chichester, 2001.
20. A. Messerschmidt, W. Bode and M. Cygler (eds), *Handbook of Metalloproteins, Volume 3*, John Wiley and Sons, Chichester, 2004.
21. P. Cornelius and S. L. Millar (eds), *Iron Uptake and Homeostasis in Microorganisms*, Caister Academic Press, Norwich, 2010.
22. R. R. Crichton, *Biological Inorganic Chemistry: An Introduction*, Elsevier, Amsterdam, 2007.
23. K. O. Konhauser, E. Percoits, S. Y. Lalonde, D. Papeneau, E. G. Nisbet, M. E. Barley, N. T. Arndt, K. Zahnle and B. S. Kamber, *Nature*, 2009, **458**, 750.
24. R. J. P. Williams, *Nature*, 1955, **184**, 44.
25. T. Cavalier-Smith, M. Brasier and T. M. Embley, *Philos. Trans. R. Soc., B*, 2006, **361**, 843.
26. P. G. Ridge, Y. Zhang and V. N. Gladyshev, *PloS ONE*, 2008, **3**, 1.
27. C. Andreini, L. Banci, I. Bertini and A. Resato, *Proteins Research*, 2008, **7**, 209.
28. C. Andreinic, I. Bertini, G. Cavallero, G. L. Halliday and J. M. Thornton, *J. Biol. Inorg. Chem.*, 2008, **13**, 1205.
29. L. A. Ba, M. Doering, T. Burkholz and C. Jacob, *Metallomics*, 2009, **1**, 292.
30. L. Decaria, I. Bertini and R. J. P. Williams, *Metallomics*, 2010, **2**, 706.
31. A. L. Zerkle, C. H. House and S. L. Brantley, *Am. J. Sci.*, 2005, **305**, 476.
32. C. Scott, T. W. Lyons, A. Bekker, Y. Shen, S. W. Poulton, X. Chu and A. D. Anbar, *Nature*, 2008, **452**, 456.
33. S. Nobuhiko and S. Takeshi, *Sci. Links Japan*, 1998, **48**, 477.
34. R. Chowdry, A. Hardy and C. J. Schofield, *Chem. Soc. Rev.*, 2008, **37**, 1308.
35. J. M. McCord and I. Fridovich, *Free Radical Biol. Med.*, 1988, **5**, 363.
36. A. Quigg, Z. V. Finkel, A. J. Irvin, Y. Rosenthal, T. Y. Ho, J. R. Reinfelder, O. Schofield, F. M. Morel and P. G. Falkowski, *Nature*, 2003, 291.
37. F. M. M. Morel, *Geobiology*, 2008, **6**, 318.
38. H. J. M. Bowen, *Trace Elements in Biochemistry*, Academic Press, London, 1966.
39. R. Morgan, S. Martin-Almedina, M. I. Gonzalez and M. P. Fernandez, *Biochem. Biophys. Acta*, 2002, **1742**, 133.
40. R. Hille, *Chem. Rev.*, 1996, **96**, 2757.
41. C. Andreini, L. Benci, I. Bertini, S. Elmi and A. Rosat, *Proteins: Struct., Funct., Bioinf.*, 2007, **67**, 317.
42. P. M. Vignais and B. Billard, *Chem. Rev.*, 2007, **107**, 4206.
43. A.-M. Sercenco, G. C. Krijger, M. W. H. Pinkse, P. D. E. M. Verhaer, W. R. Hagen and P.-L Hagedoorn, *J. Biol. Inorg. Chem.*, 2009, **14**, 631.
44. L. Banci, I. Bertini, S. Ciofi-Baffoni, X. Su and G. P. M. Borrelly, *J. Biol. Chem.*, 2004, **279**, 27502.

45. D. Magnani and M. Solioz in D. H. Nies and S. Silver (eds), *Bacterial Transition Metal Homeostasis*, Springer, Heidelberg, 2007, p. 259.
46. R. J. P. Williams, *Endeavour*, 1984, **8**, 65.
47. N. S. Wigginton, *Science*, 2010, **328**, 286.
48. S. G. N. Grant, *The Biochemist*, 2010, April, 32.
49. E. M. Bertrand, M. A. Saito, J. M. Rose, C. R. Riesselman, M. C. Lohan, A. F. Noble, P. E. Lee and G. R. Di Tullio, *Limnol. Oceanogr.*, 2007, **52**, 1079.
50. P. D. Tortell, *Limnol. Oceanogr.*, 2000, **45**, 744.
51. Y. Xu, L. Feng, P. D. Jeffrey, Y. Shi and F. M. M. Morel, *Nature*, 2008, **452**, 56–61.
52. D. L. Hatfield, M. J. Berry and V. N. Gladyshev, *Selenium: its Molecular Biology and Role in Human Health*, Springer, New York, 2006.
53. M. P. Rayman, *Biochem. Biophys. Acta*, 2009, **1790**, 1533.
54. S. E. Jackson-Rosario and W. T. Self, *Metallomics*, 2010, **2**, 112.
55. C. Goldblatt, T. M. Lenton and A. J. Watson, *Nature*, 2006, **443**, 683.

CHAPTER 7

The Amalgamation of the Chemical and the Genetic Approaches to Evolution

Part A. A Summary of the Chemical Approach

7.1 Introduction

The purpose of this book so far has been to describe the evolution of the chemistry, inorganic and organic, on the surface of Earth from its very beginning to today. We have treated all the chemistry, in the environment and in organisms, as belonging to one initial system that became energised by internal and external sources, overwhelmingly the Sun. This has given us a third view of evolution to put beside the very detailed analyses of descriptive relationships between organisms[1] and more recently of their genetics,[2] both of which support evolutionary patterns as in a tree. The two previous accounts are in close agreement, especially since about 0.54 Ga, the Cambrian Explosion, but neither provides a convincing description of organisms before this time as the data on which they rely are not available in sufficient quantity or quality. Generally they do not concern themselves greatly with the evolution of the environment. Our chemical analysis of evolution differs strikingly in that it starts from the very creation of Earth, 4.5 Ga, and the surface environment it provided. From 4.5 Ga we have followed the changes in this chemistry both in the environment and in organisms, treating organisms as if they could not change chemistry unless the environment chemistry changed (see Figure 7.1).

Evolution's Destiny: Co-evolving Chemistry of the Environment and Life
R. J. P. Williams and R. E. M. Rickaby

Published by the Royal Society of Chemistry, www.rsc.org

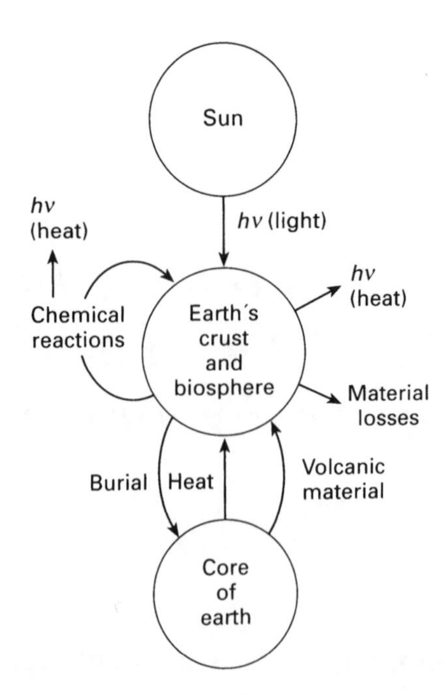

Figure 7.1 A scheme showing the interrelationship of the sun and the earth including the Earth's crust and biosphere in which light goes to heat. Several chemical steps are reversible but major ones are not, see text.

Our account falls under the heading of general systems biological studies.[3–7] The intention of Part A of this chapter is to summarise the contents of Chapters 2 to 6 so that the knowledge of the chemistry of evolution is ready to hand. In Section 7.2 we go back to the beginning of Earth's mineral chemistry history and then follow its changes to today to confirm that this knowledge of abiotic chemistry evolution fits a pattern of logical development and therefore is to a considerable degree inevitable. In Sections 7.3 to 7.6 we follow the same pattern of inquiry into the organic/inorganic system from the time life began, about 3.50 Ga until today (Table 7.1). This concludes our appraisal of evolution's chemical history. In Section 7.7 we consider the reasons why it happened and with this background we shall search for a link to the account of organism evolution given by genetics in Part 7B. This, like all accounts of evolution, rests heavily on observations and surmises of the comparative history of organisms, and on fossil findings and comparisons, using information from today's organisms (Table 7.2). We first make clear the difference between two types of chemistry with which we are concerned.

We pointed out in Chapter 1 that inorganic chemistry differed from organic chemistry in that much of inorganic chemistry relevant to this book was fast and limited by equilibria[8,9] and provided the source of all environmental and organic chemistry. It is open to a thermodynamic analysis. By contrast organic chemistry relied initially on the application of energy to environmental inorganic chemicals and then by further application of energy and control of

Table 7.1 Major Steps in the Evolution of Biological Space

Spatial Division	*Use of Space*
(1) The simplest cell of one compartment	Anaerobic reductive system 15-20 elements (prokaryotes).
(2) Cell with cytoplasm, wall and external periplasm (few vesicles)	Division of space with reductive cytoplasm and external digestion (prokaryotes, aerobes) and some oxidation. Stores of some chemicals in vesicles.
(3) Cell with cytoplasm, Nucleus and vesicular compartments. Plus (later?) organelles. No wall but elaborate outer membrane	Division of space with some reactions and signalling (Ca^{2+}), to vesicles from environments. Oxidations often in vesicles and organelles (Eukaryotes). Reductive cytoplasm.
(4) Multicellular eukaryotes with extracellular fluids and organs (later). Outer skin and shell and then inner skeleton	Differential activity within cells, see (3). Communication between cells and with the environment. Organic messengers, carriers and some activities in extracellular fluids. Biomineralisation also in (3) now. Much symbiosis.
(5) Nerves and Brain added to (4)	New electrolytic message and storage systems. Memory. New response to environment, all incorporated in (4).
(6) Mankind. External space.	Construction in external space, agriculture, machines, use of all 92 elements, electronics. "Control" over environment. Body as in (3), (4). Extensive symbiosis.

rates of transformation to the synthesis of larger molecules in chemical states far from equilibrium. These states can have long lives. The initial and the changes in organic chemistry depended heavily upon inorganic catalysts and controls. What we then showed and summarise here is that the two, the inorganic environment and organisms' chemistry, became increasingly linked in an ever more complex system.

This developing chemistry on Earth was initially mainly inorganic in character because it took about a billion years, 4.5 to 3.5 Ga, before organic chemistry could take strong hold. It is very important to realise that in a completely unknown way the initial abiotic inorganic chemistry evolved to generate organic chemistry and eventually to yield a cell, an organism, by 3.5 Ga (see Figure 7.3). All a chemist can do is no more than to marvel at the step and to provide some thoughts as to how it might have developed. We gave a probable account of what we do know of the initial inorganic chemicals and the energy supplied to them to initiate organic chemistry in an abiotic environment. We then made some speculative general proposals as to how it might have evolved before we turned to cellular chemistry. We provide first therefore a summary of the beginnings of the planet and how its inorganic surface evolved an environment for the possible evolution of the organic chemistry giving rise to life.

Table 7.2 The Major Sources of Chemical Information concerning Evolution

Markers	*Information at certain dates*
Part 7A	
Inorganic minerals Section 7.2	Sediments including radioactive elements for dating Evidence for weathering
Novel elements Section 7.3 and 7.4	Element concentration in dated sediments. Elements in the sea
Small molecules (reduced) Section 7.3 and 7.4	Origin of metabolome
Sources of energy Section 7.3	Knowledge of today's organisms
Fossils Section 7.4	Especially mineralised fossils and matrices required for their formation: Extinctions of life
Oxidation of the System Section 7.4 and 7.5	Nature of matrices for biomineralisation
Biominerals, vertebrates Section 7.4	Sediments and knowledge of today's organisms
Spatial divisions Section 7.5	Development of organelles, vesicles and multicellularity
Part 7B	
DNA/RNA/protein sequences Section 7.8	Extrapolated rates of mutation
The nature of genes Section 7.8	Analysis of genomes
DNA duplication Section 7.9	Analysis of genomes
Epigenetics Section 7.10	Modification of genes

7.2 The Reasons for the Conditions of Earth Before Life Began and its Evolution: Equilibrium, Thermodynamics and Kinetic Limitations

In this section we summarise much of the early parts of Chapter 2 on abiotic inorganic beginnings examining if they fitted a logical (unavoidable) pattern.[8] We then ask how did they give rise to organic chemistry? It is generally agreed that the formation of the Earth can be traced from its beginnings as a consequence of activities of well-described physical principles. Starting from the inexplicable but accepted Big Bang, there is a logic to the formation of the stars and the planets but we do not know how unusual is the particular nature of our planetary system which includes the Earth and the Sun. Earth is not very unusual in its physical–chemical existence as a planet, Mars is similar, but the relationship of the Earth to the Sun, its size, and the resultant rate of cooling of Earth and its atmosphere have given the surface of Earth a well-regulated temperature which could be of only modest probability. The earlier rate and degree of cooling and this subsequent, almost constant, temperature has controlled the chemistry and phases of matter on the surface after many chemicals reached close to an equilibrium condition but a few did not due to the rapid cooling and poor mixing. The nature of the equilibria is well defined by constants as described in Chapter 1. Together with Earth's particular mass these conditions ensured the retention of an atmosphere and a large but not overwhelming amount of surface water for more than 4 billion years. Liquid water is an absolute requirement for life as we know it. Its temperature range is that in which organic compounds are of sufficient chemical stability to have a long lifetime but are readily open to catalysed changes. Other features, perhaps not very likely, are the large temperature gradient from the centre to the Earth's surface, which still supplies some heat to the surface, the magnetic field, which protects the surface from very high-energy particles from the Sun, the period of rotation of the Earth and the presence of the Moon. Putting all these influences together there was formed a particular atmosphere, a layered series of surface minerals and the seas.[9] The distribution of solid and water phases included unavoidably land as a light 'continental crust' and a heavier 'ocean crust' with hydrothermal vents, movement of tectonic plates and volcanic activity because the interior remained hot (Chapter 2). These features of Earth's surface were generally to be expected but detail was not predictable. Most importantly the sea rests on the ocean crust of different chemical composition from the continental crust. Hence weathering of land had a considerable chemical consequence for the contents of the sea. By 3.5 Ga land was possibly only 50% as large as it is today but was constantly changing. The changes were especially in the first 1 billion years, when an expected but limited set of gases in the atmosphere and salts in the sea evolved. All three, gas, liquid and solid phases, flowed in different types of motion on different timescales. The ensuing flows caused further relocation of many minerals and changes of chemicals in solution in the sea and the atmosphere. All in all the initial surface of Earth could have been predicted in broad physical outline though details could not have been described. It suffered now and then major physical disturbances, as it still can do, due to bursts of meteor impacts and of volcanic activity, the timing of which were and remain unpredictable. This picture of Earth without life was one of our starting points in this book (Figure 7.1). We could well believe that a planet such as Earth could continue with this physical pattern of activity for a very long time (Chapters 1 and 2), certainly ending only with the death of the Sun, which has been predicted. So far all of this description from the Big Bang some 13 billion years ago to today was of general physical characteristics. We must now give the main details of the inorganic chemical changes.

A different way of looking at a planet such as Earth is to inquire into its chemical beginnings, starting from its initial content of inorganic chemical elements and then their early selected changes and the beginning of organic compounds.[8,9] The abundances of the elements in the universe and on Earth are known and explicable in terms of the kinetics of nuclear reactions in giant stars sequential to the Big Bang. We noted in Chapter 1, (see Figure 1.8), especially the considerable abundance of the elements H, C, N and O, the intermediate abundance in the universe and on Earth of S, P and Fe and even the moderate amounts of elements up to the atomic number of zinc.[9] Beyond zinc elements generally are only present in very small quantities. The abundances are inevitable from our knowledge of the kinetics of the formation of the nuclei of the elements. An equally important consideration for any chemical system on the surface of Earth was the availability of these, and other, elements in the surface minerals, the aqueous zone (the sea), and the atmosphere. It is known that some elements were trapped in compounds largely out of thermodynamic equilibrium in certain solid compounds as Earth cooled. Mostly, however, compounds such as water, the silicates and sulfides in the mineral surface became more or less equilibrated between solids and dilute solutions in the seas. Quite quickly the early sea had a relatively fixed high concentration of sodium, potassium, magnesium, calcium and ferrous iron cations, with chloride, phosphate and sulfide anions and a considerably lower concentration of several other metals, especially manganese and nickel. Below the surface minerals there was, and still is, a deep layer of mantle and, still deeper inside, iron and nickel formed a hot metallic core. This core and a large region of deep minerals well below the surface formed the majority of condensed matter but they have hardly evolved and they will not concern us in this book (see Figure 2.1). Meanwhile some light elements, H, C, and N were left in compounds in the atmosphere, largely H_2, H_2O, $CO_2(CH_4)$, NH_3, N_2 or in the sea. When together these small molecules were not all at equilibrium with one another and some could react in the presence of catalysts. The observed distribution of the elements in compounds in solid, liquid and gaseous phases was not far from expectation based on their abundance and inorganic chemistry properties at 20°C and related to their electronic structure as in the Periodic Table (see Figure 1.1).[9] The reason for their existence and distribution from 4.5 to 3.5 Ga could then be largely understood using the thermodynamic equilibrium constants described in Chapter 1.

Very largely all the development so far described was downhill, energetically speaking, and predictable except for the exact nature of chemicals in energy traps during cooling. Some changes of this inorganic chemistry are still occurring spontaneously. We described how quite differently the basic set of elements on or near the surface in compounds was exposed to the Sun's high-energy radiation and heated by both it and to some degree by the deep Earth (Figure 7.1). The heating caused generally predictable physical change, evaporation of water and there followed, additional to that caused by the vents and volcanic physical changes, degradative physical–chemical weathering of exposed minerals, on land by rain, aided by motion of the sea and by movements of ice. These flows produced many buried and exposed sediments all over Earth. One inorganic chemical consequence was the removal of a large percentage of the original atmosphere's CO_2 by the elements from weathering, forming Ca/Mg carbonates, another was the introduction of silica, both of which produced sediments later, and a third was the loss of H_2 and CH_4 out of the atmosphere. Many trace elements were distributed into the sea and the sediments. These largely logical irreversible evolutionary changes with time were independent of life at first but more recently weathering became partly dependent on it. (We shall discuss man's role in changing the Earth's surface towards the end of this chapter.) Once the new material from the weathering entered the sea, in understandable qualitative but uncertain quantitative amounts, we considered that all the further abiotic events there were the result of a logical progression based on well understood thermodynamic, physical–chemical equilibrium principles, described in Chapter 1. Note that elements and compounds in the sea of interest to us have fast reactions. These weathering changes, described in Chapter 2, have been continuous from Earth's formation and will continue in the future.

The action of light from the Sun also caused inevitably a separate very different, chemical change in that it energised some mixture of 'inorganic' chemicals, the chemicals which were the precursors of organic chemistry such as CO_2, into more thermodynamically unstable but kinetically stable chemical states, not at equilibrium. This led to the beginning of organic chemistry. We indicated why this specific process occurred as it aided the creation of entropy, light goes to heat. Because of the great kinetic stability of energised C–C, C–H, C–N and C–O bonds at ambient temperatures, above that of the combination of other elements, it gave the most kinetically stable compounds of a few particular forms based on these bonds. These compounds had a long life in the absence of catalysts. This is an unavoidable step. We asked next if this initial organic chemistry was bound to develop (Section 7.3). We can only begin to tackle the problem if there were very quickly produced some form of membrane-limiting vesicles,[10,11] probably made from lipids.

7.3 The Reasons for the Evolution of Organic Chemistry Before Life Began: Kinetic and Energy Controls

We shall summarise first in a little more detail all the likely initial chemical requirements in vesicles before cellular organic chemistry could evolve (Table 7.3). (The pathways in this table are maintained in all cells to today.) The containment within a membrane is necessary to stop great dilution.[10,11] We were not able to avoid much conjecture about the appearance of possible reactions in and inside the vesicle membrane because the number of cooperating events necessary to bring about chemistry close to that of a cell was large. The description we give next therefore was basic to any beginnings of reasonably kinetically stable vesicular chemistry. This chemistry has persisted in every living cell that has been observed, no matter their presumed relationship to the earliest prokaryote or the latest eukaryote. The most elementary steps, many of unknown origin, required to create and stabilise a possible initial vesicular system were:

1. The absorption and transduction of energy into kinetically stable usable forms in the vesicles by activated reactions of H_2 as a reducing agent. The chemical step is the formation of an H^+/OH^- or a charge gradient across the membrane. The second step is the formation of pyrophosphate in the

Table 7.3 Maintained Pathways Throughout Evolution

Pathway	*Example*
Syntheses and degradation of saccharides	Glycolysis (Mg)
Dicarboxylic acid reaction sequence	CO_2 incorporation in incomplete cycle. Completed later to give energy capture, i.e. Krebs cycle (Fe, Mg)
Amino-acid synthesis	Products of glycolysis and Krebs cycle + NH_3 (Fe)
Protein synthesis	Formyl initiation (Fe) and methionine initiation (Fe, Co)
DNA, RNA, syntheses	Nucleic acid pathways (Mg, Zn, Fe/S, B_{12} or Fe_2O)
Synthesis of fats	β-carbon oxidation/reduction (flavin, Fe)
Nitrogen incorporation	Formation of NH_3 (Mg, Fe, V, Mo) in symbiotic bacteria
Hydrogen reactions	H_2 as a reductant (Fe, Ni) in anaerobic archaea
Energy: electron/proton flow in membranes	Energy capture related to ATP (Fe, NADH, flavin, quinones)
Exchange of ions in membranes	Osmotic and electrolyte control (Na, K, Mg, Ca, H)
Light capture	Chlorophyll and haem (Mg, Fe)
Pumps (ATPases)	Most elements as cations or anions
Redox balance	Glutathione, Fe

form of ATP[12–15] from the gradient (Figure 7.2). It is very difficult to see how these reactions evolved.

2. The later synthesis of the necessary small metabolites by reduction using further activated H_2 directly or indirectly from H_2S and later from H_2O. They are the precursors of all large biological molecules (see above).
3. Many steps in the formation of the essential small and large metabolites are condensation reactions. They need a drying, condensation agent in water. Here the above pyrophosphate is the only possibility and its reactions are aided by complexes with Mg^{2+} (see Chapter 5). The pyrophosphate has considerable kinetic stability against hydrolysis.
4. The creation of Na^+, K^+ and Cl^- gradients across the vesicle, later to be called the cell, membrane to ensure osmotic/electrical kinetic stability of a vesicle in the sea (Section 5.10). Energy was required to reject much Na^+ and Cl^- and was provided by ATP at membrane pumps.
5. The creation of selective Mg^{2+}, Ca^{2+} and HPO_4^{2-} gradients across the same membrane for the same but also extra reasons such as to preserve chemical kinetic stability of required molecules in vesicle solutions (Sections 5.1 and 5.4). The divalent cations, especially calcium, at the concentration in the sea would have precipitated or associated organic molecules in the vesicles and had to be rejected. Energy was again required and was provided by ATP and pumps. Mg^{2+} was largely retained in the vesicle at levels not far from that in the sea while phosphate was pumped in.

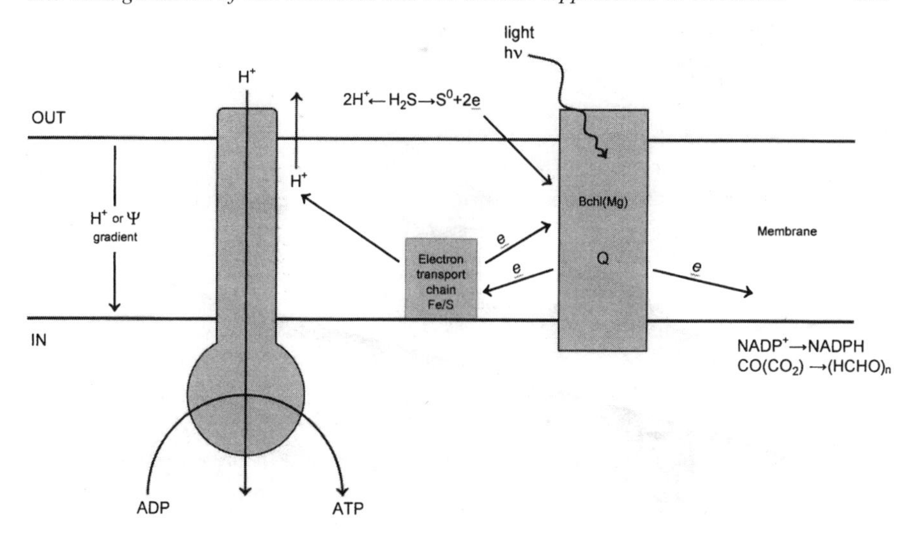

Figure 7.2 The basic nature of what was and remains the pathway of energy transduction. What was required was a way of separating oxidising and reducing units on opposite sides of a membrane leading to the production of a proton or charge gradient. The gradient formation required chains of catalysts (shown here as Fe/S proteins; Quinones (Q)) to form an initial charge separation using energised inorganic chemicals from the formation of Earth or light. The splitting of bound H into H^+ and *e* was followed by both reductive chemistry on one side of the membrane, and oxidative chemistry on the other and by a use of the gradient in a separate neutralising process to make internal pyrophosphate compounds.[15] The pyrophosphate then drove other gradients or internal condensation reactions. Some such transduction of energy is an essential part of the beginnings of life and has remained essential ever since.

6. Many of these steps required proteins of fixed sequence. They could be produced and be producible only if there was a code. This leads to a dilemma as to the origin of the code of DNA/RNA before or after proteins.
7. The extra inclusion of ions able to catalyse both redox and acid/base reactions of organic molecules, notably at first Fe^{2+} and Mg^{2+}, and perhaps some Mn^{2+}, Ni^{2+} and Zn^{2+} (Section 6.1). They were the only catalysts obtainable from the early sea. This very small set of inorganic ions was essential but considerably limited by the nature of the sea and hence in the original chemistry of life. The essential nature of the presence of these very fine metal ions follows from the inability of organic chemicals to catalyse several essential reactions. Certain organic molecules when formed could have aided metal ion capture by coordination and chelation. Note that these catalysts also needed proteins to form metalloenzymes.
8. Very many different reactions as well as expression of protein had to be controlled by internal messengers that we showed were the metal ions, small metabolites and coenzymes of the above (Table 7.3; Figure 7.8).

All seven items, 2 to 8, of these chemical steps in an original enclosed volume, were essential and they needed energy capture and a mechanism to transduce it in its membrane, point 1, into a form suitable to drive them.[12,13,15] We therefore considered that perhaps the first or a very early step in the evolution of life was a mechanism for energy capture. We gave various mechanisms after membrane formation (see Section 4.4), which we shall not repeat here, nor shall we repeat the discussion of the way the energy was then transduced to synthesis using H^+ gradients and then to pyrophosphates. The essential early use of inorganic elements in organic chemical reactions was in this capture of light and electron transfer, pumping and condensation as above. The above steps, 1 to 8, are remarkable but they and the elements involved in them are the unavoidable requirements of the evolution of the chemistry of life and evolved within the limitations of the environment and kinetic stability.[16] We showed that although the reaction pathways were very complicated and of no known certain origin the involvement of functions of the different elements was necessary and unavoidable and hence predictable from our knowledge of chemistry. This includes all the elements, H, C, N, O, S, P, Mg and Fe and Na, K, Ca and Cl. The description does not arise from considerations of the constants of real cells though it closely matches them. It will become apparent that no later living cell as far back as we can go is without these features. We also observed that all the elements involved were apparently the only available ones with particular needed properties.

The whole was a controlled, reduced, energised concentration condition which we took to be autocatalytic and partly cyclic and under feedback management in its precursor vesicles (Section 4.4). There has only ever been one selected major set of C, N, O, H pathways leading to the large molecule products internal to all cells and the selective roles of the metals ions despite the huge variety of prokaryotes and later eukaryotes which have evolved. Finally we did not understand either how the system came to be reproductive based on DNA/RNA sequences much though that has been a major reason for its survival. We gave few details of the fully formed pathways because the original cell chemistry to which our discussion had led had to be related to the first cell of which we have knowledge. All of it is given in detail in biochemistry textbooks. Given the unique nature of these requirements before or as a cell appeared within the relatively short time such complexity arose we believed that no other system of persistent chemicals could have been synthesised to achieve the observed end. In such a case we could believe that it was unavoidable before reproduction just as Earth's inorganic chemistry was unavoidable.

However for lack of detailed knowledge of the steps of the evolution of this anaerobic chemical system, from what we have asserted were unavoidable steps, we had to abandon the idea that life, in an initial reductive form, is necessarily a logical product in the sequence of evolution from the Big Bang.[5,7] Much though we may wish to believe it to be so we shall have to seek extra chemical evidence for its evolution. Clearly life could be a chance one-off, not a logical, complex, kinetic product (even in the universe), a possibility which we look at again in

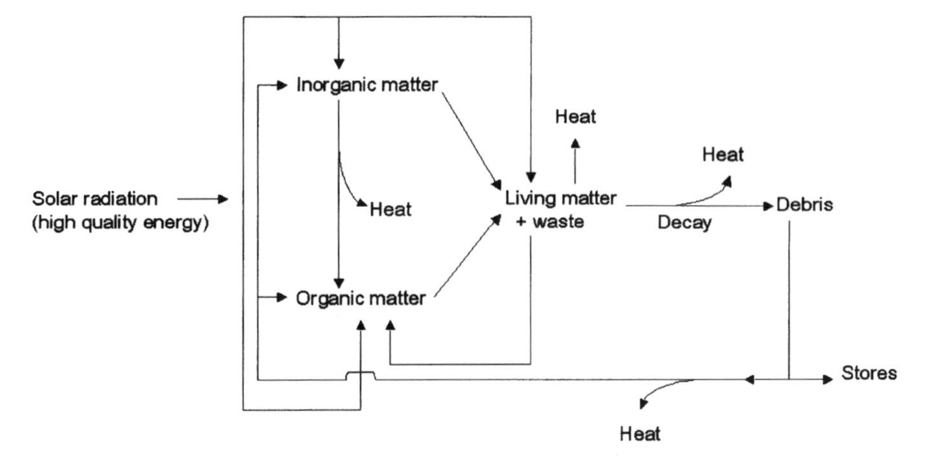

Figure 7.3 A scheme of the way material evolution is driven by energy degradation acting to give environmental change, waste and debris, itself due to organisms. We have not included purely inorganic changes included under weathering in Chapter 2.

Chapter 7B. It is however a system of chemistry, limited by element availability and possible energy capture. It could then well be the inevitable most stable, energised, kinetic system when it is in fact a logical product. It is this seeming inevitability which led us to the title of the book, 'Evolution's Destiny'. It is all driven ultimately by the degradation of light to heat, Figure 7.3.

A very different quest was to see if the further evolution from the manner of arrival of the earliest cellular organism to life was in a systematic sequence.[13] This is the major question of life's evolution from an anaerobic cell. We stated our firm conviction that the major novel chemistry involved close to 3.0 Ga was largely oxidative in both the environment and in organisms and was in stages. This is in direct contrast to the development of reduction in early cells, much of which had to have been established at their very origin and maintained by rejecting chemical oxidation. Once again we stress that our account so far was based on logically essential chemical requirements and not only from knowledge of the original chemistry of cells. We survey the experimental evidence for all of the further evolution including that of the chemistry of life and the environment in Section 7.5 but we needed to outline the evidence for the steps first.

7.4 The Direct and Indirect, Deduced, Evidence for Evolution of the System: Environment and Organisms

We gave two kinds of experimentally based information concerning chemical evolution after 3.5 Ga, and included the release of oxygen from cells around 3.0 Ga (Table 7.4). The first direct evidence for this evolution lay in the

Table 7.4 The Connected Role (with Feedback) of Inorganic Elements in Primitive Homeostasis

Function	*Required Elements*
Osmotic controls, structures	Na^+, K^+, Cl^-
Acid-base catalysis	Mg^{2+}, (Zn^{2+})
Acid-base balance	H^+, HPO_4^{2-} (ATP), Mg^{2+}
Redox catalysis and balance	H, Fe, S, Se, Ni, Co
Protein synthesis and control	(CHNOS), Fe, Mg^{2+}, (Zn^{2+} ?)
Nucleotide synthesis DNA/RNA	(CHNOP), Fe, Mg^{2+}

• NB All element uptake is connected to MgATP, or to ion gradients

changes of minerals and their trace element content, very largely on the Earth's surface, especially in sediments from the sea,[17] and in the period after 3.0 Ga (see Chapter 2). Some of the changes in the minerals were shown to be due to instability (they have positive, physical or chemical excess-free energies in a thermodynamic sense), for example some of the original Earth was initially in a kinetic trap from too rapid cooling, *e.g.* FeS with atmospheric H_2S. This cooling trap resulted too in volcanoes and hydrothermal vents, which we treated with weathering in Chapter 2. Both affected the chemistry of the ocean and also gave rise to sediments. Many such events are directly open to chemical analysis and dating, based on both straightforward analysis and on isotope fractionation data of dated sediments and of trace elements in them. The second lay in the finding of fossils, including chemical fossils (see top three entries of Table 7.2 and Chapter 3).[18,19]

The environment was changed, by the release of oxygen, to more oxidative conditions which became the main novel reaction in causing further evolution, as described below. The rise in oxidising conditions was largely due to the release of oxidising chemicals by organisms which needed reduced chemicals internally and were energised by the Sun. The evidence was supplemented by calculation of the progressive change of redox potential, as the known oxygen increased, leading to a clear prediction of the order of elements released in the environment by oxidation in time. We averaged the changes over the whole ocean for much of the further description of the environment. In this way we deduced the probable average inorganic chemistry of the ocean and atmosphere including that of metal and non-metal ions and molecules. The observations and the calculations are in general agreement despite the poor mixing of the ocean over a long period from about 2.0 to 0.75 Ga. These environmental changes at equilibrium then made much of the novel organic chemistry possible by feedback to organisms of oxidised elements invaluable for catalysis. We stress that the steps in both had a strong inevitable part.

Fossils, and fossil molecules,[18–20] found in sediments, were due to the energised organisation of chemicals by organisms (Chapter 3), and a reminder of them were given in Table 7.2. They too were very different as oxygen increased. The fossils were dated directly or indirectly from the sediments in which they were found, often using isotope fractionation. We gave reasons for the changing nature of the life forms which gave rise to changing fossils from imprints of soft bodies to genuine fossils of hard biominerals. They were linked to the rise in oxygen.

The further kind of chemical evidence for evolution we described were indirect or inferred data from the comparative examination of modern organisms to deduce chemistry, largely organic, of earlier organisms.[21–23] All of this chemistry is energised in a thermodynamic sense. We have looked at the organic chemistry together with essential involvement with it of inorganic catalysts and controls in Chapters 4 to 6. With these experimental approaches we described next the observed development of cells from anaerobic conditions. We wished to highlight the contribution of inorganic elements in stabilising these and all subsequent cells (see Table 7.4). Throughout the next sections of the later evolution of cells we must stress that the greatest stride in evolution was the evolution of cells from abiotic chemistry. It is also the greatest chemical achievement of all time.

7.5 Anaerobic Cellular Chemistry to 3.0 Ga

Section 7.3 outlined the chemistry of the precursors to the first cells, the anaerobes. Apart from the organic chemistry there were two inorganic element requirements. One was the need for catalysts and internal controls often associated with the only metal ions present in sufficient quantity, notably Fe^{2+} and Mg^{2+}. The second was the rejection of Na^{+} and Ca^{2+}. We saw earlier in Section 7.3 that the gradients so set up provided the possibility for organism evolution. At the end of that section we arrived very approximately at the chemistry of anaerobes which appeared around 3.5 Ga. We shall only describe in this section therefore the notable changes of their chemistry from 3.50 to soon after 3.00 Ga when oxygen evolution put considerable demands on anaerobic organisms due to its oxidation of environmental chemicals. The two biggest changes in the anaerobes were the coming of the porphyrins, of coenzymes and the production of oxygen. The iron porphyrins are highly effective catalysts in electron transfer supplementing the use of the earlier Fe/S proteins, for example in the steps of energy capture and transduction. However the greatest advance was the synthesis of a variety of magnesium porphyrins, chlorophylls. They became for all time the major light-collecting, transferring and initial transducing agents in the membranes of anaerobic cells to the cells of today's green plants. In Chapter 4 we deliberately started the description of the use of light as if it were the first major energy supply for precursor systems. We showed that there were other possible earlier sources of energy but chlorophyll-linked activity was and is the only certain mechanism for this activity. Somewhat less interesting was the arrival of cobalt porphyrin in the form we know today as vitamin B_{12}. This porphyrin is used in methyl-transfer reactions. The most curious was nickel porphyrin which only appeared in the Archaea. This, together with the other differences from Eubacteria, led us to wonder if there had been more than one origin of cellular life from its precursor chemistry. The nickel porphyrins are catalysts of reactions of hydrogen and hydrocarbon related chemistry. We noted that there never have been porphyrins of copper and zinc. The reason we gave was that these two metals were locked strongly in sulfides and hence not sufficiently available when porphyrin was first synthesised. It is easy to see that the arrival of the porphyrins, especially of iron and magnesium, was a major step promoting the strength of the cell evolution that followed. It also led us to search for further changes of element uses and for new elements in cells much of which derived from element changes in the environment, one of which was of the molybdenum or possibly earlier tungsten dithiolates for hydride transfer. In each of these examples as well in that of the evolution of coenzymes our knowledge of the chemistry that produced them is slight.

7.6 The Oxidation of the System

It is very important to see that the anaerobic cells we have just described and which are customarily called the Archaea and Eubacteria, though there may have been others, must have been a product of trapped environmental ions and small molecules in a small membrane-limited volume. Its energisation and reduction led to new organic/inorganic chemistry and cellular life. This very early life rejected oxidised material to compensate for reduction in activity. The oxidised material was oxygen, released from water, as soon as the supply of hydrogen from H_2 or H_2S was exhausted. It is essential to see too that much of the expansion and of the style of cellular life was due to the resultant change in the environment from which it had to obtain all its basic chemicals. The changes in the environment were observable from the metal ion content in sediments. We decided that we should follow first therefore the chemical changes in the environment, especially of metal ions largely based on sediment data, as it was progressively oxidised by the release of oxygen from organisms.

These environmental reactions must have predated any change of cellular chemistry as they are fast and inevitable. The oxidative changes back-reacted and at first were poisons to cells. The cells had to overcome other barriers to evolution too as major non-metal elements of organic chemistry required for their chemical paths were oxidised to less available forms. Uptake had then to be of N from N_2 instead of NH_3, of S from SO_4^{2-} instead of H_2S, H from H_2O instead of H_2 and C from CO_2 as all CO and CH_4 had been lost or oxidised to CO_2. The developing methods of uptake were aided by the use of the initial and the newly evolved metal elements in enzymes (Table 7.5). The coincidental changes of several elements[16,21–23] in many types of organisms with those in the environment were observed particularly at the two more rapid rises of free oxygen at about 3.0 to 2.5 Ga and at 0.75 to 0.5 Ga (Figure 7.4). These novel chemicals and their reactions were incompatible with reductive cytoplasmic chemistry and needed the evolution of more and more novel single prokaryote cells or cellular compartments (see Table 7.1, Figure 7.5 and Chapter 2). We

Table 7.5 Common Inorganic Element Changes in Eukaryotes of all Groups

Element	*Change or Increase of Use*
Zinc	Digestive functions
	Increased metalloproteases
	Protection from Superoxide
	Zinc fingers, messenger
Copper	Oxidative Reactions
	Protection from Superoxide
Calcium	Signalling, Vesicle Stores
	Biominerals (Animals)
Silica	Biominerals
Cobalt, Nickel	Reduced Use (Vitamin B_{12})
Iron	Fe/S proteins maintained
	Heme oxidases
	Non-heme non-Fe/S oxidases
Selenium	Anti-oxidation and reducing catalysts
Manganese	Glycosylation
	O_2-formation maintained
Molybdenum	Reductive enzymes at first
	Oxidative enzymes later

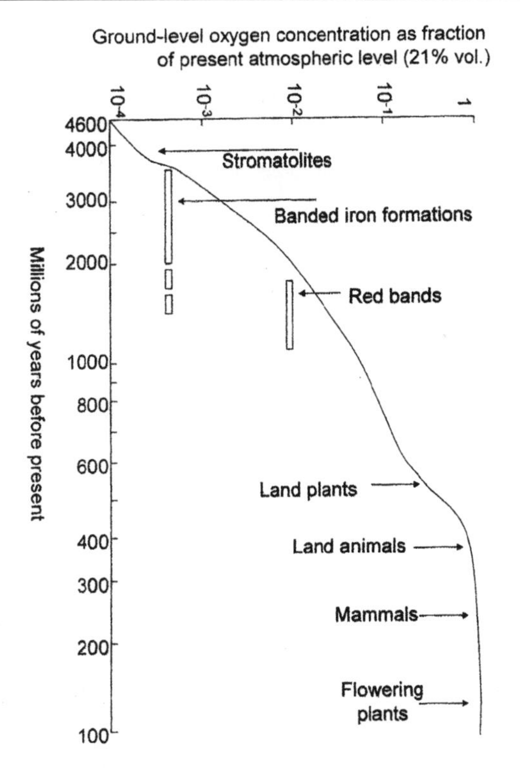

Figure 7.4 The rise of oxygen in two steps around 3.0 Ga and 0.75 Ga and the effect on some observed minerals and on living organisms. Details are given in several earlier Chapters and see Fig. 7.7.

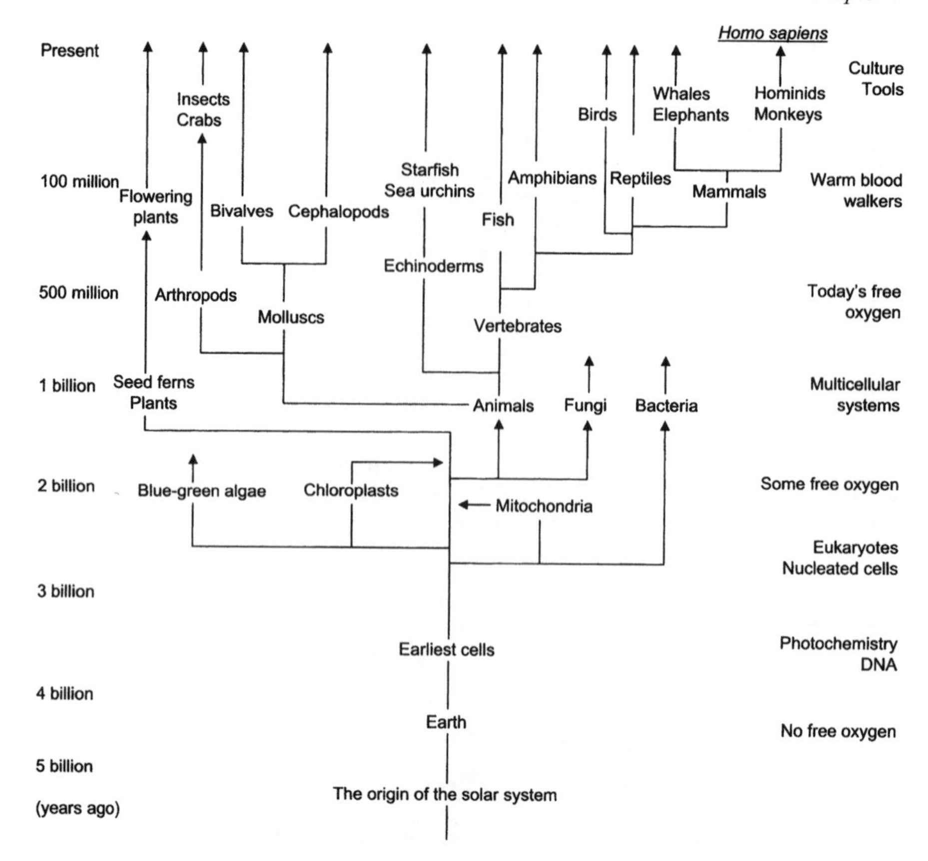

Figure 7.5 An expansion of the living matter in Fig. 7.3 to illustrate the growth of species including space division and of complexity with changes in materials available and of the atmosphere.

detail next the novel cellular chemicals and their reactions which followed in an unavoidable sequence of redox potential. This sequence coincided as the earlier oxidised elements rose in quantity in the two periods from when aerobic prokaryotes evolved slightly after 3.0 Ga, single-cell arose around 2.5 Ga and then multicellular eukaryotes evolved from 0.75 Ga onwards, Figure 7.5.

7.7 Summary of the Evolution of the Oxidative Chemistry of the Elements

Very few of the elements from the main groups of the Periodic Table, that is Na, K, Mg, Ca, B, Si, P and halogens could be oxidised. Those that were oxidised were C, N, S and Se. These four non-metals can be seen to parallel in their oxidation state rises and their timing to those of the changes in transition metal oxidations. Both followed in the environment the order of their redox potentials.[16,22,23]

We followed the progressive changes of all the elements as oxygen increased in the environment. Iron was clearly the most used element in oxidation much as it was and remains the dominant element in catalysed reductive chemistry. The first change of metals was the appearance of more Mo, Co, Ni and Zn augmented by small quantities of Cu in the first rise of oxygen around 2.5 Ga while Mn, Fe and Se became less available. Iron and manganese were lost in mineral deposits by oxidation to oxides while some sulfur, their previous combined anion, became oxidised to a higher valence state water-soluble anion. These tendencies increased in the second stage of oxidation due to the further solution of metal sulfides, giving especially an increased presence of Cu and Zn. There was also precipitation of the oxides of Mn and Fe. There was little change in the period between the two steps while the redox potential in the sea was held relatively constant by the sulfide-iron buffer (Section 2.8).

All the changes in chemistry at first required cells to have protection from them. We described in Section 4.4 how cells evolved controlled ways of removing oxygen and heavy metals probably before they could begin using them (see also Sections 7.5 and 7.6). Particularly dangerous were the reduction products of oxygen itself. We saw that the enzymes for superoxide removal were based on Fe, Mn and Ni in prokaryotes but as metal availability changed the major enzyme of eukaryotes contained Cu and Zn. We did not consider that there were any major changes in internal protective inorganic chemistry of the system afterwards.

The environmental changes were parallel in time with changes in inorganic chemistry of cells as more and more partially oxidised organic compounds were produced (Section 7.5). The metal ions Fe (somewhat), Cu and Zn (considerably) were found to be increasingly used by organisms after 2.5 Ga. Copper externally and zinc internally to the cytoplasm were more noticeably valuable after 0.75 Ga, both as catalysts and controls. During the period from 2.5 to 0.4 Ga the losses of hydrogen and methane from the environment meant that the value of nickel and cobalt used by organisms was reduced. This reduction occurred despite the increase in availability of these two metals as their sulfides were oxidised. After the second rise in oxygen the redox potential in the sea rose by 0.4 Ga to close to that of today at +0.4 V. Details of all these uses in enzymes were given in Chapter 6.

The major changes in the structure and controls of cells also followed from the oxidation of the environment.

- The separation largely of the oxidising steps from the earlier reductions by maintaining reduction in the cytoplasm and oxidation reactions in newly formed compartments and extracellular space and vesicles. First much oxidative metabolism was found in the periplasm of prokaryotes and subsequently in the new single-cell eukaryotes (see Table 7.1) which had many compartments.
- The increasing internal cooperativity that was between these increasing number of cell-derived compartments (some from bacteria including the two organelles).

- All these processes (Tables 7.4 and 7.6) were controlled by exchange and by messengers. Note was taken especially of Fe^{2+} and Mg^{2+} functions. The calcium messenger system connected the outside/vesicle/cytoplasm[24–27] evolved in the earliest eukaryotes. Notice that this is one realisation of possible functional value which derived from the unavoidable rejection of Ca^{2+} from the cytoplasm from the precursors of life (see the beginning of this section). Later zinc fingers came to control hormone-linked expression.
- The later coming together of the cells to give multicellular organisms through the synthesis of connective tissue and held by containing skins. Connective tissue evolved through oxidation of cell surface proteins by

Table 7.6 Main Biochemical Functions of the chemical elements today

General Function	*Elements*	*Chemical Form*	*Examples of Specific Functions*
Structural Functions (biological compounds and support structures)	H, O, C, N, P, Si,, B, F, Ca, (Mg), (Zn)	Combined in organic compounds or sparingly soluble inorganic compounds	Components of biological molecules; formation of tissues, membranes, skeletons, shells, teeth, internal structures, etc.
Electrochemical functions	H, Na, K, Cl, (Mg), (Ca), HPO_4^{3-}	As free ions	Transmission of messages (nerves); production of metabolic energy
Mechanical Effects	Ca, (Mg), HPO_4^{3-}	As free ions exchanging with bound ions	Triggering of muscle contraction; lysis of vesicles
Acid-base catalysis	Zn, (Ni), (Fe), (Mn), -OH, SH	Combined in enzymes	Food digestion (Zn); hydrolysis of urea (Ni); phosphate removal in acid media
Redox catalysis	Fe, Cu, Mn, Mo, S, Se, (Co), (Ni), (V)	Combined in enzymes	Reactions with oxygen (Fe, Cu); oxygen production (Mn); nitrogen fixation (Mo); inhibition of lipid peroxidation (Se); reduction of nucleotides (Co); reactions with H_2 (Ni); bromoperoxidase activity (V)
Signalling to DNA (via transcription factors)	H, O, C, N, P, S, Mg, Ca, Zn, Fe, Cu	As small molecules or ions	Binding agents to transcription factors
Various specific functions	Mg	In chlorophyll	Light harvesting in photosynthesis
	Fe, Cu	In proteins	Transport of oxygen
	I	Special covalent compounds	Hormonal action
	Fe, Ca, Si	Mineral compounds	Sensors (magnetic, gravitational)
	O_2, N_2, CO_2	Gases	Flotation (fish)

copper metalloenzymes but it had to be broken by hydrolysis now and then to allow growth. Hydrolysis was due to new or greatly increased numbers of zinc metalloenzymes.[31,32] Calcium rejected into the space between cells in the multicellular organism to build connected structures.

- Cell-to-cell information transfer arose in many different multicellular organisms using small partly oxidised organic molecules going from a donor to an acceptor cell aided by calcium messengers.[24–27] Their appearance was apparently simultaneous in different classes of organisms, in the extreme in both plants and animals.
- The development of biominerals, shells and later bones gave structures greater strength. Much of the biomineral chemistry arose in vesicles into which calcium or silicon were rejected. Biominerals evolved in many phylla simultaneously.
- Increasing advanced symbiosis of cells external to one another. Organised combinations arose of several types of what had previously been thought to have been separate species into a unit. This will force us to consider what is the common use of the word 'species'.
- Finally there evolved the nerve system linking distant cells quickly from which the brain developed, again close to simultaneously in different classes of species, *e.g.* in the octopus and nematode worms.[28–31] The essential elements here are Na, K and Cl in carrying current and Ca in the release and response to organic messengers. Here we refer again to the absolute necessity to reject sodium ions before any organic chemistry could evolve. This rejection opened up the possibilities realised in the nerves and brain.
- The continuous increase in energy used drove the whole biological development.

All of the steps of organic chemical changes were described in more detail in Chapter 4 and were related to the increasing use of space by organisms (see Table 7.1). We shall not give details here but note events outside the cytoplasm:

1. The changes of compounds in membranes including cholesterol.
2. The increase of oxidation of side chains of proteins, including –SH to –S–S–, leading to external crosslinking of them.
3. Increase in oxidation of phenols to form polymers. Together with (2) this produced external organic chemicals of great strength.
4. The use of crosslinking to give connective tissue with the use of sulfation in sugar to give open meshes.
5. The oxidation of small organic molecules in vesicles to give messengers on release.
6. Both crosslinking at times and messengers generally had to be hydrolysed by external zinc proteases and peptidases respectively.[32] The first to allow growth.

For a full description the reader is referred to textbooks on biochemistry. The dependence of organic chemistry advances in organisms on the availability of inorganic chemicals was made clear in Chapters 5 and 6.

In conclusion of this brief summary of the changes of both organic and inorganic chemistry we saw a distinct parallel between them dominated by the rise of oxygen and of the oxidation potential of the elements in the sea, followed closely by the evolution of organisms (Figure 7.4). We also noted the link between major steps of evolution with the realisation of functions from the probabilities created by the rejection of calcium and so we repeat that the question arises as to whether the parallel evolution of organisms was a case of cause and effect with feedback, directed development, or of random processes. Random process-seeking survival of the fittest would have given a great variety of organisms not easily observed in the period before 0.75 Ga but clear later. Directed development appears to arise from our survey of their almost simultaneous chemistry changes (Figure 7.5). We can only handle this problem if we have clearly in mind why the organic chemistry of genes itself evolved (Chapter 7B).

7.8 Summary of Why the Chemistry of the Environment/Organism System Arose and Evolved

We wish now to establish some of the fundamental factors behind all of the chemistry of evolution and then to summarise to what degree all this evolution was inevitable, as the title of the book suggests. The undeniable fact is that all material consists of a combination of the chemical elements. Any change, evolution, must then be analysable in terms of novel element combinations. These statements apply equally to the geochemical and biological objects. We have already described the beginnings of Earth and the settling of the inorganic environment largely as an inevitable downhill thermodynamic process approaching equilibrium. We then asked why did organic chemistry which required energy (Section 7.2) start? We could describe this most readily from the secure knowledge that it was due largely to the action of light from the Sun, although certain other energy-donating chemicals were probably available even earlier. At first the capture of light was in the environment by the water and most inorganic parts of the sea but the sea contents were not capable of retaining light before it degrades to heat. This retention was possible by light energisation of the reduction of CO_2 to give organic chemicals unavoidably formed with hydrogen donors (Sections 7.3 and 7.4). The first step had to be the synthesis of an enclosed volume made from lipids formed by reduction of CO_2 through formaldehyde with polymerisation. We gave evidence for the possibly unique way of capturing and transducing energy for this step. The further absorption of light caused reaction between the organic molecules and this increased as many molecules of higher molecular weight absorbed light at lower frequency. The increase in light absorption will continue to the limit of the ability of organic compounds to capture it. The answer as to why this process happened is therefore that organic chemicals that formed were of considerable kinetic stability in energised states. We could only see some possibilities of how the organic chemistry developed further in the general

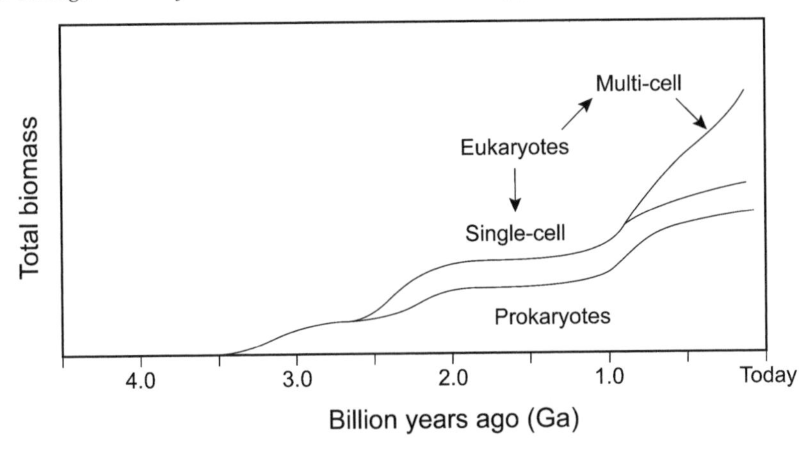

Figure 7.6 The growth in three different large groups of organisms from 0.5 Ga to today. Note the periods of rapid and of slow increase corresponding with the oxygen increases.

direction of anaerobic cells but we saw how cellular activities arose as a coordinated whole. However the very fact that cellular synthesis and activity were found in one system of polymers, only DNA, RNA and proteins strongly suggested that there was no alternative. The system became one of several types of living anaerobic cells, demanding not only the organic elements but a considerable number of energised inorganic elements. We noted that the basic chemistry of these cells arose between 4.0 and 3.5 Ga so very quickly, probably in less than half a billion years, implying that it could well have been unavoidable. It took much longer, more than 3 billion years, for human beings to evolve together with a huge range of animals, plants, fungi and unicellular prokaryotes. We gave a logical explanation as to why this second phase of evolution occurred in broad chemical terms and why its evolution was so slow.

A striking feature which allowed the existence of organic chemistry in an evolved volume was the unavoidable rejection of sodium and calcium ions. This rejection created possibilities for signalling realised later through the use of gradients in messenger systems. In particular calcium became the messenger of outside to inside signalling for all eukaryotes and maybe they could not have arisen without such signalling. It is clear that the signals are of the very essence of cells and later were absolutely required for the evolution of all mechanical movements all the way up to the highest animals. The possibility of value of the unavoidable sodium gradient was realised in the development of nerves and the brain. It appears to the authors that very little of the evolution from prokaryotes could have come about without the rejection of these two cations.

We did not discuss further why anaerobic cells came into existence after the initial action of light, but we concentrated on progressive essential chemistry (Table 7.7) related to changing complexity, Figure 7.6. We referred first to many inorganic processes in cells which were shown to be fast and go to equilibrium. Equilibrated contents did arise after compartmental energised

Table 7.7 Times of Major Evolutionary Chemical Steps

Date (Ga)	*Evolutionary step*
4.5	Earth formed
4.5 → 3.5	Weathering. Much CO_2 becomes carbonate, Volcanoes
	Sea content of Na, K, Ca, Mg, Cl develops
	Some continents, precursor chemistry
3.5 → 2.0	Weathering. Changing land and sea, volcanoes
	Prokaryotes undergo development using inorganic ions and organic chemistry
	Mo for nitrogen fixation; O-atom transfer
	Mn for oxygen production
	Fe for electron transfer, O_2 reactions, etc
	Zn for a few reactions
	Low level of copper
	Sulfate production buffered with sulfide (2.7 Ga)
	Ni and Co rise as catalysts but fall later
	Cell organic chemistry mainly reductive: some oxidative later
2.0 → 0.75	Weathering. Changing Land and Sea
	Unicellular eukaryotes
	O_2 increases slowly as sulfide removed
	NO_3 increases
	Oxidative chemistry in cells increasing slowly, iron decreases
	Several trace metals rise slowly
	Ca messages
0.75 → today	Weathering. Changing Land and Sea
	Multicellular organisms
	Extensive oxidative chemistry aided by Cu, Fe. Connective tissue. Biomineralisation, Bone
	Organic messages
	Na/K message. Nerves and brains
Today	Evolution changes its character: human activity

transfer from or to the environment across membranes. They included the internal balanced concentration between bound and free forms of monovalent and divalent states of metal ions and their redox states including Na^+, K^+, Cl^- (no binding), Mg^{2+}, Ca^{2+}, Cu^{2+}, Mn^{2+}, Co^{2+}, (Ni^{2+}, Cu^{2+}, and Zn^{2+}) in any given isolated compartment. The last three only equilibrated between bound states. We assumed that both the environment and organisms must follow equilibrium conditions in their compartments from about 4.5 Ga to today in so far as these fast-reacting ions are concerned. Some were important as stabilisers of a cell's osmotic and electrolytic content, Na^+, K^+ and Cl^-, catalysts, controls or messengers. We showed earlier that each metal had a very selective availability and chemical capability and hence they were essential in particular functional ways for the kinetic stability of cells. Their evolution entirely conformed to expectation from general chemistry, much of which is maintained in all cells: no other elements can replace them. This gave a very easy explanation of why they were so used in cells. As we have already explained evolution of the functions of non-metal organic chemistry in cells was partly restricted by these metal ion equilibria. The forms in which non-metals appeared in cells were not simply related to outside concentrations by energised pumping but were controlled by directed catalysis and energy to give products of temporary kinetic stability.

All these energised unstable gradients and synthesised organic molecules decayed with time to their initial environment conditions in a cycle of environmental chemicals – uptake, synthesis, degradation, Figure 7.3. The light energy used to create gradients and synthesis was degraded to heat. The conversion light to heat increased the more light was captured and used as the organisms grew in numbers and size in the sea and on land. The driving force of life is then the entropy gain of light going to heat via chemical intermediates. The remaining puzzle is how the evolutionary process was sustained by reproduction.

There is one further logical development, almost disconnected from natural evolution, which is the development of chemistry and physics by man. This is a development making use of material, environments and energy not previously available to organisms. The question arises as to whether or not this development will conflict with the natural processes. If it conflicts with present organisms we shall need to discover if biological systems can utilise the products of this chemistry which would mean new animals and plants.

7.9 Added Note on a Novel Genetic Analysis Related to Chemical Development

We have not used genetic analysis in this description of the evolution of biological chemistry but a very recent study has made such an analysis using genomic data possible in principle.[33] The effort is to include the beginnings of inorganic and organic biological chemistry. This was accomplished by analysing sequences of DNA related to the proteins of organic chemistry, averaged over some 4,000 species. The results indicate that from life's beginning carbon dioxide, formate and acetate with ammonia and hydrogen sulfide were involved. In contrast methane, nitrogen and its oxides were not present and sulfate binding only occurred and increased after 3.0 Ga. Nitrate only became bound significantly after 1.5 Ga. The orders but not the dating details agree with our account. Of course our concern is not with gross evolution of organisms but we have concentrated on that of particular groups of species, chemotypes.

A parallel analysis of the origin of the binding of trace metal ions by proteins, again grossly averaged over 4,000 species, showed that from the earliest times iron with some zinc and manganese were present in organisms.[33] Copper, molybdenum, cobalt and nickel were absent. The first to be found strongly was iron but surprisingly by this analysis, zinc binding was also considerable. Manganese binding decreased later and copper, nickel and molybdenum were not present in quantity until close to 1.5 Ga. Most of these findings agree with the general descriptions we have given apart from the exact datings.

Part B. The Connections Between the Chemical, the Biological and the Genetic Approaches to Evolution

The first section of Part B is an introduction to the nature of genes and inheritance and a discussion of ways in which genes may be altered to suit the changes of the environment. We shall point to several difficulties in present analyses. Only later do we come to grips with the problem of seeing how changes in genes depend upon environmental chemical changes.

7.10 The Nature of Genes: Gains and Losses of Genes and Inheritance

The essence of continuity and propagation of changes in organisms lies in reproduction, but it does not require sexual combination which only arose some 1 billion years ago. We can do no more than give our impression of the immense subject of reproduction and genetics[2–7,31–41] in this book while we present our view of its connection with our overall topic, the chemistry of evolution. One view is that reproduction is simply definable by the description of DNA doubling (with cell division) and evolution is related directly to DNA changes, mutations, when bases in a sequence are changed. Now we have shown that evolution is led by environmental change which does not appear in the definition. We need therefore to relate environmental and organism chemical change to reproduction with changes of DNA in a different definition. We start from our knowledge of the inherited active components in a cell. All cells depend ultimately on light for energy and then material intake, 'food'. This energy is then used to connect environmental chemicals to the chemicals in a cell. Division must therefore include the reproduction of all the machinery necessary for each cell to survive including means of obtaining and using energy and material. In Chapters 4 to 6 we showed that this conversion involved first a range of small inorganic molecules and ions from the environment, which are energised in uptake and initial synthesis in controlled concentrations. Further later syntheses from them include those of all metabolites, proteins, RNA, saccharides, lipids and DNA itself. It is this whole set of chemicals and their activities which must be reproduced. We can consider these activities in a sequence. DNA is the coded molecule which is to be run into the reproductive machinery to make *a single copy* of itself, DNA, called replication, and also into other machinery to make controlled numbers, multiple copies, of coded RNA, some of which are then run through other machines, ribosomes, which make controlled multiple copies of sequenced proteins. It is the controlled multiple copies which characterise a cell not just the singular DNA. The catalytic proteins, enzymes, often combined with metal ions which interact with substrates. In the cell the enzymes then make small saccharides and lipids from the small substrates. The lipids form a membrane enclosing a limited volume of liquid, the cytosol. Each one of the small units, ions or small molecules, either directly or after combination into particular compounds, metabolites, inside the cytosol, back-interacts with the proteins, enzymes and pumps to control numerically their input and outward transfer. The input units often act on proteins, called transcription factors, which switch on or off many protein syntheses including part of the machinery which feeds the DNA tape into the DNA replication machinery. There are additional RNAs produced from the coded DNA such as transfer t-RNA and RNA which back-interacts with DNA, i-RNA. The translated code must include proteins which contain metal ions for the transduction of 'external energy' to activate all the steps. Reproduction is of a homogeneous condition.

All of these components of the original cell, including all the machinery, are in an intensive feedback system and must be reproduced, inherited, on division and we must relate changes in them to changes in DNA. The reproducing cells must also have a definite external inherited environment supplying the essential small molecules and energy. Adaptation to a new environment requires DNA changes or at least changes of expression. There is no doubt therefore that the environment, non-metals and metal ions, interact with all the internal activity of cells, using external energy sources. The whole organism can be described as a sum of *gene units of inheritance*, not just DNA. We look then at sections of the code (DNA) as coding for particular RNAs and proteins and giving rise to particular activities of the cell. The inheritance also must include all the other required components for these activities from an environment. We shall call the combination of both internal and external chemicals in studies of biological evolution as an essential part of a *cellular gene*. This common, informal use of the word gene, *i.e.* cellular gene, must not be confused with the use of 'gene' as an exclusive one-to-one relationship between DNA and inheritance.[23] We shall call this sense of a gene, which is often used in popular biology, *the DNA-gene*. It is not related to the environment directly nor to many characteristics of an organism as they depend on the environment. We can now ask, 'Are the DNA changes random, allowing cellular activity to catch up with and benefit from environmental change but without its direct or indirect influence on DNA, consistent with the idea of a DNA-gene?' Alternatively we can ask, 'Can the environment drive inheritable change in which genes are referred to as 'cellular genes'. Note that a DNA or RNA virus cannot reproduce in isolation so that codes themselves are not living genes much though they can reproduce in cells inside an organism. Clearly cellular machinery and conditions are required. No protein can be produced without DNA of a cell and no reproducible changes of cellular genes can occur without the cell's DNA changes. We shall turn to further very important components of this machinery included in the cellular DNA under the heading of epigenetics (Section 7.10).

We must consider therefore the way in which changes of the coded tape of DNA can be made so as to change expression of proteins in genes in organisms using either description of genes. The first way is the conventional effect of random damage to the whole DNA which causes single-base changes by chance and which is reflected in changes in multiple copies of RNA and proteins. This could well be related to but not caused by an environmental factor. The protein change, say in an enzyme, is then persistent if it gives a selected improved survival value to the cells.[34,35]

An example of the conventional way in which the non-metal environment might be said to influence the evolution of organisms in prokaryotes to unicellular eukaryotes is apparent when the relative rates of reaction of a natural inhibitor to those of a substrate of an enzyme change, see Section 4.13. We take the case of the protein called RuBisCO, ribulose-1,5-bisphosphate carboxylase/oxygenase. It has two competing molecules for binding to the active site, CO_2 and O_2, of which the CO_2 reaction for carbon incorporation is essential. Now the ratio of CO_2 to O_2 in the atmosphere has changed dramatically over the 3.5 billion years of life (see Figure 2.5). Of concern here is the ratio of the response to CO_2 as opposed to O_2, after cyanobacteria produced O_2, say after 2.5 Ga, and its subsequent gross increase to today while CO_2 declined very severely from 3.5 Ga and even from 600 Ma toward 200 Ma. The study of RuBisCO's specificity factor (τ) of CO_2 relative to O_2 is shown in Figure 4.20 for a number of organisms from 2.5 Ga to close to today. The general trend is for a large change in activity in RuBisCO in favour of CO_2 with some red algae as a peculiar exception. Here the suggestion is that random positive mutation has been utilised to drive change to overcome unfavourable environmental CO_2/O_2 ratios. Any conclusion also must take into account the total production of the enzyme and other factors involved in C/N accumulation. A thorough study will be complicated, needing a check on other environmental factors and an analysis of DNA sequences. This case is given as an example of an easily understood relationship of protein change to DNA-genes through mutation but we shall tackle the problem in all its complexity later in this chapter.

There is little doubt that this mechanism, often called the survival of the fittest, has been active and that is demonstrated by many DNA/RNA protein sequence studies. It is easy to see that it fits quite well into a picture of a connected set of slow changes in organisms in one branch of a tree of evolution. It is very general and likely to be true in the evolution of organisms which have very similar chemical natures with no changes in chemical elements except in quantity. A problem arises when the rates of DNA change are often many times greater than normally expected, with rapid changes in the proteins expressed, including new elements and proteins. However if an initial DNA of a given protein is to generate a novel chemistry, as in a greater step in evolution, then to maintain the essential activity of the cell the original unmutated protein must still be expressed. This requires an increase in the DNA with maintenance of the critical original DNA. In passing we observe that there is no way that the mechanism of random mutation alone could achieve the evolution. Looking back to the precursor beginning of life there is no mechanism of production of proteins which could have depended on DNA

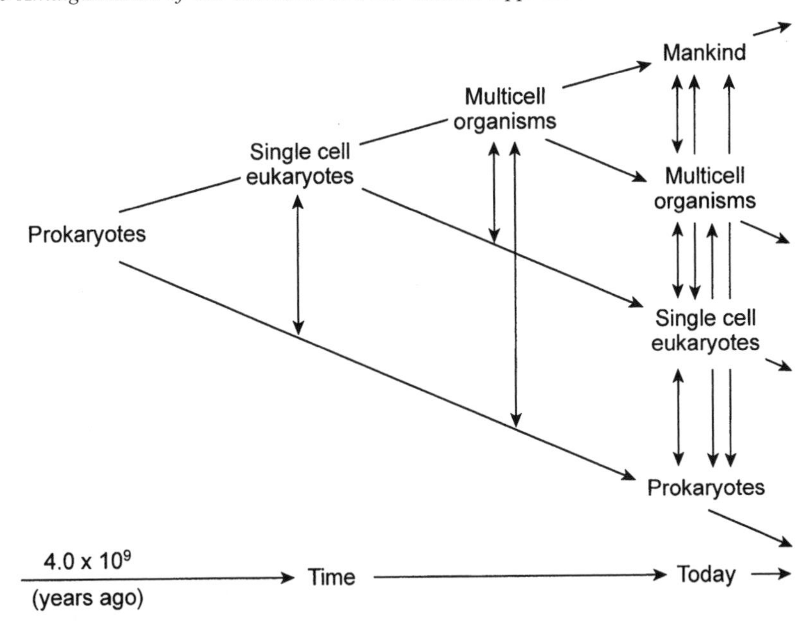

Figure 7.7 The probable gene transfers which have occurred. Those between the earliest organisms classes are well demonstrated. Transfer between the higher groups is uncertain but possible arguments for it are given in the text within a discussion of cooperativity rather than competition.

as DNA did not exist. Whatever that mechanism was, could it not operate again? It took far less than 1 billion years to evolve the first cell but the cytoplasmic activity has changed little, although cells have existed for 3.5 billion years.

A quite different possibility for evolution after DNA modification in one organism is either by transfer of genes from it to another directly through contact or through carriers such as phages or viruses (Figure 7.7). This transfer is often discussed under the heading of reticulate evolution.[42–47] In fact the DNA of man is known to contain a large quantity of transferred viral DNA but we do not know how much novel genetic information has been introduced in this way as there is much 'junk' DNA (90%) in human and other eukaryotes (but see below for a discussion of 'junk'). We all know how virulence spreads between people but we also know, through the work of Barbara McClintock, that there are some units of DNA, transposons, which enable transfer of an isolated DNA activity internally, so that transfer between cells has a possible starting point. Transposons allow shuffling of sections of DNA so giving rise to variety. We know that this step is active in the human immune system but does not lead directly to novel evolution. The mechanism of DNA transfer through contact was also aided by the movement between organisms of small circular pieces of DNA, plasmids, which is known in many types of cells (see Figure 7.8).[43] Now the plasmids have a special interest for us as they are capable of entering the main DNA of bacteria by conjugation and can carry

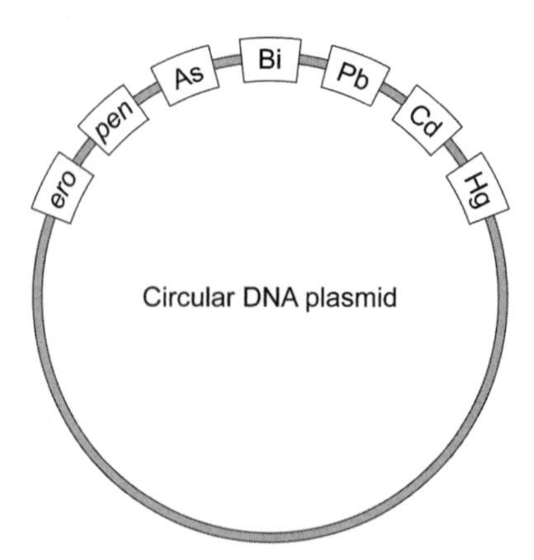

Figure 7.8 The DNA of a plasmid is found to carry resistance DNA-genes to both man-made antibiotics and poisonous metal ions in the environment.

DNA-genes to and from the DNA. This may well be possible in some eukaryotes such as yeast too. They can certainly be integrated into human DNA.[46] Bacteria are able to expand their information in plasmids relatively easily, including extra DNA. This is shown by their capacity to develop genes against newly introduced adverse factors in the environment such as protection against poisonous metals, Cd, Hg, Pb and antibiotics (see Figure 7.8). It is through this ability, with their independent DNA replication, plus DNA (gene) transfer that resistance to, for example, drugs, can spread widely amongst bacteria. Thus plasmids, like viruses, could have aided evolution considerably. Possible early examples are the introduction of useful metallo-enzymes many for oxygen reactions in many phylla simultaneously (Figure 7.9). The obvious question is how did the resistant genes arise in the plasmids or viruses in the first place, as they require additional sections of DNA to be created? This activity has no clear connection to mutation. Next we remind the reader of the relative ease of transfer of sections of DNA between cells in the creation of eukaryotes,[47] before we refer again to increases in DNA. No eukaryotes exists without mitochondria even if in some cells the organelles are degraded to hydrogenomes.

As described in several sections already the nature of organelles, mitochondria and different chloroplasts arose through acceptance of bacteria probably by some form of primeval eukaryote. The result was the creation of modern eukaryotes and happened with or due to a change of the environment on the introduction of O_2. The acceptance occurs with extremely large DNA-gene transfer from the bacteria to the larger cell DNA.[43,47] The advance in complexity in what was presumably some kind of basic eukaryote cell, with a special membrane, vesicular and nuclear properties with this organelle 'cooperativity', led to gain in overall efficiency of energy capture from light or oxidative phosphorylation.[48] (In passing, a grave weakness in all descriptions of evolution is the lack of knowledge of these large cells which

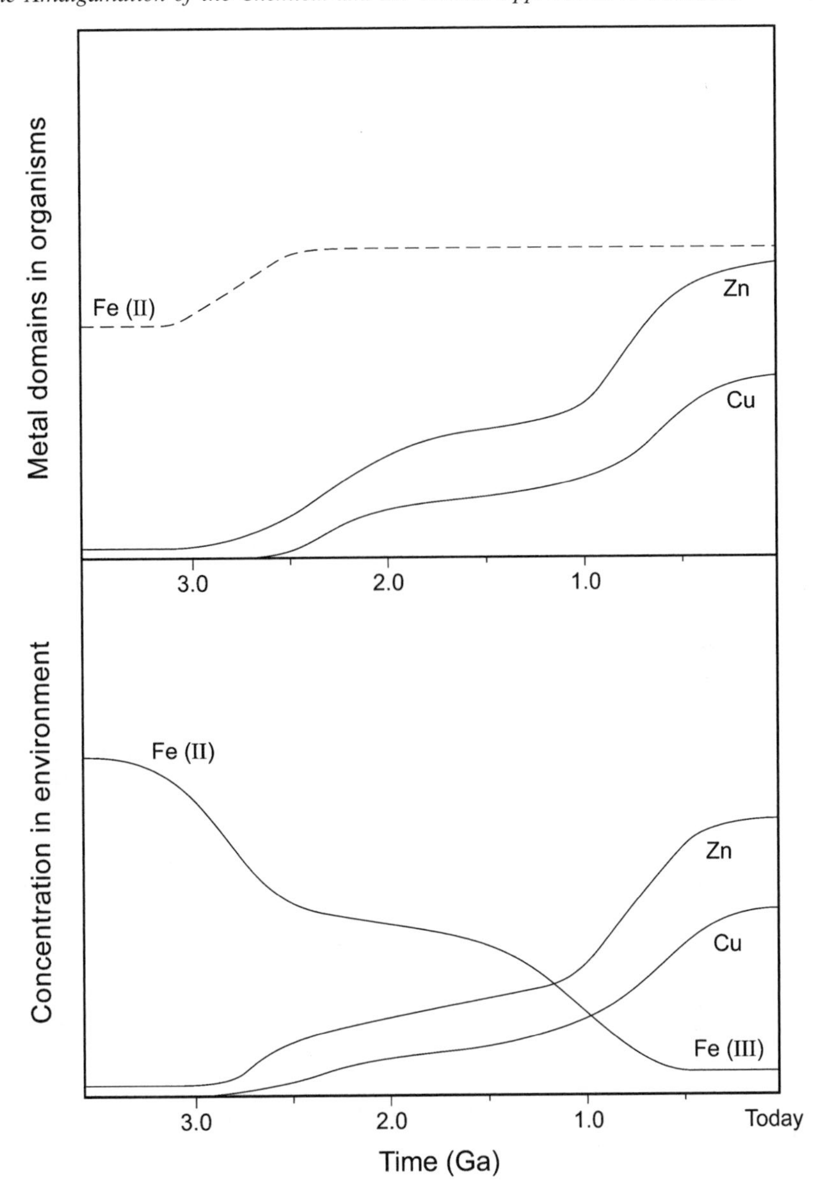

Figure 7.9 The parallel development during evolution of element availability and of protein domains of organisms.[22] Much recent DNA/RNA sequence evidence confirms this pattern. The bottom figure is the same as Fig. 2.16.

incorporated the bacteria, Figure 7.10) Other cellular genes possessed by the captured bacteria may have been transferred too, perhaps zinc and copper functions. There is later also coalescence of two eukaryotes in photosynthesising cells. This means that at somewhat similar times in different ways it has paid branches of life to become so dependent on other branches that they exist

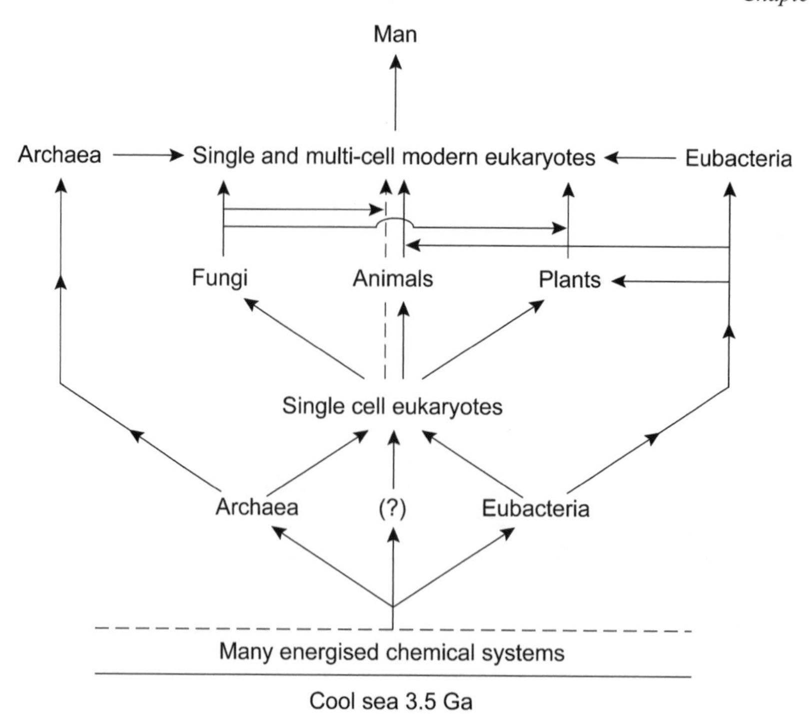

Figure 7.10 The relative early development during evolution of organisms, showing all the uncertainty of the original organisms and the arrival of eukaryotes.

only because they assist one another mutually for the sake of a now blended, survival strength. (The coalescence of genes is still present in the coalescence of sperm with an egg in sexual reproduction.) No matter how the coalescence happened, different fusions led to the several different major lines within classes of animals, algae (plants), and fungi at almost the same time (Table 7.10) (perhaps around 2.5 to 2.00 Ga). Plants with different chloroplasts, originating from several different bacteria with different photosynthesising molecules and the same mitochondria, show that the origin of eukaryotes, such as brown, red and green algae, arose in different ways. (We also suggested that coalescence of vesicles might have led to the origin of cells in Section 4.7.) In fact coalescence or combined existence is one part of symbiosis that is a dominant feature of later evolution.[47] In conclusion it is clear that the coalescence of two DNAs did not require any mutation in DNA but was in response to environmental change. It was closely associated with the rise in oxygen and with it a novel environment and a novel membrane with molecules, cholesterol, needing oxygen for its synthesis. The new structure gave access to a new energy source, light, in some eukaryotes and potentially to a wide range of oxidised environmental factors. It did require expansion of DNA and the complementary enzymes to handle oxygen of the new environment in

Table 7.8 Closely Timed Evolutionary Steps across many Phyla

Time (Ga)	*Evolutionary Step*
3.5	Appearance of Prokaryotes
2.5	Aerobic metabolism (Eukaryotes)
	Mitochondrial incorporation
	Calcium Signalling
	Zing Fingers
0.75-0.50	Multicellular eukaryotes
	Biomineralisation
	Increase in Zn, Cu, Se Enzymes
	Zn Transcription Factor Increase
	Organic Messengers (Oxidised)
0.40-0.35	Nervous system
	Na/K Messengers

additional ways (see Sections 7.10 and 7.11). It also required new proteins, some for exchange of substrates between compartments, *i.e.* trafficking. Now coalescence could well have introduced two types of an essential functional protein (gene) in the product. This, like other forms of duplication (see below), would allow evolution of new functions from one of them without loss or mutation of the essential activity in the other.

Here we reintroduce the opposite subject that of loss of genes in general.[42–47] A very surprising feature of evolution is that many of the genes lost by later complex organisms are those absolutely essential for life of all organisms. Remarkably the first things lost from the main DNA of the eukaryote cells was the need to produce chemical energy, ATP, and certain pathways such as the reverse Krebs cycle, fatty acid oxidation and nitrogen fixation. Their DNA is found only in the DNA from organelles, incorporated prokaryotes. What could be a more remarkable indication of one route to escape the difficulty of management of the chemical complexity introduced both by use of oxygen and of its environmental products? A possible rationale is that it is much easier to produce bioenergetic gradients in the smallest enclosed volume. Later losses in many later eukaryotes concern such essential syntheses as those of coenzymes, often called vitamins, amino acids, sugar and fatty acids essential for man, the loss of ability to reduce SO_4^{2-} and NO_3^- as sources of S and N, and of proteins containing certain metal ions. Other notable genes which are also lost are those which concerned the use of substrates such as hydrogen and methane and the requirement of much uptake of some metal ions involved, for example much Ni and Co. This is a clear indication that the genes (and their DNA) which are lost (or not expressed) are those which provide essential products obtainable from the chemicals of the environment, that is from simpler organisms which must make them or from losses of environmental chemicals. But why do this? The losses are extensive and selective. The result is that symbiosis, internal or attached rather than the remote association of simpler with more complex organisms, became common, beginning with the arrival of eukaryotes some 2 or more billion years ago. Whatever the explanation it has to result from a feedback to a part of the DNA to make it redundant. The feedback has to be from the information that a synthesis is not required as the supply is adequate from its cellular environment. This is the same puzzle that arose in the creation of new genes for new chemicals. Clearly the DNA is sensitive in selective ways to its environment whether it is an external feature or a symbiont. The loss of redundant genes is of course an advantage in that it reduces complexity. A feature of it is that loss of certain genes is common across large groups of organisms and is of course a loss of a particular activity and often with a complete loss of local pieces of DNA.

Loss of a cellular inherited factor does not necessarily involve loss of a DNA-gene. There are ways in which a DNA-gene can be silenced by the binding of a small RNA which is included under epigenetics in Section 7.10. However there are parasitic organisms such as trypanosomes which have lost large sections of their DNA, selectively, and have a small DNA. They only survive in the host's environment.

In each novel evolutionary step, except when gene coalescence or loss is involved, we see that we need extra proteins. A further way of introducing new genes *de novo* is by modifying 'junk' DNA, sometimes referred to as including pseudo-genes (DNA).[49] They are sequences of DNA which are inherited but have no apparent function. Recent discussion has centred on the conversion of these pseudo-genes (DNA) into active expressed genes. This finding gives a possible reason for maintaining some junk DNA in eukaryotes in that the junk allows modification without damage to the DNA of the genes which were already required to be expressed. Junk DNA may have been introduced by 'coalescence' with viruses. Alternatively it may well be that much 'junk' DNA is introduced by duplication of genes. It is not present in Eubacteria but there is some in Archaea. Once 'junk' DNA is present then only mutation is required to give a novel activity.

7.11 DNA Gene Duplication: A Possible Resolution of the Problem of Gene/Environment Interaction

Throughout the above discussion of DNA there is the overriding evolutionary need for an increase in the DNA-gene complement in order to introduce novel functions, Figure 7.8. Apart from coalescence the greatest probability for increase is duplication which is known to be extremely common and rapid, often forming a considerable part of the genome and 'junk' DNA.[49] It is clearly a prior requirement of enzyme complement increase if somewhat different DNA-genes are to evolve from a parent by later mutation.[49–54] In many genomes, including that of *Homo sapiens*, at least one-third, 15,000, of the total number of DNA genes, 40,000, have arisen from duplication. It is seen immediately that the duplication is not of the whole DNA although this can occur and accounts for the high multiplicity of chromosomes and for division of classes of some groups of organisms. Now the observed duplication of a local segment of DNA will just produce identical DNA-genes and amplify activity. There are two further possibilities. The initial DNA segment can have been of an expressed protein with two activities both of which can be separately improved by mutation after duplication. Alternatively the duplicate can be inactive away from a promoter and become junk DNA or a pseudo-gene which through mutation eventually finds a new valuable use. We saw throughout this book that such new uses generally appeared after a change in the environment, which must have been stressful. We examine the possibility that stress is responsible for duplication later (see Section 7.10). Duplication may also be advantageous, apart from in the evolution of different organisms, in multicellular cells of an organism because in each differentiated cell the differences in expression and hence metabolic activity could make a particular optimal rate of one enzyme the best for a particular organ and for cooperative purposes.[55–58] This we see in the well-known case of iso-enzymes associated with different organs.

Our major interest however is in the evolution of new organisms in stages with the rise of oxygen (Chapter 7A). We have observed in Chapters 4 to 6 that changes in the environment were accompanied by expansions in the proteome. We drew attention especially to metalloproteins with similar functions.[59–60] Knowledge of the proteins duplicated could lead to conclusions as to the cause of duplications.

We decided therefore to inspect the numbers of duplicated proteins which bind the metal ions Zn, Cu or Fe in proteomes of some 800 different organisms, including both prokaryotes and a series of eukaryotes.[13,59,60] The numbers of possible proteins were derived from the inspection of the genomes. In this we were greatly helped by Bertini[59] and Dupont[60] who developed suitable bioinformatic programs. The results are quite striking in that in all organisms only a few proteins show multiple duplication. Of the classes of 30 Zn, 20 Cu, 120 Fe and 50 haem proteins in multicellular eukaryotes only about five in each metal ion group had more than 10 copies usually representing very similar or related functions (Table 7.9 and Figure 7.11).[61,62] Moreover the

Table 7.9 Local DNA (Gene) Duplication

	Numbers of very Similar Proteins					
	Peroxidase	*P450*	*Cu oxidases*	*Fe oxidases*	*Zn fingers*	*Calmodulin*
Organism		*Heme*	*EC:1*	*20G-FeII*		
Eubacteria	1	1	3	1?	5	0
Yeast	1	3	9	1	373	4
Arabidopsis	95	268	111	116	1631	80
Homo Sapiens	0	70	21	9	2984	160

Note Proteome size from top ~3500, 6,000, 32,615, 37,742.

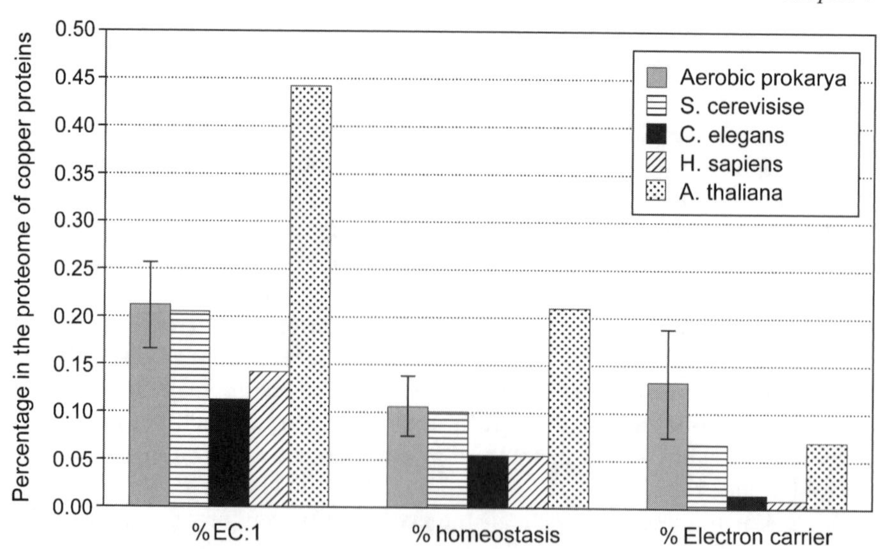

Figure 7.11 A comparison of the percentages in the proteome of some copper proteins in five different organisms, EC:1, Oxidases, the other two proteins are homeostatic and electron transfer proteins containing

number of copies of a particular metalloenzyme varied greatly in different organisms. There were very few in prokaryotes,[55] but many oxidases (Cu, Fe and haem) and transcription factors (Zn) in multicellular organisms (Table 7.9). Looking at Table 7.9 and Figure 7.11 in more detail we note that in the plant, arabidopsis, the numbers of copper, haem and non-haem oxidative enzymes all exceed 50 copies and each duplication is much larger than in a complex animal such as *Homo sapiens.*[62] The oxidases of a plant act on a wide range of related organic compounds such as phenols, many for production of crosslinking proteins used in synthesising structures such as bark and other compounds, but importantly for defensive purposes against toxins. In a plant damage, seen as a wound, is managed by enclosing a wound by the growth of 'plastics' such as lignin, synthesised from such phenols, in a protuberance, a gall, made by oxidation, using these enzymes. Modern plants have seeds which are more open to a variety of environments and hence could be expected to show greater duplication of protective enzymes than animal genes. The suggestion from these observations is that stress causes gene multiplication and, when followed by mutation, is a major cause of evolution. Oxygen and the oxidised metal elements, and other molecules produced by the reactions of these two, are known to be the source of major stresses. A further example is the development in plants and animals of proteins for calcium and silicon biomineral synthesis (Table 7.9 and Chapter 5). These proteins were shown to require oxidative crosslinking of an external protein matrix for Ca and Si binding by metal (Cu) enzymes which must have appeared in several forms only in classes of animals. Protection is a response to stress.

A striking second difference between plants and animals is in the degree of duplication of certain transcription factors relative to that of oxidative enzymes. The hormones and other organic messengers to these transcription factors are often compounds requiring oxygen and copper and iron metalloenzymes for their synthesis. The particular transcription factors which we have analysed are the zinc fingers, because here again we have a link, through zinc, to the environment (see Table 7.9).[16,61] These fingers are often the acceptors for hormonal signal messages. Quite differently from the oxidative enzymes it is now the animal not the plant proteins which are more greatly duplicated (see Table 7.9). This is also the case for the calcium-signalling proteins (Chapter 5).[61,62] While, as stated, many of the chemical agents which apparently lead to duplication of protective proteins are linked to oxygen and its environmental products, in the cases of zinc and calcium proteins this cannot be the case. It appears that their duplication is related to the greater need of animals to coordinate response from outside to inside cells (Ca) or between cells (Zn receptors for hormonal organic messenger molecules). Animals need to respond to external events, stress and internal growth (stresses), respectively. Growth itself was undoubtedly aided by the oxidative action of oxygen aided by iron and copper increase but growth also requires external hydrolysis. We have noted that many zinc hydrolases are also needed in growth. Human DNA has over 500 copies of such hydrolases.

We have concentrated so far on the duplications of certain metalloproteins because they are the easiest to link to stresses coming directly from environmental change, which is mostly of metal ions. The DNA sequences to these proteins are far from the only ones which have been extensively duplicated. One noteworthy duplication is in the Hox series of proteins known to control spatial development.[53,56] Here duplication gives proteins that are related to different body part development. It has also been noted that duplication of crystallins,[57] assisted the ability to use different regions of light absorption by the eye. This could arise by adaptation since the wider use of light of different wavelengths. There are many other examples of duplication acting to aid variation in function.[58–64] For example isoenzymes of RNAase are also common in different tissues and heat shock proteins have hundreds of copies in human DNA.

A very interesting case is the effect of certain drugs in humans. In particular the drug methotrexate causes fast, local multiplication of the enzyme dihydrofolate reductase[65] and is correlated with mutations in the P-53 zinc protein which acts strongly in cancer cells. A very interesting second case is that of the response to damage by hydroxyurea.[66] Here the drug inhibits the enzyme ribonucleotide reductase, an Fe/S protein. In both cases the sequence was drug resistance possibly by considerable duplication and mutation. We shall refer to this study and the references in it to other examples in Sections 7.10 to 7.12, where we describe the mechanism of duplication. Finally foreign metal ions were observed to introduce new genes, much as did drugs in bacterial plasmids (Figure 7.8) and, as stated, this required gene duplication.

We then are led by combining our knowledge of the very similar directed local effects of defence against drugs and of metal ions locally in plasmid DNA and of those metal ions in multiplied gene content in higher organisms to the possibility that the changes of chemicals in the environment directly affected the genome of eukaryotes during evolution. Thus evolution, in major part of novel chemotypes, could well be an inevitable chemical process in the periods in which they occurred, not due to a random search. At the same time the multiplication of these proteins could lead by differential mutation to a great variety of similar proteins in similar organisms.

It is equally important to observe those proteins which do not undergo duplication. Amongst oxidases generally the enzymes which have been duplicated are those described above and we add electron-transfer proteins as well as related proteins required for uptake and transfer of the required metal ions.[55] It has been observed recently that from early in evolution only some of the proteins of bioenergetics were duplicated. Those which were not extensively duplicated include proteins of the bioenergetic production of proton gradients. These enzymes of oxidative phosphorylation have only very few known forms, using either oxygen or nitric oxide in the case of cytochrome oxidase (Section 6.7). There are also only a small number of duplicates of the photosynthesis apparatus. We have found that amongst zinc enzymes multiple duplication is much greater amongst hydrolases,[61] where there are many similar organic substrates attacked by the activation of the same second substrate, water, than of the almost singular nature of isomerases where the substrates are specific and not many such reactions are required. A parallel example of evolution of a non-metal enzyme without duplication is that of CO_2 uptake by plants using the enzyme RuBisCO, as described earlier. The simple conclusion is that those enzymes with one possible substrate and only one or two known inhibitors are increased in selectivity by random mutation and do not duplicate greatly. Those classes of enzymes with many possible substrates are open to further changes in the environment. The duplicated enzyme must be protected from inhibitors. It is known that each duplicated enzyme is very selective amongst possible substrates. Of course duplication followed by different mutations is essential. It is the two together, the first of which could clearly be more important in the evolution of large changes.

Evolution is then linked to DNA but clearly dependent on the changes in environmental or internal pressures causing duplication.

7.12 Epigenetics and the Mechanism of Duplication

We have been led from chemical evidence to the probable conclusion that the major cause of chemical evolution could be not so much random chance exploration of the possibilities of responses to environmental change, but is the effect of the environmental change more directly on DNA via selective local duplication followed by mutation in the duplicated DNA.[66–68] It was further proposed that DNA changes were caused by stress due to the environmental chemical change. If this is the case we need clear examples of stress directly influencing the genes. To gain greater insight into this possibility we turn to the analysis of the human immune system.[69,70] Here the DNA is modified in special cells which are not inherited but refine the sensitivity of protective proteins to environmental stress, that is antigens, arising from invasion by a foreign bacterium or a foreign chemical.[63,70] The defence is the multiplication with different local mutations of antibodies and of uptake systems in several very similar cells so that the damaging invasion is detected, refined and destroyed by phagocytes in one of the cells, T-cells. The antibodies are rapidly refined in their binding to the antigen by changes in the genetic codes of the specialist cells which lead predominantly to changes in the binding region of the antibody, called the variable region, and not in the rest of the protein, called the constant region, which maintains the antibody fold. This is clearly due to local DNA changes by duplication with mutation. Very few changes are observed in the rest of the DNA so that mutation is strongly directed. The duplication of the antibodies is considerable. How this arises has been intensely studied and is often included under epigenetics (see below). A major route is through the action of transposons, described earlier but now different in different cells, so as to rearrange DNA fragments of the antibodies to give novel proteins and a variety of antibodies in different specialised cells but this does not require duplication. The duplication and transposition of protein fragments are not random but are extremely locally selective. This immune system is restricted to certain specialist cells and it does not affect evolution in man. However transposons are known to have existed at 0.5 Ga and are known in most organisms, even in invertebrates, where they are inherited, adding one more facet to the sudden changes following the Cambrian Explosion. However transposon activity is not the only mechanism of increase in DNA due to local duplication.

For a successful understanding of the role of gene duplication dependent on environmental factors in the advance of evolution we need to find the mechanisms by which it can occur reproductively and the underlying causes of the mechanism. Here we know that the role of reverse (retro) transcription is essential in which mRNA generates a new stretch of DNA after mismatching has caused the two DNA strands to remain apart. The central point is the mismatching of strands[67] which will arise if one of the strands is chemically modified. The effects we wish to describe must be localised and not random in the DNA.

There are a number of ways in which the DNA bases and their attendant proteins, histones, are now known to be modified internally and which alter expression, included under the title of epigenetics, some of which are inherited.[68,69] The direct changes of bases, most important here, are first methylation of cytosine, often adjacent to guanosine, and second, of deamination of cytidine to uridine which is then recognised as thymidine. Both cause DNA chain breaks and base changes, local general mutation.[69–72]

The action of deaminases is considered to be one route by which diversity of bases appears in immunological production of antibodies.[71–76] The change is described under the heading of somatic hypermutation. It is known to give rise to specialist cell evolution if it is coupled to gene duplication in the specialist cells of the immune system where one or more of the duplicates are deaminated. While the immune system itself is not inherited from one generation to the next in man, the deamination system is found in very early eukaryotes and the cytosine deamination is already present in bacteria.[61–71] Clearly the equivalent to the basic part of the immune system 'evolution' is duplication in specialist cells in humans. It is present also much earlier in organisms where it could be dependent on environmental changes. This again amounts to selected, directed DNA change and in lower organisms this apparatus is certainly inherited.[72–76]

Apart from this immunological activity we have looked for DNA changes which come about in man by the presence of drugs which we mentioned earlier. The particular drugs in which we are interested are methatrexate[65] and hydroxyurea.[66] It was found that these drugs caused inhibition of particular enzymes and it is these enzymes which are reproduced. The effect of the inhibition has then been shown to be disruption of one DNA replication at points where there are the promoters of the particular enzyme synthesis. A plausible explanation is that at least in some cases the disruption of DNA strand moves it away from binding to the other strand, making access for duplication which aids the recreation of double-stranded DNA later in the sequence. (An intriguing additional point is that the cytosine deaminases are zinc-dependent enzymes.)

The third type of modification is of chemical changes of the histone proteins by at least four chemical methods.[67–74] Moreover selective binding of the small RNAs is guided to the sites at which these modifications occur.[74–76] The small RNA inhibits expression.

If we combine epigenetics, much learnt from the immune system, with gene multiplication we see that different cell types in evolution can be made by different gene modification brought about by epigenetic factors and internal chemical change or mutation after duplication and ultimately linked to the environmental changes.[68] If this is the major change of DNA then considerably more experimental evidence is required to prove it.

7.13 The Definition of Species and Symbiosis

Let us now put aside our efforts to understand how reproduction arose and how DNA can be modified in line with the observed changes in organisms and associated environmental chemistry and summarise our view of the use of the word 'species'. We could define an idealised organism system as one of 'DNA species', a simple sequence of bases, as the DNA increased in size from prokaryotes to eukaryotes, and it would then define a connected tree of evolution based on a clear definition of species. Complications arise increasingly from the finding of plasmids, more than one DNA, duplicating independently and transferable in bacteria at least. The idealised species model becomes more difficult again because symbiotic organelles with their own replicating DNA became the norm in single-cell eukaryotes.[43] The definition of a unique 'DNA species' as identifying an organism is less and less tenable as the developing organisms are seen to be dependent more and more on many different kinds of symbionts, both inside and outside organisms, from organelles to parasites inside the skin of multicellular organisms or attached externally to them. The observed increasing loss of essential parts of DNA and genes and dependence on chemicals from outside reflects again this cooperativity of chemistry of different DNA species. The human DNA sequence does not relate to humans as an organism because this organism depends on the cells' mitochondrial DNA and all the DNAs of the essential symbiotic 'organisms' living within the human body or outside it. As all advanced eukaryotes are cooperative organisms consisting of 'organisms', they are not then DNA species. Thus a 'tree' of evolution has a multitude of interlinks between branches of organisms, for example fungi, with prokaryotes and with plants (see Figure 7.7). The word 'species', like the word 'gene', is confused by different common and specialist usages.

Part C. Concluding Perspectives

The first section of Part C is a final summary of all of Parts A and B. A considerable part of it repeats material in earlier sections but we want the reader to have a clear vision of our proposal of an integrated view of joint environmental and organism chemistry with many unavoidable features. In the second section we attempt to give a perspective on the present state of evolution with man's influence upon it and the future prospects of the system, organisms/environment. The last section describes the well-known approach to contemporary problems using the concept of Gaia, which we shall show to have a number of problems.

7.14 Final Summary of Chemical Evolution with Reproduction

In Section 7.6 we summarised the evolution of the inorganic/organic chemistry observed in the environment/organism system from 4.5 Ga. In Section 7.7 we gave a possible explanation of why these changes occurred from the point of view of chemistry and physics. We shall not repeat any details here but give the principles which governed the evolution of this limited chemistry. The main point we made was the physical and chemical inevitability of the steps after the Big Bang, an incomprehensible event about 13 billion years ago, leading to the creation of the elements and after 8.5 billion years, of Earth itself. From 4.5 Ga to 3.5 Ga the evolution of Earth's inorganic environmental chemistry was again explicable or at least interpretable. At the end of this time a slowly evolving physical–chemical system of minerals, the seas and the atmosphere had arisen due to weathering and suffering many physical impacts by meteorites and internal disturbances. It had moved by then close to a state of a well-controlled temperature at the surface. The weathering of the minerals and now to a lesser degree the physical disturbances have continued to this day and have been a major part of environmental change, shores, soils and elements in the sea, leading to sediments. All remain as probably inevitable logical processes going to equilibrium with fluctuations.

Quite differently we had to accept that the energised organic chemistry in an aqueous solution, the sea, grew from a few initial, energised, inorganic chemicals to form a cell by 3.5 Ga without any limitations except those imposed by the limitations of kinetics and of the environment.[9,16] The energy from the Sun became the dominant requisite life energy source very shortly after 3.5 Ga, if not before. The synthesis of the molecular constructs for energy capture and transduction to a useful form to drive chemical reactions is very difficult to conceive. The initial steps of this energised synthesis of this and all other organic molecules had to start and continue from oxides of carbon and a source of hydrogen and lead through steps of synthesis to the molecules observed in all cells even to today. All such organic chemicals or any other chemicals concentrated within a membrane, had to be long-lasting due to kinetic barriers but only stable for a period. Their inevitable degradation then required continual reproduction and this was managed by coded nucleotide molecules, either DNA or RNA. The degradation is eventually to the original thermodynamically stable starting materials with the production of heat. This overall energy flow of light to heat is the drive of an entropy gain as is general in many happenings in the universe and is in accord with the Second Law of Thermodynamics. Note that only one major group of large organic molecules evolved, suggesting that it is through particular kinetic barriers that only they exist. They could well be then an inevitable product.

From the beginning it was necessary to have one group of inorganic ions to balance the physical–chemical properties of cells with those of the sea. We noted that Na^+ and Ca^{2+} were rejected, which gave gradients across the cell membrane and that these gradients had possible significances which were realised later and were of extreme consequence in evolution. A second group of inorganic ions was necessary to assist change of organic compounds by catalysis. Some of both groups of elements also provided fast diffusing messengers to help to control and so balance the many pathways of the organic compounds. Initially there were very few such ions and the available Fe^{2+} and Mg^{2+} were dominantly important. Unlike the organic compounds almost all of the free inorganic ions, used here or later, were in fast exchange either with bound states or between bound states and between oxidation states. Therefore they were also in equilibrium in any compartment of the sea or in a vesicle (cell). It is this fast exchange to equilibrium which is an unavoidable feature of aqueous inorganic ion chemistry. It is open to thermodynamic analysis, Chapter 1. This gave predictability to changes of the environment and later it also gave a degree of predictability to cellular life[17] through the impact of these external changes (see Table 7.8). Further processes have been in an inevitable chemical direction driven by energy uptake. We turned to the unique nature of the first anaerobic cells knowing that it is impossible for us to appreciate how this complicated system of chemistry arose so quickly. We can assert with complete confidence, however, that it arose as a product of energised inorganic chemistry and then energised reduced organic chemistry in an enclosure at 3.5 Ga. This necessitated the release of some oxidised less stable chemicals, overwhelmingly oxygen later into the environment. It is only from this time, 3.5 Ga, that we had to attempt to relate the chemical changes to those of genes as the role of DNA (RNA) is quite uncertain until the first cells evolved.

The oxygen and the oxidised chemicals in the environment gave rise to the second most important change for the future of evolution. Environmentally oxidised compounds immediately began to back-react with the organic molecules of cells. Many of the back-reacting chemicals were inorganic ions which gave rise to new catalysts using oxygen and then novel oxidative chemistry.

A strong confirmation of the inevitable nature of this energised (cell) chemistry is that the free metal concentrations of the cytoplasm of all organisms, the metallomes, are fixed for all types of organism, anaerobic or aerobic. These concentrations are then part of the singular general nature of the chemistry of all life much as it was eventually based on one coded set of large molecules, DNA (RNA), and the production of another set of large molecules, proteins, saccharides and lipids. We concluded that the major chemical features of life probably evolved once in cells because any alternative sets of reactions must have been much less kinetically stable. The inorganic/organic chemistry increased until the first cells arose.

We asked next the easier question as to the manner in which the joint chemistry of the environment/organisms system evolved since the formation of these first cells. It was much easier to follow the inorganic changes than those of organic chemistry or of genes as we showed and this remains the case until around 0.40 Ga. (We did not use the description of a tree of evolution based on the evolution of observed organisms as first described by Darwin under 'species'. We avoided too the word 'species' as it appears to be used differently to describe general inheritance amongst organisms and, in a less popular sense, a single DNA sequence from one simple cell type (see also the word 'gene' in section 7.10). We concentrated on the sequential oxidative chemical changes of the elements and their combinations in the environment,[9,13] and then on their effects on cell evolution. We now summarise the progress of this evolution beginning with the first anaerobic cell and followed by the creation of the aerobic cell and then the chemistry of all the following prokaryotes and eukaryotes. We note the reasons for this evolution to today.

1. As we have stated the anaerobic prokaryote cell is an amazing chemical creation. We labelled it the first and most important chemical step in the whole of evolution and it could only have late involvement with genetic DNA/RNA because the earliest precursor organic chemistry steps of life formed DNA itself.

2. The anaerobic prokaryote chemistry inside a restricted volume was reductive, needing hydrogen and came to use water as a source with the rejection of oxygen. This resulted in the change of the environment from 3.0 Ga to more oxidised conditions which back-interacted with the cells. As a consequence of this interaction the anaerobic cell developed first protection from oxygen and later gave rise to aerobic prokaryotes. It is here that we have to ask how did the changes in chemistry link to the changes in DNA? At this time DNA fragments could exchange between different cells.
3. Not long afterwards the eukaryote cell appeared. This we dubbed the second most important chemical change in evolution. This rapid change occurred in step with or just following the necessary major evolution of compartments to separate the unavoidable oxidative from the reductive chemistry. This and subsequent events led us to suppose that there was a systematic development of the combined inorganic/organic chemistry in cells led by inorganic environmental change. Outstanding novelties were the new membrane, the capture of bacteria to give organelles and the calcium messenger system. This realisation of a function for the calcium gradient, established in the first prokaryotes, was essential for eukaryote life. Further features were the small extension of zinc functions, the introduction of low level of copper catalysts and the number of compartments including the mitochondria and the chloroplast. We related these changes to DNA changes with the direct transfer of segments from the original bacteria. There were at this stage either two or three different DNAs in one cell. This is called intimate symbiosis.
4. There followed a period of little change in organisms coinciding with a period in which there was little environmental change of oxygen or environmental chemicals, from about 2.0 to 1.0 Ga.
5. Shortly following 1.0 Ga the oxygen began to rise again and there were more extensive unavoidable changes in environmental chemistry following reversible thermodynamics. In parallel, and we say consequently, the oxidative chemistry of cells could again evolve to produce the multicellular eukaryotes. We took this to be the third very important chemical change in evolution. The DNA had expanded some ten times by this stage. Moreover symbiosis was now with organisms not directly in the same cell.
6. We noted that the different types of inorganic/organic chemistry which arose demanded greater numbers of compartments from those in the earliest eukaryote cells to those of the multicellular organisms. We see this as maintaining largely the essential cytoplasmic reductive chemistry with an increase in general complexity, especially outside cells based on oxidative chemistry. Several new enzymes were based on copper oxidative enzymes and zinc hydrolytic enzymes.
7. From the beginning of this period the many cells of one organism were contained within a 'skin'. The extracellular and vesicle oxidative chemistry gave rise to the crosslinking of connective structure which limited some

movement and created organs of differentiated cells. It was necessary for the connective tissue to grow with the increase of numbers of cells. The copper enzymes helped in this synthesis and that of many new organic messengers for communicating between cells and organs. At the same time the hydrolytic zinc proteins arose to cut connective tissue, making space for growth. There was also a series of zinc transcription factors, zinc fingers which responded to several hormones collectively via dissociable zinc. Zinc therefore acted as an overriding growth hormone. Some of the connective tissue gave rise to biomineral formation. However at the beginning of this period separate differentiated cells appeared with new organic messengers of two kinds passing between them. The same DNA was now transcribed in different ways in separate differentiated cells and also in different organs.

8. The synthesis of connective tissue and messengers for control of the whole organism with an outer skin now made possible the evolution of huge plants and animals. The animals developed fast Na^+/K^+ exchange currents in nerves. This led to the evolution of the brain. Both the nerves and the brain were a direct realisation of the potential of the Na^+/K^+ gradients seen in the earliest cells and essential for life.
9. Between about 0.60 and 0.40 Ga, including the Cambrian Explosion, the oxygen ceased to rise as did changes in environmental chemistry to today. We stated that this has meant no change in cellular chemistry could occur. DNA could only change by mutation, not by chemical novelty.
10. During the whole progress of life from 3.5 Ga to today there has clearly been a huge number of random variations within the chemotypic classes of prokaryotes and eukaryote chemistry. We consider that from about 0.40 Ga to today the number of variants (species?) was the only change and it increased rapidly.
11. There has been very considerable evolution of one of the last novel chemical changes at about 0.50 to 0.40 Ga that gave rise to the brain of man. This fourth and possibly the last big development gave rise very recently to an understanding of physics and chemistry. This development in turn has created the deployment of all the chemical elements of the Periodic Table. This is then the last possible chemical element change associated with the environment/organism system. It has led to the modern environment of life. It has also allowed control of all other forms of life. This self-determination using discovered chemistry does not give internal reproduction of any objects created by it. It has no connection with previous novel chemical developments from 3.5 Ga to 0.5 Ga, much though it is clearly parallel to them, or with variations of organisms from 0.4 Ga to a million years ago (see Section 1.13). It is not linked to any chemical genetic changes after self-consciousness appeared in man close to 100,000 years ago.

Only at this point did we attempt to link the clearly inevitable overall direction of environmental and, we say therefore much of inorganic and organic chemistry in cells (Figure 7.12), to the genome, DNA, changes (Sections 7.8 and 7.9). After describing the nature of changes in DNA itself, we

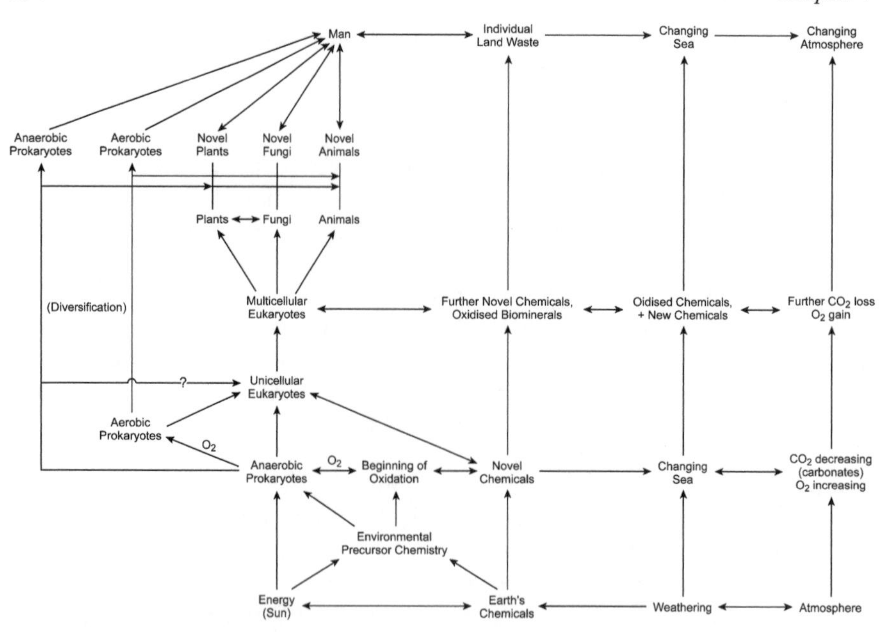

Figure 7.12 A pattern of evolution based on chemistry and incorporating the environment and organisms in one scheme. The cooperation between organisms in their observed chemical combination effectively excludes a tree-like evolution and such a description is not attempted for this cooperative chemistry of the whole system.

asked, did the 'genes' and 'species' follow chemical change of the environment by chance, survival of the fittest, or were the DNA major changes more or less directly connected to the environment? In order to answer this question we found it necessary to define the two approaches to the evolution of a cell based on different definitions of a gene (Section 7.11). One was inheritable traits, as generally used by 'old-fashioned' biologists while the second, that of molecular biology (genetics) seems to be restricted to the definition of changes in inheritance through DNA by itself. A species in the second case is defined by a cellular DNA sequence, while in the first cells are related to the whole ensemble of chemicals in it which we could connect to the cell's environment. One process, backed by much experimental evidence, is that a cell defined in either way can evolve by mutation of the DNA bases which is surely the manner of evolution of different but very closely related organisms in very small steps. The inheritance of the characteristics of the whole cell is then interpreted as due to the expression of the DNA as a code. This is especially true of the evolution of many variations between 0.4 Ga and today. Not so readily explained were the large, seemingly quick, changes in cell character required for the development of novel chemistry in very different prokaryotes and then in turn single-cell and multicellular eukaryote organisms, especially earlier and before 0.40 Ga. Note that we have linked the evolution of eukaryotes with that of the appearance of calcium gradients and that of the evolution of nerves and the brain to sodium/potassium gradients. Both had arisen as absolute requirements for cell stability from 3.5 Ga. The introduction of these two chemical features could hardly have been directly related to DNA changes. The two major periods of rapid change of organisms, the coming of single-cell and multicell eukaryotes, were associated, as we have just described, with two unavoidable separate rapid changes in oxygen and in trace elements by the oxidation of the environment. Oxygen release was unavoidable from about 3.0 Ga. We cannot relate this release to DNA developments. The chemical changes in both periods were of several elements. There was also a break of buffered chemistry from 2.0 to 1.0 Ga in the environment and in the development of organisms. This was due to the unavoidably buffered inorganic chemistry. The whole complement of oxidative chemical changes was seen to be very similar in chemistry in very different classes of organisms, plants and animals. These advances in the chemistry of organisms cannot be related to mutation alone. Before we went further we observed that there were several other ways in which DNA evolution can advance not due to mutation.[75] One different and additional style of development, seen in the evolution of eukaryotes, was for one organism to evolve from two or more sets of cells both containing DNA, RNA and proteins. Here there were two or more DNA sequences which became operational in one organism. The first example of this novel evolution bears no relationship to mutation but in chemistry it closely coincides with environmental changes around 2.5 to 2.0 Ga and apparently occurs in many types of organism at the same time. We have termed this coalescence. As complexity increased organisms came to be based upon more and more of such cooperative DNA species and each of the dependent DNAs evolved separately in one organism. At the same time organisms distant from one another exchanged essential chemicals. This forced us to look again at possible changes in DNA/RNA not by mutation. Various mechanisms we noted including by gene exchange (loss and gain), by viral infection[76] and by transposition of segments of DNA. We turned finally to the very extensive duplication of either whole genomes or of segments of DNA representing genes. We proposed the possibility that the observed local duplication followed by mutational activity could happen by directed chemical changes to local DNA regions or its associated proteins, histones, caused by the stresses in the cell's metabolism strongly generated by oxygen itself and oxidative environmental changes. Many such stresses must have arisen from particular inorganic element changes in the environment, they are the major changes forced upon all living organisms. They could have led to many coincidental or parallel cell advances in uses of inorganic elements both as messengers and catalysts in novel structures recognisable as a need from fossils, and the use of symbiosis (Table 7.10). Hence we linked the directional changes of environmental and cellular chemistry to the multiplicity of DNA changes before mutation. As an aid to our thinking we examined the immune system of man and of organisms of much earlier origin in order to find cases of and mechanisms of duplication.

Table 7.10 Biological Evidence and Changes in the Environment in Evolution.

Time (Ma)	*Biological Evidence*	*Interpretation*	*Glucose Use*	*Geological Evidence*	*Oxygen per cent PAL*	*Loss of Element*	*Gain of Element*	*Little Change of Element*
400	Large fishes, First land plants	Bone, Vascular structure			100			
550	Cambrian Fauna	Shelly metazoans, absorption through external shell			10			
670	Ediacaran fauna	Metazoans, collagen			7			
1400	Cells larger in diameter	Eukaryotic cells, mitosis uses actomyosin, organelles			>1			
2000	Enlarged, thick-walled cells at intervals on algal filaments	Oxygen tolerating bacteria, protection against photo-oxidation			1			
2800	Stromatolites, filamentous chains	Resemble living cyanobacteria			0.1			
3500		First evidence of cells			<0.01			
3800	Rhythmically banded rocks, depletion of ^{13}C	Microbial organisms (?) Biological Activity (?)			$\ll 0.01$			

It is thought that an antigen effects certain essential protein activity in the cell. The cell response has to be to produce more of this protein so opening the DNA strands. To recover matching of strands one strand can be duplicated. Duplication is then clearly a response to stress. The DNA bases exposed can be mutated more rapidly after duplication by at least one zinc enzyme, cytosine deaminase. A variety of such products of multiplication of differently mutated proteins appears in different immune cells which then release the changed antibodies and remove the antigen. Mutations in the duplicated protein are several times faster than in the non-duplicated DNA. Man's immune response only occurs in specialised cells, T-cells, and is not found in his reproductive cells. However similar proteins to those in the immune cells have been found from the analysis of the DNA of earlier organisms.

A much stronger, general argument for duplication as a response to stress has been observed in human cells responding to drugs. The first example we mentioned above was the response to methatrexate. Other examples are now known and a good summary is included in a paper on the effect of the drug hydroxyurea. This paper also describes the way in which local effects on DNA arise in the DNA.[66] These duplications are followed by rapid mutagenic changes. The result is the increased protection of the cells.

Returning more directly to the analysis of the possible effect of the environment on DNA we analysed the duplication of metal ion-dependent proteins in a wide range of organisms. We observed the number of duplications, especially of oxidases in minerals and plants. They occurred long before the immune systems arose. These oxidases include those which generate protective complexes and remove poisons. We considered that they arose through stress due to the compounds in the environment. There were other examples which we related to physical or chemical stress. The stress of the evolution of oxygen and of its dependent environmental changes we suggest is the general cause of large and quite sudden changes in organisms at the times when these stresses arose. Our conclusion was that the large change in evolution within DNA was due to changes in the environmental chemistry (see Figure 7.12). There is then a considerable degree of inevitability in all features of the organism chemistry in evolution as chemistry was forced to change by environmental chemistry.

In our view our chemical description in no way conflicts with the Darwinian description of a tree of evolution[7,14] along branches so long as that description relates to closely related organisms in limited groups, that is along a spreading branch. We noted that this explanation is entirely consistent with findings from molecular biology (genetics) in the way it interprets single DNA mutational sequences (DNA species). In fact we see no chemical or physical logic or evidence of the vast numbers of closely linked variants of organisms within the groups of chemotypes, which we then accept as being due to random selection of fitness. As an example of the present time very small differences exist, for example between varieties of grazing animals occupying slightly different physical–chemical environments. It seems to us that these descriptions were particularly suitable and sufficient to explain the vast majority of developments after 0.4 Ga when chemical development was virtually complete as oxygen increases stopped. On the other hand the earlier chemistry we have outlined forced us to look for alternative explanations to mutation for the clearly observed earlier quite sudden main 'branching and expansion of the tree' at certain times. They were in our opinion fundamentally based on novel equilibria in the environment which had then to be connected to DNA change.

A second and largely early development of organisms which we have described arose in the form of chemical interactions between the branches of the tree through combinations of organisms in multicellular organisms, symbiosis, that is the necessity for the strength of the total chemistry of life. It is included under symbiosis and is more effective in association, in eukaryotes, than in prokaryotes which exchange chemicals through external solutions. We noted that increases in symbiosis occurred at times of increase in the complexity of organisms linked to chemical environmental changes. This does not fit a Darwinian tree.

We conclude by stating that we have provided a convincing explanation of the chemical influence of the environmental changes on the inorganic and organic chemistry observed in this evolution of the different groups of organisms, chemotypes. We have then tried to find a link between the observed chemistry and the necessary changes of the DNA to reproduce this chemistry. A special limitation was the lack of a mechanism for the very beginning of life, knowledge of which could well assist in undertaking all evolutionary developments but it cannot be related to mutation. We turned to the value of known common duplication of sections of DNA in all organisms in aiding the coupling of the environment and DNA. We correlated the local duplications and subsequent mutation with the introduction of novel functions, especially those of newly available metals for metalloenzymes. Very illuminating evidence was from recent data on drug resistance, where the response is very fast and has been shown to be due to replication with mutation. We then likened this effect of drugs to that of newly introduced metal ions taking both to be stressful poisons. We concluded that duplication followed by mutation in DNA could well be the reason for the apparent linking of DNA change to the environment at least until around 0.4 Ga.

With the full development of multicellular organisms by 0.4 Ga inevitable changes of chemistry in the environment or in organisms ceased. Throughout the 3 billion years of the inevitable changes of chemistry random exploration undeniably also helped the evolution of the chemistry of organisms. Undoubtedly stresses arose other than by novel chemical change, for example in periods of extinction, and recovery could well have been by duplication and mutation. Generally advance in evolution could only be by such random trial and error search for fitness, utilising all the previous unavoidable chemical advances before 0.40 Ga. One of these was the introduction of nerves and then the brain. By random exploration of both these features man's brain evolved in an outstanding way. (Remember that they arose from utilisation of the sodium gradient.) Quite suddenly and very recently man developed with an ability to change the environment through the use of an understanding of chemistry and

Table 7.11 The New Parts of the Possible Future System

1) Tapping of New Energy Sources. Atomic, Deep Earth, Water, Wind and Wave Flows, Power Transmission
2) Use of All Elements of the Periodic Table, see Fig. 1.
3) New Structural Materials: Inorganic Metals, Minerals and Organic Plastics
4) Use of All Space: not connected to Cells
5) New Catalysed Reactions: Industrial Chemistry
6) Use of New Conditions of Temperature and Pressure on the Surface: Industrial Metal Production
7) New Message Systems: Electronics
8) New Transport Networks on Land and in Air
9) Protective Organic Treatments: Synthesised Medicines
10) Dominance over other Organisms: Animals and Plants, Use of Genetics
11) Control of the Population

physics. A completely new stage of evolution of chemistry began which was not related to random searching. We do not know where this will lead evolution. Looking back at all of evolution through the eyes of a chemist, remember that all evolution is of chemistry, we see that in large measure much of its outline was in essence inevitable, justifying the title of this book, 'Evolution's Destiny'.

We are well aware that throughout evolution the major changes were of organic chemicals. However their very great number and complexity have not been examined by quantitative analysis which means that we cannot include anything but the simplest outline of them.

7.15 The Chemical System and Mankind Today and its Future

It may be useful to consider here the present condition of mankind and our environment, assuming our view of the ecosystem, and then to give possible prospects for Man's further evolution (Table 7.11). Man's activity is now one major driving force on the whole of life's evolution and changes in much of Earth's surface and seas. His activity is of a physical–chemical nature similar to that of all evolution. As it brings physical–chemical change it is comparable to evolution before 0.5 Ga, not by chance but directed action. The timescale of man's development of new chemical and physical activities in the last 1,000 years is a major evolutionary problem for it is millions of times faster than we can expect for changes to be successful in complex organisms. The environment for all life is being quickly man-made by changes due to agriculture and industry which are ongoing and have been far-reaching in the last 200 years. Such an environmental change is not too different in timescale to that of some previous changes which led to extinctions (see below). Extinction may be happening now to some parts of biodiversity.

It is appropriate to concentrate first on atmospheric conditions today, as apparently the main short-term concern (for man) today is global warming.

We accept that although there are changes in the Sun's energy emission the cause of the present atmospheric changes lies with man's engagement with material well-being and wealth from industry and agricultural activity. One solution appears to be to stop our dependence on fossil fuel. There are alternative ways of generating energy to fossil fuels but they now seem unlikely to be adequate for man's needs in the next 30 years. However in discussing evolution generally we must keep a sense of proportion and compare the scale of these problems over long periods with those of Earth's history. In Chapter 2 we noted huge successive physical changes of the environment due to large temperature changes, Snowball Earth and groups of volcanoes. Vast chemical changes also arose from production of oxygen around 2.5 to 2.0 and 1.0 to 0.4 Ga, which then grossly increased 'pollution' of the sea by trace metal ions. Again there has been a fall of orders of magnitude in carbon dioxide and more recently within the last half a billion years there have been many considerable fluctuations in both oxygen and carbon dioxide (Figures 2.11 and 2.12) and consequently of temperature. In later chapters we have shown that despite massive losses, extinctions, organisms of one kind or another have survived or adapted or new organisms have evolved through these huge changes. There is no doubt that many, perhaps most, species can survive the worst effects we can imagine of our own relatively small-scale activities. The planet's surface system as a whole, including organisms, therefore cannot be in catastrophic danger from small changes of temperature introduced by man and they are very unlikely to affect evolution in creating novelty.

An equally difficult problem is that access to and use of more energy, demanded as quickly as possible to obtain chemicals and construction with them will introduce quite other risks. This increase in energy use, lying behind all other changes, has been of course the drive of all evolution but in the past energy use has only increased gradually. It is clear that man is seeking ways of increasing available energy by every means possible – from fossil fuels, from deep Earth, from nuclear reactors, from the Sun, from wind and rain and water flow, and so on, and as fast as possible. Apparently we must have growth in our activities to maintain our 'economies'. However the demand of our economies is for more and more material using this energy and this will put a strain on the total chemical resources on or near the surface. While it could also lead to pollution by trace elements, sooner or later there will come limitations on one chemical source or another due to such demands. If we are to avoid grave problems there has to be a general acceptance that there is a need to reduce exploitation of material as well as energy use. We add that it is our very success in technology and chemistry, which, like all previous physical and chemical changes in the environment feeds back to organism development, but inevitably generates danger. The obvious example is the introduction of new chemicals to kill off unwanted plants or bacteria which threaten health. The downside is that lower organisms mutate quickly, become insensitive to the chemical and then a greater human hazard.

There are also dreams for a 'science' future either on Earth or on other planets but the probability of achieving them on an extensive effective scale in the near future, say in the next 100 years, is so small that we put them aside. The dreams include use of DNA to create novel organisms, the very extensive use of robots, and several other scenarios from science fiction.

We are also becoming increasingly aware that we cannot circumvent natural disasters other than those caused by global warming caused naturally or by man. The scale of recent disasters due to earthquakes is very large locally but is hardly noticeable globally. There are much greater dangers from other chaotic events, for example from a major volcanic eruption. We have escaped such an activity for nearly 200 years but it can only be a matter of a similar time within which such a devastating volcanic eruption occurs again. Such a giant volcano would give rise to less of the Sun's energy reaching the Earth's surface. Even if only 2 or 3 years are experienced without much sunshine globally, due to such volcanoes, the present human population cannot survive. The relative short-term future is certain to bring such a 'disaster'. Again protection giving escape from a disaster due to impact from an asteroid cannot be totally dismissed but we certainly cannot control an asteroid. It, in a similar manner to even greater disasters such as have occurred before, will not destroy life, but will cause general loss of life as in periods of extinction.

In thinking about our problems we must appreciate our peculiar position in evolution. In our account of early evolution from 3.5 to 0.5 Ga we showed that, at each change of the available chemicals in the environment organisms responded by developing novel chemistry and management systems. Differently we considered that since about 0.5 Ga chemical change ceased but this did not stop a different type of evolution. Utilising to the full the last large chemical changes led to the generation of a huge variety of organisms based on the same chemistry but refining the use of many chemicals and most importantly of the novel external messenger system, electrolytics with control over many events and of a new messenger system in management. This was the nerve cells in animals from which the brain evolved. Say about 100,000 years ago man's brain began to have properties, different from that of the generality of animals, when man became a novel self-aware or self-conscious species. Quite quickly awareness in man developed until he became able to construct simple ways of communicating while controlling some chemicals from the environment. This extraordinary ability accelerated greatly so that he developed a more and more sophisticated ability to handle the world around himself without causing a major environmental impact. Now about 500 years ago man began even to appreciate and later to understand the physics and chemistry around him. Man could then develop a more and more complex life maintained by advances in physics, giving new communication systems as well constructs and chemistry using many novel elements. This evolution was completely different from that in all other periods from 0.5 Ga to 0.01 Ga in that it was completely external to any biological development but in some respect it did not differ from that between 3.5 and 0.5 Ga. There was now the risk that the size of activity, the population, will would affect the environment as described earlier.

Looking at this period from 3.5 Ga to 0.5 Ga evolution we observed that organisms were driven by external chemical change which back-interacted with life. The external changes were a consequence of the actions of organisms themselves, producing oxygen, much as is the case today of a changed external environment due to man's activities. Organisms adjusted their life in many ways but in particular by employing novel communicating systems they managed to come to terms with their new lifestyle with new organisation at any one time. Today man uses external chemicals extensively not so very differently from the more primitive use of chemicals. But through the fast change and large scale of technical progress he has not been able to reach quickly a balanced condition with feedback to himself and other organisms in this newly created environment. There is a determined way in which the explanation of physics and chemistry and its use by man has given him his environment much as there were determined environmental changes for earlier organisms created by them. The earlier organism could only evolve slowly too. We can learn from an understanding of the manner in which the earlier organisms came to terms with their environment by chemical change matching environmental chemical change brought about by organisms themselves. We must adjust ourselves to our new environment and that includes all of us. It will mean that 'fast forward' progress is not possible. We need slower change while we use our intellectual powers to adjust ourselves to this new environment, as it is very unlikely that we can change our physical chemical makeup. This implies that we must make a change in mental attitudes, remembering that it was mental effort which has brought us to our present physical–chemical condition. We need to organise society and communication systems so as to evolve maintainable conditions. This will leave little room for outright competition between individuals and especially between races or nations. Competition has often been thought to be the basis of the essence of evolution, the so-called 'survival of the fittest', but we have shown that at all times cooperation was essential.

The conclusion is inevitable. There are serious long- and short-term risks for Earth's surface and atmosphere but they cannot be called dangers to the planet much though they are a danger for man. There is even a long-term possible future for life as it is, that is, say, for a few thousand years under man's control, if we are lucky, but it does not seem to be possible to maintain our present attitudes to energy and materials for so long. We have to be more aware of the changes of the environment which generate many risks for us, one of which is that of being loaded with what we ourselves have created in the last 200 years. This change in the environment is not chemically compatible with life as in earlier periods where life adjusted itself chemically. Man cannot change his chemical nature. It may well be that the only realistic calculation must be based on an intellectual effort to overcome the sum of the risks. Maybe we must realise that human life can exist only with a much smaller population than today and with quite another attitude to technology.

A different approach to the future of mankind and of the environment/organism system is described next under the title Gaia, which appears to state that all is well apart from global warming, which is avoidable in the relatively immediate future. Unfortunately in many people's minds it confuses the short-term considerations of man with long-term considerations of the man/environment interaction and some underlying ideas in Gaia may be incorrect.

7.16 A Note on Gaia

Lovelock[77] has proposed that there could be at present (or perhaps more likely there was some 200 years ago) a balanced reduction of carbon dioxide in organisms by photosynthesis using hydrogen from water and the oxidation of the reduced carbon back to carbon dioxide by all living things using the necessary simultaneous release by life of oxygen to the environment. With him we could consider that the overall reactions have resulted in a flowing cyclic steady state in which oxygen cannot be further increased by life because to do so would only increase the oxidation of life lowering oxygen production and returning the whole cycle back to this balanced state. The increase of carbon dioxide is equally self-balancing for this gas causes global warming and thence increase in the rate of weathering and of oxidation, so lowering the biomass. As the carbon dioxide came from degradation of organic matter which diminishes, the system is again self-correcting. The previous history of the planet with life from 3.5 Ga can then be treated as a drive towards this present cyclic steady state, Gaia. In our book we have described that, from the time of Earth's formation, the development to the present condition has been led from an initial state of excess of carbon dioxide and methane and no oxygen by weathering and then by the splitting of energised water into reduced carbon material in life and to an oxygen-rich condition. Much CO_2 was absorbed as $CaCO_3$. The oxygen removed the methane and introduced a gradual increase in the carbon/oxygen cycle. It is agreed that the reactions are driven by energy mainly from the Sun. Life evolved therefore as increasingly reduced chemicals in cells with oxygen in the atmosphere. It was in large part the oxygen which caused evolution of the environment, that is of many elements in the environment. They helped evolution of organisms because the environmental changes back-reacted with the living organisms. They were then forced to utilise effectively what they could manage of the oxidation of chemicals in what we see as life's evolution. Reduction increased, mainly in plants while oxidation increased, largely by animals. In principle this could have given a Gaia-like cycle as stated by Lovelock. In fact there was a period some 200 years ago when this view of the state of the planet might have held approximately. However this neglects the real condition of the system, which relates to man's activity. Man today is interfering with this 'Gaia' state by generating excess carbon dioxide by burning fuels, *i.e.* reduced carbon with oxygen, and Earth will suffer the consequences.

Unfortunately a period arose long ago and it left a disturbing effect on the possibility of Gaia based on today's carbon dioxide. Some 300 to 400 million years ago (0.4 to 0.3 Ga) plant life, reduction of CO_2, evolved a chemical very resistant to degradation, oxidation generally and by animals, lignin. As we mentioned in Chapter 6 lignin was produced through catalysis by the newly available copper in enzymes, starting just before this period. As a result there was a large increase in plant life which on death escaped oxidation and became quite quickly buried as reduced carbon. Under pressure it became a hidden store of oil and coal (C/H compounds) undermining any possible true balance of CO_2 and H_2O in the whole ecosystem. In a fascinating turn of fortune man has found a way of excavating the buried carbon from the stores and using it in large quantities as a fuel. In so doing man is in fact helping to return the total balance of CO_2 to a state closer to that of 0.4 Ga. This would be near to the state of a possible true 'Gaia'. What we term pollution by CO_2 is in fact our contribution to moving toward this true 'Gaia', although it may turn out not to be to our liking. It is undoubtedly damaging to man that he should become his own worst enemy in this process, but is this a real concern of evolution? Is man not doing only what should have occurred naturally and possibly would have done so eventually through fluctuations of Earth bringing the carbon back to the surface? Now it may be that by returning to the true atmospheric conditions of 0.4 Ga we shall cause very substantial change in the conditions in parts of the Earth. Much of what we cherish may be lost due to the climate change. However this is not a disaster for the planet and it may be that it will rectify what have been excesses of man both in use of resources and in population. The life that then arises may become a true Gaia, an organism/environment system, but there are further problems caused by man which can affect evolution via changes in reaction rates. This is due to his exploitation of many materials which could act as catalysts in organic chemistry.

One major different problem is that many other elements are involved apart from carbon, oxygen and hydrogen in these reactions and have changed in availability by man's industry. While some are catalytic, others are inhibitory and only some are reversible. The energised organic chemistry/oxygenated atmosphere which is influenced by these elements is therefore only a part of the problem. We have to remember especially those irreversible energised events which change elements and while organisms cannot correct for these changes they may even help to increase some of them. One is physical weathering due to energised water and wind flow and a second is general inorganic chemical changes, for example by oxidation of sulfides. The sea is now a sulfate and not a sulfide compartment. These flows are not reversible by organisms but strongly interactive with them and an essential part of evolution. We cannot know how the changes in elements in the sea by man's increase of mineral turnover will affect the present state in the long term. It is in this progression that it is very difficult to see how Gaia can exist except on a short timescale, the length dependent on the rapidity of change. It is not easy to see how the nature of man can adapt to these changes as described in Section 7.13.

Looking back over the whole of evolution of the chemical system the condition of man today may be just a minor misdirected short-term fluctuation of evolution's inevitability in a very long process. Man has escaped from the restrictions of 'natural evolution' but cannot be really interested in the reality of long-term evolution, only in his own fate which is a short-term problem. We must not confuse the short-term concerns of hundreds of years with the analysis of evolution over hundreds of millions of years. We do however have a choice as to how we behave. We can try to maintain the present state of Earth following the idea of a short-term 'Gaia', which will mean changing our lifestyle, or we can push ahead with our activities optimising the opportunities we uncover for our own well-being. In the second case we can use the long-term evolution of evolution to see the possible risks we shall be taking, for there can be no Gaia while we want growth.

References

1. C. Darwin, *On the Origin of Species by Natural Selection*, 6th edn, John Murray, London, 1862.
2. T. Cavalier-Smith, M. Brasier and T. M. Embley, *Philos. Trans. R. Soc., B*, 2006, **361**, 843.
3. P. A. Corning, *Kybernetics*, 2001, **30**, 1272.
4. U. Alim, *An Introduction to Systems Biology*, Chapman and Hall/CRC, London, 2006.
5. C. Cunningham, *Evolution*, W. M. B. Earlmans, Grand Rapids, , USA (and references therein).
6. C. R.Woese, *Proc. Natl. Acad. Sci.*, 1998, **95**, 6854.
7. J. Barrow, S. Conway Morris, S. Freeland and C. Harper (eds), *Fitness of the Cosmos for Life: Biochemistry and Fine Tuning*, Cambridge University Press, Cambridge.
8. H. Blatt and R. J. Tracy, *Petrology: Igneous, Sedimentary and Metamorphic*, Freeman, New York, 1994.
9. J. J. R. Fraústo da Silva and R. J. P. Williams, *The Biological Chemistry of the Elements*, Oxford University Press, Oxford, 2001, Chapters 1 and 2.
10. W. Martin and M. J. Russell, *Philos. Trans. R. Soc. London, Ser. B*, 2003, 388, 5985.
11. N. Mulkidjdjanian and M. Gaperin, *Proc. Natl. Acad. Sci.*, 2010, **107**, E137.
12. R. J. P. Williams, *J. Theor. Biol.*, 1961, **1**, 1.
13. P. Mitchell, *Nature*, 1961, **191**, 144.
14. D. G. Nicholls and S. Ferguson, *Bioenergetics 2*, Academic Press, San Diego, 1992.
15. D. Stock, A. G. W. Leslie and J. E. Walker, *Science*, 1999, **286**, 1700.
16. R. J. P. Williams and J. J. R. Fraústo da Silva, *The Chemistry of Evolution*, Elsevier, Amsterdam, 2006.
17. C. Klein, *Am. Mineral*, 2005, **90**, 1473.
18. M. A. Fedonkin, J. G. Gehling, K. Grey, G. M. Narbonne, P. Vickers-Rich, *The Rise of the Animals: Evolution and Diversification of the Kingdom Animalia*, John Hopkins University Press, Baltimore, 2007.
19. M. Brasier, *Darwin's Lost World*, Oxford University Press, Oxford, 2009.

20. J. R. Waldbauer, L. S. Sherman, D. Y. Summer and R. E. Summons, *Precambrian Res.*, 2009, **169**, 28.
21. A. D. Anbar and A. H. Knoll, *Science*, 2002, **297**, 1137.
22. C. L. Dupont, S. Yang, P. Palenik and P. E. Bourne, *Proc. Nat. Acad. Sci. USA*, 2006, **103**, 17822 (and see reference 53).
23. Y. Zhang and N. V. Gladyshev, *Chem. Rev.*, 2009, **109**, 4828 (and references therein).
24. K. Hoelm and M. E. Nicpon, *Human Anatomy and Physiology*, 7th edn, Benjamin Cummings, San Francisco, 2007.
25. F. M. Harold, *The Way of the Cell*, Oxford University Press, Oxford, 2001.
26. E. Carafoli and C. Klee, *Calcium as a Cell Regulator*, Oxford University Press, New York, 1998.
27. C. Toyoshima and G. Inesi, *Annu. Rev. Biochem.*, 2004, **73**, 269.
28. D. Krogh, *A Brief Guide to Biochemistry and Physiology*, Prentice and Hall, New York, 2007.
29. Y. Jiang, A. Lee, J. Chen, V. Ruta, M. Cadane, B. T. Chait and R. Mackinnon, *Nature*, 2003, **423**, 33.
30. D. Purves, G. J. Augustus, D. Fitzpatrick, W. E. Hall, A. S. L'Amantia, J. O. McNamara and L. E. White, *Neuroscience*, 4th edn, Sinauer Associates, Sunderlands, , MA, 2008.
31. R. A. Colvin, W. R. Holmes, C. P. Fontaine and W. Maret, *Metallomics*, 2010, **2**, 306.
32. C. Andreini, L. Banci, I. Bertini and A. Rosato, *Proteome Res.*, 2008, 7, 209.
33. A. L. David and E. J. Alm, *Nature*, 2011, **469**, 93.
34. R. Dawkins, *The Selfish Gene*, Oxford University Press, Oxford, 2006.
35. J. M. Smith and R. Szathmary, *The Major Transitions in Evolution*, W. H. Freeman, San Francisco, 1995.
36. S. A. Kauffman, *The Origin of Order*, Oxford University Press, Oxford, 1993.
37. A. Lazcano and S. L. Miller, *J. Mol. Evol.*, 1994, **39**, 546.
38. L. E. Orgel, *Origins of Life: a Review of Facts and Speculations*, Elsevier, Amsterdam, 1998.
39. C. de Duve, *Life Evolving: Molecules, Mind and Meaning*, Oxford University Press, New York, 2002.
40. S. J. Gould, *The Structure of Evolutionary Theory*, Harvard University Press, Cambridge, , MA, 1995.
41. D. J. Depew and B. Weber, *Darwinism Evolving*, MIT Press, Cambridge, , MA, 1995.
42. S. Conway Morris, *Philos. Trans. R. Soc., B*, 2006, 361.
43. L. Margulis, *Symbiotic Universe*, Basic Books, New York, 1998.
44. M. L. Arnold and N. D. Forgarty, *Int. J. Mol. Sci.*, 2009, **10**, 3836.
45. A. Moat, J. W. Foster and M. P. Specktor, *Microbial Physiology*, Wiley-Liss, New York, 2002.
46. J. Spring, *J. Struct. Funct. Genomics*, 2004, **3**, 19.
47. J. C. Herron, *Evolutionary Analysis*, 4th edn, Freeman-Scott, New York, 2008.
48. N. Lane and W. Martin, *Nature*, 2010, **467**, 929.

49. S. Ohno, *Evolution by Gene Duplication*, Springer Verlag, Berlin, 1970.
50. M. Linch and J. S. Conery, *Science*, 2000, **290**, 1151.
51. J. S. Taylor and J. Raes, *Annu. Rev. Genet.*, 2004, **38**, 615.
52. I. Zhang, *Trends Ecol. Evol.*, 2003, **18**, 292.
53. E. V. Koonin, *Nucleic Acids Res.*, 2009, **37**, 1011.
54. A. L. Hughes, *Proc. Natl. Acad. Sci. USA*, 2005, **102**, 8791.
55. J. M. McClintock, R. Carlson, D. M. Mann and V. E. Prince, *Development*, 2001, **128**, 2471.
56. J. Platigorsky and G. Wistow, *Science*, 2008, **252**, 1078.
57. P. Deschamps, H. Moreau, A. Z. Worden, D. Douvillee and S. G. Ball, *Genetics*, 2008, **178**, 2373.
58. T. M. Embley and W. Martin, *Nature*, 2006, **440**, 623.
59. C. Andreini, I. Bertini, G. Cavallero, G. M. Holliday and J. M. Thornton, *Bioinformation*, 2009, **25**, 2088.
60. C. L. Dupont, A. Butcher, R. E. Valas, P. E. Bourne, G. Caetano-Anolles, *Proc. Natl. Acad. Sci. USA*, 2010, **107**, 10567.
61. L. Decaria, I. Bertini and R. J. P. Williams, *Metallomics*, 2010, **2**, 706.
62. L. Decaria, I. Bertini and R. J. P. Williams, *Metallomics*, 2011, **3**, 56.
63. M. H. Serves, A. R.W. Kerr, T. J. McCormack and M. Riley, *Biol. Direct*, 2009, **4**, 46.
64. B. Conrad and E. Antonarakis, *Annu. Rev. Genomic Hum. Genet.*, 2007, **8**, 17.
65. J. Mellor, *The Biochemist*, 2010, **32**, 14 (and references therein). See also eight following articles, pages 18–33.
66. A. Petronis, *Nature*, 2010, **465**, 721.
67. E. Jablonka and M. Lamb, *Evolution in Four Dimensions: Genetics, Epigenetics, Behavioural and Symbolic Variation in the History of Life*, MIT Press, Cambridge, , MA, 2005.
68. E. J. Richards, *Nat. Rev. Genet.*, 2006, **7**, 395.
69. E. Jablonka and G. Roaz, *Q. Rev. Biol.*, 2009, **84**, 131.
70. J. Travers, *Science*, 2009, **324**, 580 (and references therein).
71. R. Chatwin, S. N. Wantakal and S. Roa, *Trends Genet.*, 2010, **26**, 443.
72. M. E. Feder, *J. Exp. Biol.*, 2007, **210**, 1653.
73. A. M. Nedeleu, *Proc. R. Soc. London, Ser. B*, 2007, **272**, 1935.
74. I. Karasov, P. W. Messer and D. A. Petrov, *PloS Genet.*, 2010, **6**, e1000987.
75. T. F. MacKay, *Philos. Trans. R. Soc. London, Ser. B*, 2010, **365**, 1229.
76. S. W. Ding, *Nat. Rev. Immunol.* 2010, **10**, 632.
77. J. Lovelock, *Homage to Gaia*, 2000, Oxford University Press, Oxford.

Subject Index

Note : page numbers in *italic* refer to figures and tables.